Solid-State NMR in MATERIALS SCIENCE

Principles and Applications

Solid-State NMR in MATERIALS SCIENCE

Principles and Applications

Vladimir I. Bakhmutov

CRC Press
Taylor & Francis Group
Boca Raton London New York

CRC Press is an imprint of the
Taylor & Francis Group, an **informa** business

CRC Press
Taylor & Francis Group
6000 Broken Sound Parkway NW, Suite 300
Boca Raton, FL 33487-2742

Printed in the United States of America on acid-free paper
Version Date: 20110803

International Standard Book Number: 978-1-4398-6963-5 (Hardback)

Library of Congress Cataloging-in-Publication Data

Bakhmutov, Vladimir I.
Solid-state NMR in materials science : principles and applications / Vladimir I. Bakhmutov.
p. cm.
Summary: "Solid-state NMR is an extraordinarily powerful method that is widely applied in materials chemistry and is constantly being developed through application of new techniques and approaches. As the most exciting direction of modem materials science, the controlled design of new materials requires constant application of new methods. Although young scientists are familiar with solution NMR, the solid-state version of this method often remains a black box, so this book describes experimental problems in solid-state NMR while avoiding unnecessary theoretical details. The text covers amorphous and diamagnetic materials, paramagnetic effects, the porous system, and more"-- Provided by publisher.
Summary: "Nuclear Magnetic Resonance (NMR), discovered as physical phenomenon by Bloch and Purcell in 1945, rapidly became the theoretical basis for of a powerful and most popular analytical method, which is widely applied in different fields of modern fundamental and applied science, and also medicine and industry. This circumstance is easy explained by capability of NMR to solve rapidly a large number of practical tasks which can be connected not only with structural descriptions of systems on macro and/or molecular levels but also identifications of their various dynamic behavior often responsible for mechanical properties of materials"-- Provided by publisher.
Includes bibliographical references and index.
ISBN 978-1-4398-6963-5 (hardback)
1. Materials--Testing. 2. Nuclear magnetic resonance spectroscopy. 3. Solid state chemistry. I. Title.

TA417.36.B35 2011
620.1'1278--dc23 2011027262

Visit the Taylor & Francis Web site at
http://www.taylorandfrancis.com

and the CRC Press Web site at
http://www.crcpress.com

Contents

Preface xi
Author xv

Chapter 1 General Principles of Pulsed NMR and NMR Techniques 1

1.1 Early History of NMR in the Solid State 2
1.2 Nuclei in the External Magnetic Field 3
1.3 Effects of Radio Frequency Pulses 9
1.4 From Macroscopic Magnetization to Nuclear Relaxation and Shapes of NMR Signals 12
1.4.1 Common Representations of Molecular Motions and Their Influence on NMR 14
1.4.2 Contribution of Molecular Mobility to Spin–Spin and Spin–Lattice Nuclear Relaxation 17
1.5 Parameters Characterizing Nuclei in the External Magnetic Field 20
1.5.1 Chemical Shift 20
1.5.2 Spin–Spin Coupling 22
1.5.3 Quadrupolar Coupling 24
1.6 Introducing NMR Equipment and NMR Experiments 26
1.6.1 Magnets and NMR Probes 28
1.6.2 Homogeneity of the External Magnetic Field and Digital Resolution in NMR Spectra 30
1.6.3 Dead Time of NMR Spectrometers: Intensity of Signals and Distortion of Baselines in NMR Spectra 30
1.7 Main Principles of Two-Dimensional NMR 32
1.8 Enhancement of Sensitivity in NMR 34
References 35

Chapter 2 On Amorphous Materials as Potential Objects in Solid-State NMR Studies: Common and Key Problems 37

2.1 Hybrid Materials 38
2.2 Polymer Solids 40
2.3 Systems Based on Nanostructures 41
2.4 Wood and Wood Components 42
2.5 Glasses 44
2.6 Zeolites 45

2.7 Design of Porous and Layered Molecular Systems 47
2.7.1 General Problems in Structural Characterizations of Amorphous Porous Materials 48
2.7.2 Some Key Structural Issues in Characterizations of Amorphous Materials Modified by Metal Ions and Methods for Structure Solution 50
References 54

Chapter 3 Features of Solid-State NMR: Diamagnetic Materials 57

3.1 From Isotropic Chemical Shifts to Magnetic Shielding Tensors 57
3.1.1 Visualization of Chemical Shift Tensors and Determination of Their Orientations 62
3.2 Nuclei with a Spin of 1/2 66
3.3 Quadrupolar Nuclei in Solids 70
3.3.1 Two Categories of Quadrupolar Nuclei 72
3.4 Detection of NMR Signals in Solids: Common Aspects 73
3.4.1 Wide-Line NMR 73
3.4.2 Direct Excitation 74
3.4.3 Cross-Polarization 76
3.4.4 Cross-Polarization with Quadrupolar Nuclei 79
3.5 High-Power Decoupling 81
3.6 Multi-Quantum NMR Experiments 83
3.7 Dipolar Recoupling NMR Techniques 85
3.8 Sideband Manipulations 90
3.9 Zero-Field Solid-State NMR Experiments 92
3.10 Solid-State NMR Imaging 94
References 96

Chapter 4 Practice of Multinuclear NMR in Diamagnetic Solids: General Information and Examples of Applications 99

4.1 Structural Information by Increasing the Spectral Resolution in Multinuclear Solid-State NMR 99
4.1.1 Spectral Resolution for Nonquadrupolar Nuclei 100
4.1.2 Signal Assignments 102
4.1.3 Resolving Quadrupolar Nuclei 104
4.2 Probing Proximities of Nuclei and Measurements of Internuclear Distances in Diamagnetic Solids 106
4.3 ^{31}P Solid-State NMR 111
4.4 Solid-State NMR of Halogen Nuclei and Oxygen 116
4.5 Solid-State NMR of Oxide Materials: Nuclei ^{51}V, ^{93}Nb, and ^{181}Ta 123
4.6 Solid-State ^{13}C NMR: Wood and Wood Products 127
4.7 ^{2}H NMR in Solids at Deuterium Natural Abundance 129

4.8 Between Solution- and Solid-State NMR: Nanoparticles and Suspensions ... 131
References ... 137

Chapter 5 Dynamics in Diamagnetic Materials from Solid-State NMR: Methods, Measurements, and Analysis ... 139

5.1 Variable-Temperature Solid-State NMR Experiments and a Line-Shape Analysis ... 141
5.1.1 Dynamics in Polymeric Materials from a Line-Shape Analysis ... 143
5.2 Molecular Mobility from the Solid-State NMR Relaxation ... 145
5.2.1 Measurements of Relaxation Times and Errors of the Measurements ... 145
5.2.2 How Molecular Motions Affect Nuclear Relaxation ... 149
5.2.2.1 Correlation Time Distribution versus Single Correlation Time ... 151
5.2.3 Relaxation Approaches: Methodology and Examples ... 153
5.3 One- and Two-Dimensional Exchange Solid-State NMR Spectroscopy ... 156
5.4 Molecular Mobility from the Cross-Polarization NMR Experiments ... 158
5.5 Molecular Mobility from Magnetic Relaxation Dispersion Experiments ... 160
References ... 163

Chapter 6 Strategies in Solid-State Multinuclear NMR: Studies of Diamagnetic Porous Materials ... 165

6.1 Porosity of Materials from NMR ... 165
6.1.1 NMR Cryoporometry ... 168
6.1.2 NMR Relaxometry ... 171
6.1.3 NMR Behavior of Gases in Pore Spaces ... 173
6.2 Structural Features of Silica Lattice and Surface by Solid-State ^{1}H ^{29}Si, ^{27}Al, and ^{17}O NMR Spectra ... 176
6.2.1 Distribution of Diamagnetic Metal Ions from Solid-State NMR ... 181
6.3 Molecular Mobility in Diamagnetic Porous Materials ... 186
References ... 189

Chapter 7 Paramagnetic Effects in Solid-State NMR ... 191

7.1 Theoretical Aspects ... 193
7.1.1 Chemical Shifts and Magnetic Shielding Tensors in the Presence of Unpaired Electrons ... 193

7.2 Nuclear Relaxation in the Presence of Unpaired Electrons....197
7.2.1 Spin Diffusion....198
7.2.2 Nuclear Relaxation via Nucleus–Electron Dipolar and Contact Interactions....199
7.3 Practical Consequences from the Theory of Paramagnetic Effects....200
7.4 ^{1}H, ^{31}P, ^{13}C, ^{2}H, ^{29}Si, ^{7}Li, and ^{6}Li NMR Spectra of Paramagnetic Solids....205
7.5 ^{51}V and ^{55}Mn NMR Spectra of Paramagnetic Solids....210
7.5.1 Analyzing the Nature of Metal Ions by Their Direct Observation in Solid-State NMR Spectra....212
7.6 Special NMR Techniques for Observations of "Invisible" Target Nuclei....215
7.7 Relaxation Measurements and Relaxation Times in Paramagnetic Solids....217
7.7.1 Solid-State T_1 Measurements....224
7.7.2 Spin–Spin Relaxation Times and Anomalies in Solid-State T_2 Measurements....225
References....228

Chapter 8 Strategy in NMR Studies of Amorphous Porous Paramagnetic Materials....231

8.1 Intense Sideband Patterns in MAS NMR Spectra of Paramagnetic Amorphous Materials and Their Analysis....231
8.2 Direct Detection of the Nuclei Closest to Paramagnetic Ions in Porous Materials by the Hahn-Echo Mapping NMR Experiments....236
8.3 NMR Relaxation Approaches to Structure of Porous Amorphous Paramagnetic Silica-Based Materials: From Experiments to Models and Interpretations....240
8.3.1 Factors Affecting Relaxation Times in Porous Solids....241
8.3.2 Spin–Lattice Relaxation Times in Porous Silica-Based Materials and Concentrations of Paramagnetic Ions....242
8.3.3 ^{29}Si Spin–Lattice and Spin–Spin Relaxation in Static and Spinning Porous Amorphous Paramagnetic Silica-Based Materials....243
8.3.4 T_1 Criteria for Locations of Paramagnetic Ions: Relaxation of Isotropic Resonances....247
8.3.5 T_1 Criteria Based on Relaxation of Sideband Patterns in MAS Spectra of Porous Paramagnetic Materials....251
8.3.6 BMS Effects and NMR Relaxation of Sideband Patterns in MAS Spectra of Porous Paramagnetic Materials....254

8.4 Protocol for Quantification of "Invisible Nuclei" in MAS NMR ... 255
8.5 Concluding Remarks ... 256
References ... 257

Index ... 259

Preface

Nuclear magnetic resonance (NMR), discovered as a physical phenomenon by Bloch and Purcell in 1945, has rapidly become the theoretical basis for a powerful analytical method that is widely applied in different fields of modern fundamental and applied science as well as in medicine and industry. This is easily explained not only by the capability of NMR to solve a large number of practical tasks rapidly, which can be connected with structural descriptions of systems at the macro and/or molecular levels, but also by its capability to identify various types of dynamic behavior often responsible for the mechanical properties of materials.

The development of NMR techniques for solution and solid-state studies is impressive. The areas of NMR applications are constantly expanding from atomic and low-temperature physics (where they are used to monitor multiphoton transitions and probing distortions of magnetic fields in superfluids, respectively) to geology (where they are used to search the pathways of cation diffusion) to archeology (where natural destruction of wood is the focus of investigators).

It is impossible to overestimate the role of NMR in fundamental and applied chemistry where the objects of study are molecular systems varying from simple organic or inorganic compounds to complex biomolecules, such as proteins, and NMR data are collected by one-, two-, and three-dimensional experiments performed on different nuclei.

In 1995, the NMR community celebrated the 50th anniversary of the discovery of NMR in condensed phase emphasizing the strong impact of NMR on the development of different fields of science. While the first NMR experiments were carried out in the gaseous phase by Rabi and coworkers in 1937 to confirm experimentally the ideas of quantum mechanics, at the present time, NMR is an indispensable tool for chemists, biochemists, and biologists. It can be added that different aspects of NMR applications in solutions, liquids, and solids can now be found in the *Encyclopedia of Nuclear Magnetic Resonance* and/or in the *Encyclopedia of Applied Spectroscopy*, which is a remarkable feat by itself for the NMR community.

Generally, chemists apply regular NMR spectroscopy in solutions and/or liquids to characterize, for example, new or currently synthesized compounds and to control chemical reactions. However, even in this context, chemical manipulations can lead to systems that are insoluble in conventional solvents. On the other hand, the principal interests of chemists can be concentrated on the properties and structure of solid molecular systems and composites, which are the focus of materials science. Since the nature, properties, and complexity of new materials change constantly from crystalline to amorphous systems (or systems containing both the phases) and from diamagnetic to paramagnetic materials, solid-state NMR techniques and methodical approaches also change from single-pulse NMR at magic angle spinning (MAS) to various polarization transfers, multiquantum manipulations, and nuclear relaxation measurements performed in static and spinning samples. In fact, according to Dybowski, the literature on solid-state NMR increased by more than 3300 original articles as recently as 2005–2009. This ocean of NMR literature is too large to be studied by chemists, particularly young researchers or beginners in materials science.

Moreover, while in the past NMR spectroscopy and relaxation experiments preformed to collect frequency-dependent and time-dependent data, respectively, were two separate areas of research; at the present time, both the domains are equally important for chemists working with complex materials and composites. This is particularly so in the controlled design of new layered and/or porous molecular systems with three-dimensional porous networks, which constitute the most exciting direction of modern materials science. This field needs constant usage of methods capable of a structural analysis of *amorphous heterogeneous* solids and their dynamics. Solid-state NMR and NMR relaxation are ideal tools for solving such tasks; however, they require in-depth knowledge of both techniques. Therefore, this book can serve as a useful reference for researchers as it describes the strategy of solid-state NMR experiments, including theoretical and experimental details, and also considers important technical problems.

This book has been written for the reader who is not familiar with the applications of solid-state NMR. Therefore, it begins by covering the general physical principles of pulsed NMR, by including elements of the theory and practice in the registration of NMR signals, and by explaining different NMR equipment (Chapter 1). This chapter describes the main spectral characteristics, such as chemical shift and spin–spin and quadrupole coupling, and considers NMR sensitivity and its enhancement, including principles of two-dimensional NMR. Chapter 2 introduces some important classes of materials that are potential objects for NMR studies. It formulates key problems of structure and dynamics of these materials and demonstrates the different physical methods used for their resolution.

Chapter 3 describes the important features of solid-state NMR data collected in diamagnetic systems, from isotropic chemical shifts to three-dimensional magnetic shielding in NMR spectra, from nuclei with spins of half to quadrupolar nuclei, often creating problems in their registration. The elements of detection by cross-polarization and multiquantum experiments are also considered. The chapter concludes with a description of sideband manipulations in MAS NMR experiments and briefly touches upon zero-field experiments.

Chapter 4 considers the practice of solid-state NMR applied to diamagnetic systems. It focuses on different practical important questions, from increasing the spectral resolution to principles in the measurement of internuclear distances in solids. General data on solid-state NMR of ^{2}H, ^{31}P, ^{35,37}Cl, $^{79/81}$Br, ^{127}I, ^{17}O, ^{51}V, ^{93}Nb, and ^{181}Ta nuclei needed for materials scientists are also provided. The chapter concludes by considering the NMR of nanoparticles and suspensions.

Dynamic solid-state NMR is presented in Chapter 5. The chapter introduces the relaxation theory and relaxation time measurements as well as the principles of a line-shape analysis and an exchange NMR spectroscopy for probing molecular mobility in diamagnetic materials.

Chapter 6 describes the strategy of solid-state experiments in different nuclei performed on amorphous porous diamagnetic materials and covers specific NMR approaches to studying distributions of diamagnetic metal ions through silica-based systems, illustrated by typical examples. NMR porosity experiments are also presented in this chapter.

Chapter 7 is devoted to the theory of paramagnetic effects in solid-state NMR, taking into account chemical shifts and magnetic shielding tensors in the presence of unpaired electrons as a function of nuclear nature. The chapter introduces the theory of nuclear relaxation in the presence of paramagnetic centers in solids, demonstrating the concept of spin–diffusion and electron–nucleus interactions and formulating their relative effectiveness. The theoretical and practical aspects of NMR in metals, specifically, are presented in this chapter.

As a consequence of the theory of paramagnetic effects, Chapter 7 discusses the ^{1,2}H, ^{31}P, ^{13}C, ^{27}Al, ^{29}Si, and ^{6,7}Li NMR data collected for static and spinning paramagnetic molecular systems. It demonstrates the specific NMR technique developed for the spectral observation of "invisible" nuclei, which are located close to the paramagnetic centers. Attempts to directly observe the nuclei in paramagnetic centers are also considered. The chapter concludes with T_1, T_2 relaxation time measurements and presents some anomalies in the nuclear relaxation behavior of paramagnetic systems.

Chapter 8 discusses the strategy of solid-state NMR experiments performed in different nuclei for some amorphous silica-based paramagnetic materials. These systems are used as examples, revealing possible applications for other complex materials. This chapter provides an analysis of intense sideband patterns typically observed in MAS NMR spectra of paramagnetic systems, discusses the factors affecting relaxation times, and describes the Hahn echo mapping experiments for direct observation of target nuclei in such materials. Finally, it formulates solid-state NMR spectral and relaxation criteria for localization of paramagnetic metal ions in the matrix of materials.

This book has been written in simple language and consists of clear illustrations, numerous examples, and detailed bibliographies. We hope that it will be useful not only for young and experienced researchers in materials science, but also for university and college students.

Author

Professor Vladimir I. Bakhmutov has worked in different fields of chemical physics and is the author of more than 300 scientific publications, including topic reviews and books. Among them, two of his books, *Practical Nuclear Magnetic Resonance Relaxation for Chemists* (Wiley & Sons, New York, 2005) and *Ejemplos prácticos del uso de la RMN en la Química* (CINVESTAV, Mexico, 2006), could be of interest to NMR users.

1 General Principles of Pulsed NMR and NMR Techniques

As has been emphasized in the literature, NMR is capable of performing various tasks rapidly for different objects from homogeneous systems, gases, solutions, and liquids to heterogeneous and strongly amorphous solids. Hence, it is widely used in molecular physics, chemistry, biochemistry, biology, pharmacy, geophysics, materials science, and archaeology. Moreover, nowadays, NMR is the diagnostic method used in veterinary science and medicine particularly in clinical research of the human brain by magnetic resonance imaging (MRI) [1].

A large variety of techniques are applied in NMR spectrometry and MRI. Further, technology is constantly being evolved for studying increasingly complex objects. In other words, new scientific fields demand new NMR techniques, the development of which, in turn, affects these fields. For example, in the past, quick successes in radio frequency technology resulted in the appearance of the first commercial NMR spectrometers in 1953, which strongly impacted research in chemistry, biochemistry, and biology. Then, the synthesis and study of more complex molecular systems, such as biochemical molecular objects and/or complex molecular aggregates, required more powerful NMR spectrometers, with better spectral resolution and sensitivity, modifications in magnet technology for creation of stronger magnetic fields, and development of predominantly new methodological spectroscopic approaches. Thus, the modern commercial NMR spectrometers represent multiple-pulse devices, capable of experiments in solutions and the solid state. They can operate at very high magnetic fields up to 21.14 T and detect even the so-called rare nuclei. In addition, the development of computer techniques and increasing speed of computers, performing very fast Fourier transformation (FT), have led to routine use of the one- (ID), two- (2D), or three-dimensional (3D) NMR applied widely in solutions and the solid state.

This book is devoted to solid-state NMR. In spite of the appearance of the first commercial NMR spectrometers as early as the 1950s, solid-state NMR was not very popular among chemists due to relatively low spectral resolution of the spectrometers and their low sensitivity, which is quite critical for solids. Only in the 1980s did the expansion of solid-state NMR techniques become very rapid.

1.1 EARLY HISTORY OF NMR IN THE SOLID STATE

The history of NMR, very interesting and sometimes dramatic, has been reviewed by Andrew and Szczesniak [2]. They note that the first NMR experiments were successfully performed in gases (1937) to experimentally verify theoretical concepts in quantum mechanics by accurately measuring nuclear magnetic moments. However, an attempt by Gorter in 1936 to observe an NMR signal in the condensed phase (that could lead to a new analytical method) was unsuccessful. Gorter attempted to detect resonances of Li nuclei in crystalline LiF and protons in crystalline potassium alum. The second attempt was reported by the same research group 6 years later, where similar experiments were performed for ^{19}F nuclei in crystal solid KF. Again no signal was observed due to a singularly long relaxation time of ^{19}F nuclei in these very pure crystals. The important conclusion was that a long relaxation time can cause a saturation effect and resonance lines can become "invisible." In other words, behavior of target nuclei in NMR experiments is time dependent and expressed as infinitely long relaxation times, particularly in rigid solids free from paramagnetic impurities, and NMR signals are not detected. This important relationship between relaxation times and results observed in NMR experiments is better recognized by theoretical considerations of the relaxation phenomena in Chapter 5.

The first successful observation of NMR signals in the condensed phase was reported by Purcell and Bloch on probing protons in paraffin wax and water, due to their relatively short relaxation times. Finally, Bloembergen, Purcell, and Pound carried out careful investigations of nuclear relaxation and formulated the theory, now known as the BPP theory, which is valid for all states of matter: solids, liquids, and gases. The authors introduced the dipole–dipole mechanism explaining nuclear relaxation via molecular motions characterized quantitatively by correlation times, depending on the nature of objects. It is interesting that after this classic work, other treatments of relaxation processes led to only small changes in the relaxation theory.

One of the first solid objects was ice, showing the principal difference between NMR in solids and liquids: line-width of the ^{1}H resonance in liquid water is very small (significantly <1 Hz) while the resonance of ice is dramatically broadened up to ~10^5 Hz. This effect related to strong magnetic dipolar proton–proton interactions is the second factor that complicates observation of NMR signals in solids. It became obvious that detection of NMR signals in liquid water and ice requires different technical solutions.

It is significant that the first simple, noncommercial NMR spectrometer was created in 1946, capable of resonance observations in liquids and solids as well as measurement of relaxation times necessary for detailed understanding of the role of molecular motions in the NMR phenomenon. In 1947 extremely low temperature relaxation measurements were performed in solid H_2 at a temperature of 1 K to reconfirm the importance of molecular motions in the dipole–dipole relaxation process. In spite of these successes, the relaxation mechanism operating in such crystals as CaF_2 remained unclear. Only in 1949 did Bloembergen explore the nature of this mechanism via energy diffusion from nuclear spins to centers containing unpaired electrons, due to their dipolar coupling [3]. The cycle of these classic works was especially important because it became clear that the NMR spectrum in solids, namely, widths of the resonances and their shapes, are dictated by magnetic dipolar nuclear interactions, while relaxation of nuclei occurs due to a time dependence of these interactions.

In the context of practical applications of NMR, important for chemists, the aforementioned statement can be reformulated as the NMR spectra in solids contain information about the number of nuclei in structurally different molecules and/or groups and their mutual dispositions, while nuclear relaxation time measurements at different temperatures open the way to understanding the dynamic behavior of these groups and molecules from isotropic and anisotropic reorientations to phase transitions. These two unique features allow us to consider solid-state NMR as a powerful analytical method, including NMR techniques based on magic angle spinning, multiple pulse sequences, cross-polarization, double and triple resonance, nuclear Overhauser effect (NOE), and experiments in zero fields [2], which are considered in this book.

1.2 NUCLEI IN THE EXTERNAL MAGNETIC FIELD

Generally the structural chemistry of materials requires determination of the number of molecules in structural units and their symmetry in the lattice of materials. Solid-state NMR gives a solution for this general task. Moreover some NMR approaches, as an alternative of the x-ray diffraction method, can even provide accurate internuclear distances or bond angles, for example, angles Si–O–Si in glasses that are determined on the basis of relationships between the line dispositions in the NMR spectra and the structural parameters. All these points are directly related to definitions of chemical shifts, line shapes, line widths, and spectral resolution, as fundamental properties of nuclear behavior in the external magnetic field.

In spite of the great variety of NMR experiments and NMR techniques [4–8], the physical meaning of NMR can be reformulated in a simple form as follows: *nuclei are placed into the external magnetic field and excited by a radio frequency irradiation; then, the resulting absorbed energy is registered as NMR signals.* These signals form the NMR spectra recorded usually as plots in the axes "line intensity/resonance frequency."

Dispositions of resonance lines in the NMR spectra are characterized by chemical shifts generally measured in ppm (Figure 1.1). The fine splitting of the NMR signals due to quadrupole coupling or spin–spin coupling (Figure 1.2) is measured in Hz and depends on electronic environments of nuclei. Finally, integral intensities of the signals are proportional to the number of the nuclei, when the nuclei are excited directly and relaxed completely. That is why the NMR spectra are directly related to molecular structures.

It should be emphasized that the spin–spin coupling constants regularly used by chemists to elucidate molecular structures in solution are also important for the solid-state NMR spectra. It is obvious, however, that they should be well resolved (e.g., the *high resolution* can be reached in magic angle spinning [MAS] NMR spectra). Figure 1.1 shows clearly the presence of spin–spin coupling constant $^{1}J(^{13}C\text{-}^{199}Hg)$ in the $^{13}C\{^{1}H\}$ CP MAS NMR spectrum of the mercury complex in Scheme 1.1. This complex forms in the solid state a network typical of porous materials capable of gas absorption [9]. As the scheme shows, the ^{13}C resonance appearing at ~160 ppm

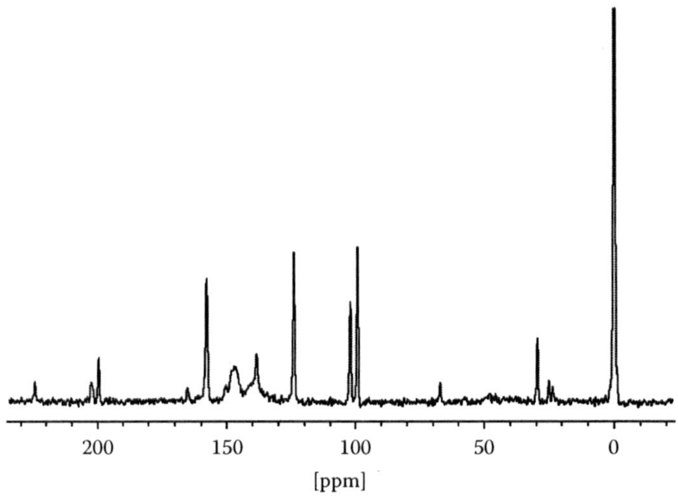

FIGURE 1.1 The solid-state ^{13}C CP MAS NMR spectrum of a porous material created by a network of the compounds in Scheme 1.1. The spectrum was recorded at a spinning rate of 10 kHz with a standard 4 mm NMR rotor. The mixing time for cross-polarization was 1500 μs at a 90° (^{1}H) pulse of 4 μs. ^{1}H Decoupling was provided by a tppm–pulse sequence operating at ^{1}H frequency. The relaxation delay was 5 s.

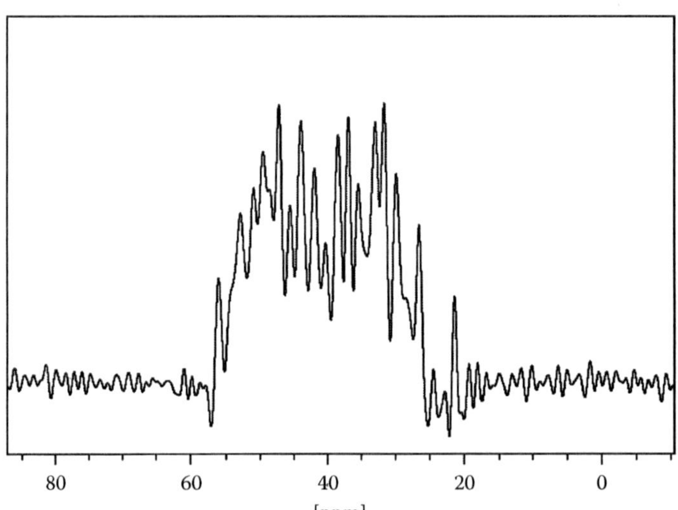

FIGURE 1.2 The solid-state single-pulse ^{19}F MAS NMR spectrum of compound $Fe(II)_5(bptz)_4(SbF_6)_{10}(CH_3CN)_{10}$, recorded at a spinning rate of 12 kHz with a standard 4 mm NMR rotor, a 60°-(^{19}F) pulse and relaxation delay of 6 s. The high-resolution spectrum was obtained due to the Gauss treatment provided by standard software to increase the spectral resolution. Therefore, the initial NMR data were collected with a relatively large memory size of 16k and at a large number of scans to reach a good signal-to-noise ratio.

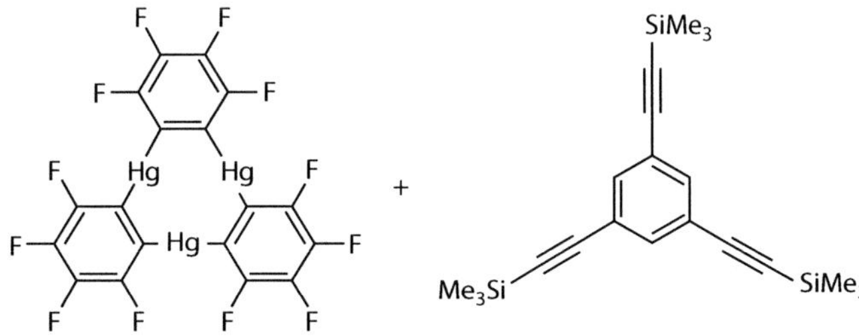

SCHEME 1.1 The molecular system based on the mercury complex forming in the soild state a network typical of porous materials capable of gas absorption.

is accompanied by two satellites separated by 1483 Hz. First, observations of the J constants simplify spectral assignments that play a major role in structural analysis. Second, for example, accurate measurements of these $^1J(^{13}C\text{-}^{199}Hg)$ magnitudes can potentially show the changing electronic density on carbon and mercury atoms in chemical modifying systems of this family.

The second intriguing example is represented in Figure 1.2, where the ^{19}F MAS NMR spectrum of solid compound $Fe(II)_5(bptz)_4(SbF_6)_{10}(CH_3CN)_{10}$ reported by Campos-Fernández et al. [10] exhibits a ^{19}F resonance consisting of 20 lines, at least. A portion of the lines corresponds to the $^1J(^{19}F\text{-}^{121}Sb)$ constant of 1940–1950 Hz, while the second part shows the $^1J(^{19}F\text{-}^{123}Sb)$ splitting of 1033 Hz. Since the isotopic shift ($^{19}F\text{-}^{121,123}Sb$) is not detectable in the solid state due to relatively low spectral resolution, the number of the observed lines proves the presence of three nonequivalent ions $[SbF_6]^-$ in this solid compound, one of which can be encapsulated in the bptz cycle in accordance with a single crystal x-ray structure. In addition, the absence of a $J(^{19}F\text{-}^{19}F)$ splitting demonstrates equivalency of ^{19}F nuclei. This equivalency, in turn, proves fast isotropic reorientations of $[SbF_6]^-$ ions.

The practical importance of chemical shift and spin–spin coupling is obvious from the earlier examples. Both of them originate from the behavior of nuclei placed into the external magnetic field. This behavior is generally described in terms of nuclear spins, represented via wave functions, ψ, similar to electron wave functions. These ψ functions depend on the spatial and spin coordinates and can be used to calculate energy of spin states via Schrödinger's equation, $\hat{H}\psi = E\psi$, where $\hat{H}$ is an operator (*Hamiltonian*) that acts on the wave function. Then, spin operator $\hat{I}$ with components $\hat{I}_X$, $\hat{I}_Y$, and $\hat{I}_Z$ corresponds to the spin coordinates of the wave functions [11]. This quantum mechanical formalism finally leads to energy levels whose appearance in the external magnetic field explains the phenomenon of NMR. A semiclassical description of the NMR effect is also possible, where a central role belongs to behavior of the net magnetization created in a sample that is placed in the external magnetic field. Both of these approaches are partially used in the following.

According to the aforementioned formalism (see, e.g., Abragam [4]), any nucleus is magnetically active when it possesses a *nonzero angular momentum*, P. It is significant that most of the elements in the periodic table have magnetically active isotopes with different natural magnetic abundance. The rule is simple: a nucleus has a zero spin when numbers of protons and neutrons are equal.

The angular momentum P is responsible for appearance of *nuclear magnetic moment*, μ, which is expressed via,

$$\mu = \gamma P \tag{1.1}$$

In turn, nuclear momentum μ is capable of responding to the presence of the external magnetic field, noted in the following as B_0. Coefficient γ in Equation 1.1, representing the nuclear magnetogyric ratio is one of the fundamental magnetic nuclear constants dependent on the nuclear nature. Table 1.1 lists some magnetic properties for the nuclei that are most popular as target nuclei for materials chemists. First, as the table shows, the γ values change in a very large diapason illustrating, thus, a very large diapason of resonance frequencies. Second, the γ constant has a sign that plays an important role in some NMR experiments; for example, at indirect detection of nuclei. Finally, it should be emphasized that sensitivity, as a very important parameter in NMR experiments, fundamentally depends on the nuclear nature, that is, on magnetogyric constants of nuclei and their natural abundance. For example, sensitivity related to ^{1}H nuclei reduces by factor of ~10^3, 10^4, and 10^6 in going to ^{117}Sn, ^{13}C, and ^{2}H nuclei, respectively. This circumstance sometimes creates significant problems in detection of NMR signals, particularly in the solid state where

TABLE 1.1
NMR Properties of Selected Nuclei

Nucleus	Spin I	Natural Abundance (%)	NMR Frequency, ν_0, in MHz at B_0 = 2.3488 T	γ (10^7 rad/Ts)
^{1}H	1/2	99.98	100	26.752
^{2}H	1	0.016	15.35	4.107
^{11}B	3/2	80.42	32.08	8.584
^{13}C	1/2	1.108	25.14	6.728
^{14}N	1	99.63	7.22	1.934
^{15}N	−1/2	0.366	10.14	−2.712
^{29}Si	−1/2	4.67	19.87	−5.319
^{19}F	1/2	100	94.08	25.18
^{31}P	1/2	100	40.48	10.84
^{17}O	−5/2	0.037	13.557	−3.628
^{93}Nb	9/2	100	24.549	6.567
^{51}V	7/2	99.75	26.3	7.036
^{59}Co	7/2	100	23.7	6.347
^{109}Ag	−1/2	48.2	4.65	−1.083
^{117}Sn	1/2	7.61	35.63	−9.578
^{119}Sn	1/2	8.58	37.29	−10.02
^{129}Xe	−1/2	26.4	27.8	−7.399
^{113}Cd	−1/2	12.2	22.2	−5.933
^{199}Hg	1/2	16.84	17.91	4.815
^{205}Tl	1/2	70.5	57.63	15.589

resonances can be strongly broadened due to strong dipolar internuclear interactions and relaxation times are elongated.

As follows from Table 1.1, ^{1}H and ^{19}F nuclei are most convenient in chemical investigations by the NMR method due to their natural abundance. On the other hand, these nuclei undergo the strongest dipolar interactions, complicating their detection with high resolution and requiring a special technique for collection of the data.

The P and μ magnitudes are *quantized* and projections of the angular nuclear moment, P_Z, on the OZ coordinate axis shown in Figure 1.3 can be written as follows:

$$P_Z = \hbar m_I \qquad \hbar = \frac{h}{2\pi} \tag{1.2}$$

where

$\hbar$ is Planck's constant ($h/2\pi = 1.05457266 \times 10^{-34}$ J s)

m_I is the magnetic quantum number dependent on the nuclear spin, I, taking values from I to –I as: I, I – 1, I – 2 … –I

In turn, according to the quantum-mechanical formalism, I is a multiple of 1/2. For example, at I = 1/2 the angular and magnetic moments can be expressed by

$$P_Z = \pm\left(\frac{1}{2}\right)\hbar \qquad \mu_Z = \pm\left(\frac{1}{2}\right)\gamma\hbar \tag{1.3}$$

Hence, the μ_Z (or P_Z) projections, which are imagined in the following as vectors, take only two "permitted" spatial orientations: *parallel* or *antiparallel* relative to the OZ axis. It should be emphasized that in the absence of the external magnetic field B_0 both of the orientations are obviously energetically equivalent and thus the nuclei occupy a single energy level.

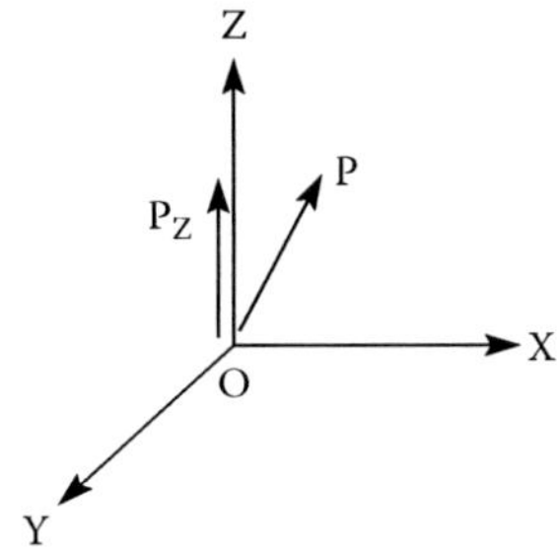

FIGURE 1.3 Projection of the nuclear angular moment P, P_Z, on the OZ axis. The external magnetic field, B_0, is generally applied along the OZ direction, while the radio frequency field affecting the P_Z projection is applied along the OX (or OY) coordinate.

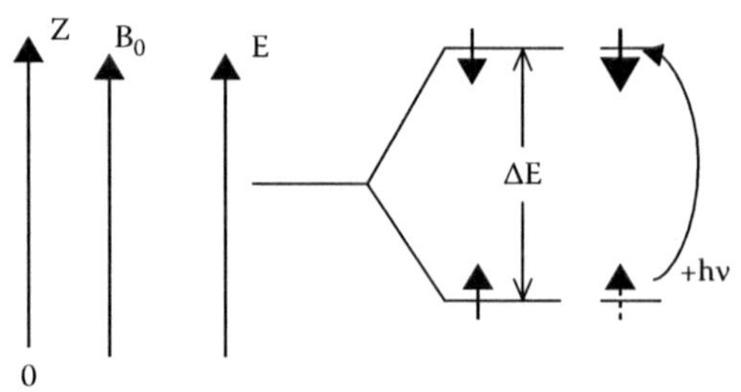

FIGURE 1.4 Zeeman's energy levels that appear for nuclear spins I = 1/2 placed into the external magnetic field, B_0, applied along the OZ coordinate.

When the strong magnetic field is applied, for example, along the OZ axis, the aforementioned parallel or antiparallel μ_Z (or P_Z) orientations become nonequivalent. Then, the initially degenerated energy level undergoes the so-called Zeeman's splitting shown in Figure 1.4 and this splitting ΔE expressed via Hamiltonian (1.4) will be proportional to the μ_Z magnitude and the strength of the applied magnetic field, B_0:

$$\hat{H} = \gamma\hbar\hat{I}B_0 \quad \text{and} \quad \Delta E = 2\mu_Z B_0 \tag{1.4}$$

Since a minimal energy difference can be formulated as *one quantum*,

$$\Delta E = h\nu \tag{1.5}$$

Equation 1.4 can be rewritten as

$$\begin{aligned} h\nu_0 &= 2\mu_Z B_0 = \gamma\hbar B_0 \\ \nu_0 &= \gamma B_0 \end{aligned} \tag{1.6}$$

formulating, thus, the fundamental conditions of the resonance phenomenon for a single uncoupled nucleus: the nucleus, characterized by the γ constant and the spin of 1/2, is placed into the external magnetic field, B_0, and undergoes a single-quantum transition (i.e., the m_I number changes from −1/2 to +1/2) between the Zeeman's energy levels from a low energy to a high energy level at radio frequency irradiation with frequency ν_0. Following from the equations, the ν_0 frequency, named Larmor's frequency, depends on the nature of the nuclei and the strength of the applied magnetic field, B_0.

For nuclei with spin ≥1, named quadrupolar nuclei, the number of energy levels obviously increases. For example, ^2H or ^{14}N nuclei with spin I = 1, produce three energy levels, corresponding to the μ_Z values of $+\gamma\hbar(1)$, 0 and $-\gamma\hbar(1)$. It should be noted that in solutions and liquids this circumstance does not complicate the situation because, on the one hand, the levels are energetically equidistant, and on the other hand the double quantum nuclear transitions (the m_I value changes from +1 to −1) are forbidden. For this reason, in spite of the presence of three energy levels in solutions, the nuclear transitions between them are registered as a single NMR resonance.

In contrast, a larger number of the energy levels for quadrupolar nuclei strongly affect the shape of NMR signals in the solid state. Here besides the strongly intense transitions from −1/2 to +1/2, other transitions can be also excited and detected. As is seen in the following, such an excitation can lead to very wide sideband patterns in the MAS NMR spectra of quadrupolar nuclei.

1.3 EFFECTS OF RADIO FREQUENCY PULSES

As the previous section shows, an NMR signal can be detected when the frequency of irradiation approaches the Larmor frequency ν_0 due to absorption of energy ΔE. Technically the continuous wave (CW) irradiation methods have been used in the earliest NMR experiments to observe *change in macroscopic magnetization* that is created in a sample in the presence of the B_0. In these terms, the sample is considered as an ensemble of nuclei that populate the Zeeman's energy levels proportionally to the factor

$$\exp\left(-\frac{\Delta E}{kT}\right) \tag{1.7}$$

where

T is the temperature
k is Boltzmann's constant (measured as 1.380658×10^{-23} J/K)

Under these conditions, the unequal populations of the energy levels in the external magnetic field produce the equilibrium macroscopic magnetization that undergoes Larmor's precession, as shown in Figure 1.5. The frequency of this precession is ν_0 or $\omega_0 = 2\pi\nu_0$ and the CW ν_0 irradiation reorients the magnetization vector that is directly registered as an NMR spectrum.

Besides the CW irradiation the resonance can be reached with the help of radio frequency pulses (RFPs). By definition, a powerful radio frequency (RF) pulse at acting frequency ω_{rf} can simultaneously excite a very large frequency band resulting

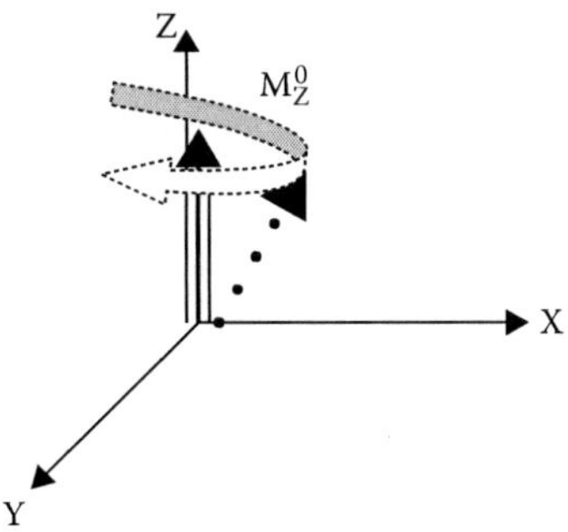

FIGURE 1.5 The macroscopic magnetization M_Z^0 is created in a sample in the presence of the external magnetic field B_0 oriented along the OZ coordinate. The interaction between the M_Z^0 and the B_0 leads to the Larmor precession with frequency ν_0 or $\omega_0 = 2\pi\nu_0$ as an equilibrium state of a nuclear system.

in appearance of a *time-dependent* oscillating field. This field contributes to the external magnetic field B_0 to produce a total field that can be expressed as

$$B_{total} = iB_1 \cos(\omega_{rf} t) + kB_0 \quad (1.8)$$

Here, the unit vector i is situated along the OX direction and the unit vector k is along the OZ direction. Under these conditions, the Hamiltonian can be transformed to the form

$$\hat{H} = -\gamma\hbar(\hat{I}_Z B_0 + \hat{I}_X B_1 \cos(\omega_{rf} t)) \quad (1.9)$$

Now, the expression contains the rotating component, and the wave function, ψ', describing spin states in the new rotating coordinate system takes a form (1.10)

$$\psi' = \exp(-i\psi_{rf} t \hat{I}_Z) \times \psi \quad (1.10)$$

where the first term represents the rotation operator. The accurate quantum-mechanical description of the spin system and the corresponding solution of Schrödinger's equation are not simple and they can be found, for example, in the literature [4,5]. The aforementioned equation is just to show that description of spin states under action of a RF pulse requires introduction of the new rotating coordinate system. Broadly, the nuclear behavior can be easy imagined by a vector model shown in Figure 1.6.

Here magnetization vector M, orientated at an equilibrium state along the OZ direction in a rotating coordinate system, undergoes reorientations by the action of RF pulses at their different durations. It should be emphasized that durations and amplitudes of RF pulses should be sufficient to rotate the magnetization vector, for example, by angle α. According to the Ernst rule [12], the pulse angle, α, is defined as

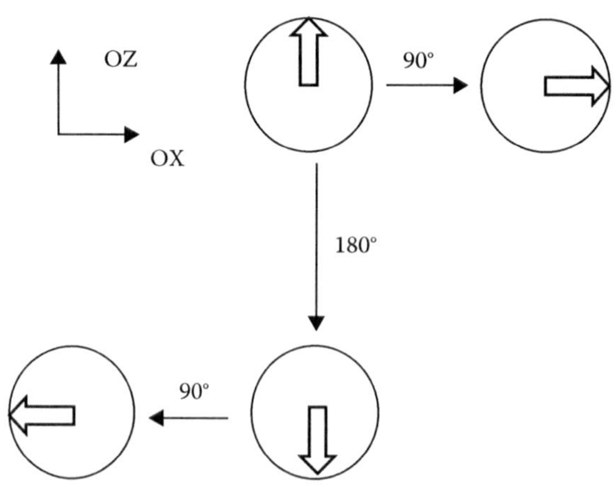

FIGURE 1.6 A simple vector representation, visualizing how the magnetization vector changes the orientation in the rotating frame by the action of RFP of 90° and 180° or their combination.

$$\alpha = \gamma B_1 t_P \tag{1.11}$$

where

B_1 is the power of the pulse
t_P is its duration

It should be noted that the RFP can be nonselective to excite large frequency ranges or selective to act in a limited frequency diapason. In addition, the shapes of the pulses can also be different, affecting the result in some advanced NMR experiments. However, this aspect is not the focus of this book and can be found in original papers or in Duer [11].

Figure 1.6 shows that after an action of a 90° pulse the new M orientation will be along the positive OX axis or along the negative OZ axis after a 180° RFP. Here, the M orientation along the positive OX axis will correspond to a *maximal* intensity of an NMR signal because the registering radio frequency coil is situated along this axis. In contrast, due to a 180° RF pulse the NMR signal will not be observed for the same reason. Finally the signal intensity will be visibly negative after a combination of 180° and 90° pulses when the time between them is not sufficient for nuclear relaxation. Generally speaking, in spite of the simplicity of these representations, they are necessary and sufficient to understand the main principles of solid-state NMR experiments and some problems that can appear in their performance.

The simplest NMR experiment with direct excitation of nuclei consists of several time sections represented in Figure 1.7: (1) an initial time delay, RD, (or the so-called relaxation delay or cycle delay), (2) action of a RFP that excites target nuclei, and (3) collection of NMR data as free induction decays (FID) during the acquisition time (AT). Sections (1)–(3) can be obviously repeated many times to accumulate the NMR data. The latter is necessary to obtain NMR spectra with better signal-to-noise

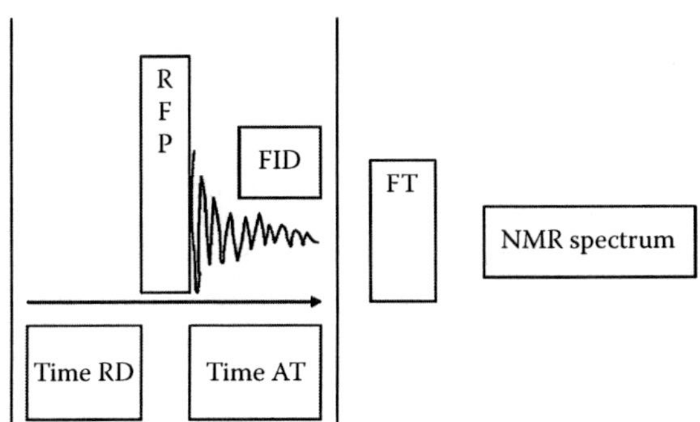

FIGURE 1.7 Schematic presentation of the simple 1D NMR experiment, containing three time sections. Time delay, RD, is necessary for full relaxation to recover the initial equilibrium state. Time RFP corresponds to the duration of the RFP. Time AT is needed to collect the NMR data. In practice, typical durations of RFP, relaxation delay RD, and AT are ~5–10 μs, 1–4 s, and 1–3 s, respectively.

ratios particularly important for rare nuclei (nuclei are named rare when their natural abundance is low). As the figure shows, the collected resulting NMR data are *time dependent* and therefore they can be expressed as function f(t). In contrast, the NMR spectra represent the *frequency-dependent* data, expressed as the F(ν) function. In other words, the FIDs require the mathematical (computer) procedure, FT,

$$F(\nu) = \int f(t)\exp(-i2\pi\nu t)dt$$
$$f(t) = \int F(\nu)\exp(+i2\pi\nu t)d\nu \qquad (1.12)$$

converting the time domain into the frequency domain and vice versa. Finally these basic experiments do not require advanced techniques but NMR spectrometers should be equipped with computers capable of fast FT.

1.4 FROM MACROSCOPIC MAGNETIZATION TO NUCLEAR RELAXATION AND SHAPES OF NMR SIGNALS

Under conditions of equilibrium, the macroscopic magnetization undergoing Larmor's precession corresponds to the maximal M_Z^0 value in the rotating framework, while the M_X^0 and M_Y^0 components are equal to zero. Irradiation of the sample leads to excitation of the spin system and the M vector reorients: M_Z^0 reduces to M_Z and the M_X and M_Y components become nonzero. After (and during) passing the resonance, the excited nuclear spins relax (when they are in principle capable of an energy exchange with their environments) to reach the initial equilibrium state.

Since the magnetization has longitudinal and transverse components, there are two principally different relaxation mechanisms. The first mechanism, named spin–lattice relaxation, represents the energetic exchange between nuclear spins and lattice. The latter can be defined as a continuum of nuclear and/or electron magnetic moments of any sort. The second mechanism, spin–spin relaxation, occurs when two (or more) spins undergo so-called flip-flop reorientations. It is significant that these reorientations thus shorten the lifetimes of the excited spins while the total energy of the system does not change. For this reason, such flips are often called energy-conserving spin transitions. In the aforementioned terms, the relaxation of macroscopic magnetization is given by Bloch's equations

$$\frac{dM_Z}{dt} = -\frac{(M_Z - M_Z^0)}{T_1}$$
$$\frac{dM_{X,Y}}{dt} = -\frac{M_{X,Y}}{T_2} \qquad (1.13)$$

The equations show the recovery of the Z (longitudinal) component with a time constant T_1 and the Y and X (transverse) components with a time constant T_2.

For NMR signals, the T_1 time is responsible for intensity of the NMR line while the T_2 time controls its width. In other words, even for uncoupled spins it is seen that the relaxation time constants directly affect detection of NMR signals: infinitely long T_1 times due to the absence of an energy exchange between excited spins and the lattice do not allow registration of NMR signals, and infinitely short T_2 times due to intense spin flips prevent observation of the signals due to their strong broadenings.

Since NMR experiments use RF pulses acting at the carrier frequency, ω, they help to observe all the time-evolving signals in the rotating frame. They can be characterized by the amplitudes C expressed, in turn, as magnitudes with their real and imaginary components:

$$C\exp(i\omega t) = C\cos(\omega t) + iC\sin(i\omega t) \tag{1.14}$$

Then, FT of term C exp(iωt) leads to the complex frequency spectrum. Here the real part will be associated with an absorption line shape (or line shape of observed NMR signals) and the imaginary part will represent a dispersive line shape.

According to Abragam's formalism [4], shape of an NMR signal can take the form of a Gauss curve

$$f(\omega) \sim (\Delta\nu(2\pi)^{1/2})^{-1}\exp\left(-\frac{(\omega-\omega_0)^2}{2(\Delta\nu)^2}\right) \tag{1.15}$$

or the Lorentz curve

$$f(\omega) = \left(\frac{\Delta\nu}{\pi}\right)\left(\frac{1}{(\Delta\nu)^2 + (\omega-\omega_0)^2}\right) \tag{1.16}$$

as shown schematically in Figure 1.8. For isolated nuclear spins line-width determined at half-height is calculated through spin–spin (T_2) times as $\Delta\nu$ (Hz) = $1/\pi T_2$ (s) or $\sim 1/T_2$ for the Lorentz and Gauss shapes, respectively.

Generally, if an investigated system is characterized by fast high-amplitude isotropic molecular motions typical of liquids, solutions, or plastic crystals or even solids with high-temperature phases, then researchers can expect Lorentz's resonance shapes. In contrast, the Gauss shapes are more probable in rigid solids. It should be noted that power spectra do not follow this rule. Such spectra can be built on the basis of both the real and imaginary parts using standard software. This is particularly useful to determine, for example, line dispositions in the spectra when the regular NMR spectra show poor signal-to-noise ratios and/or require nontrivial phase and baseline corrections.

The knowledge of the line shapes is essential for researchers not only from the theoretical point of view. First, if a resonance line obtained for an NMR standard exhibits an unusual shape, this is certainly the result of poor shimming coil regulations or inaccurate magic angle adjustments needed for the high-resolution solid-state NMR spectra. Second, as Figure 1.8 shows, the wings for the Lorentz curves

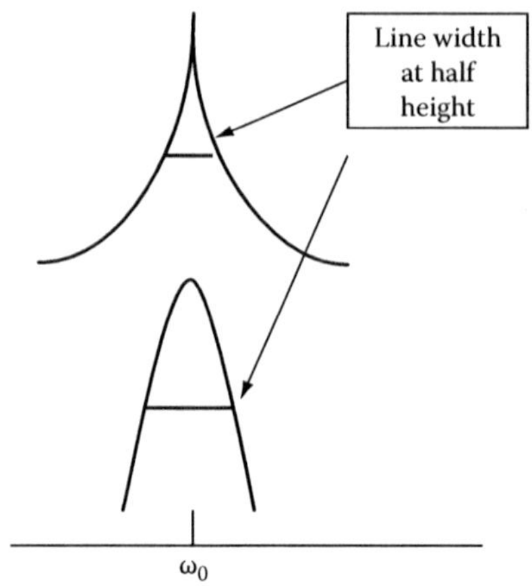

FIGURE 1.8 The Lorenz (top) and Gauss (bottom) curves are shown schematically. Generally shapes of NMR signals are described by these curves and characterized via line widths measured at half-heights in rad or Hz.

are more stretched than those of the Gauss curves. In other words, two Lorentz lines could be overlapped rather than two Gauss curves at the same chemical shift difference. The latter should be taken into account for their integration. The knowledge of line shapes is particularly important in solids at simulations of NMR signals and/or at their deconvolutions. In the absence of this knowledge, the deconvolution procedures are often performed by mixing the Gauss and Lorentz curves to reproduce the natural line shapes.

1.4.1 Common Representations of Molecular Motions and Their Influence on NMR

As we have noted in previous sections, nuclear relaxation plays a major role in observation of NMR signals. By definition, an energetic exchange between excited nuclear spins and their environment can lead to effective spin–lattice relaxation, if the lattice creates magnetic fields *fluctuating* at frequencies close to Larmor's resonance frequency, ν_0 (or ω_0) [1]. Such fields are created in the lattice by nuclear and/or electron magnetic moments physically located in molecules and they can fluctuate if the molecules undergo thermal motions.

Figure 1.9 illustrates the situation where magnetic dipoles of nuclei A and B, oriented along the direction of the applied external magnetic filed, B_0, interact due to dipole–dipole A–B coupling. The A–B interaction depends on the internuclear distance, r(A–B), and spatial orientation of the dipolar vector, r(A–B), relatively to the OZ axis. Then, the strength of a magnetic field created by the A nucleus at the location of the B nucleus also depends on the internuclear distance and the r(A–B) orientation. Following a molecular motion leading to reorientation of the r(A–B) dipolar vector, the magnetic field on the B nucleus will change. However, it is important to understand that any molecular motion, for example, a free rotation around the axis lying along the r(A–B) vector in Figure 1.10, does

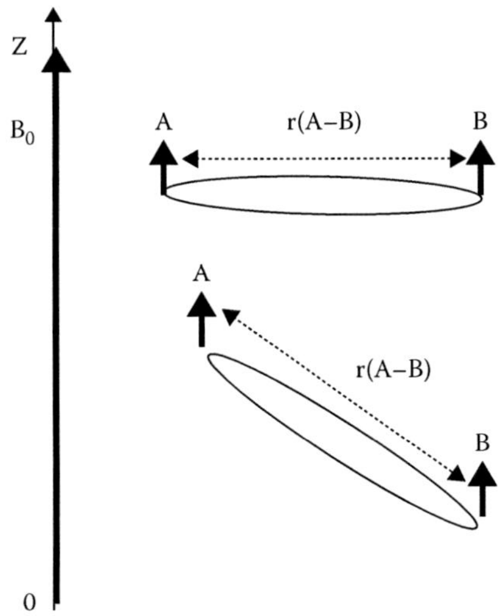

FIGURE 1.9 The representation of a molecular reorientation from top to bottom and dipolar coupling between spins of nuclei A and B. The nuclei are located in a molecule placed in the external magnetic field B_0. This molecular motion causes the reorientation of the dipolar vector with respect to the OZ direction of the external magnetic field.

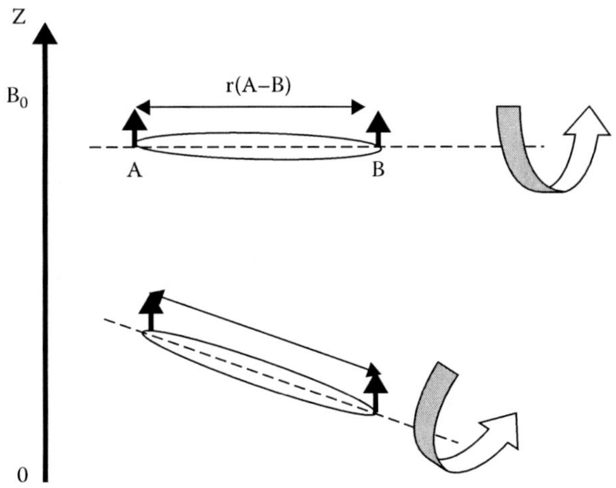

FIGURE 1.10 A free rotation around the A–B axis does not result in a space orientation of the dipolar A–B vector and therefore it does not affect the dipolar nuclear relaxation. At the same time, nuclear relaxation occurs due to a molecular jump from top to bottom.

not change its spatial orientation and hence it does not contribute to nuclear dipole–dipole relaxation.

Obviously the effects of molecular motions will be stronger when their frequencies (or their reverse correlation times) are closer to Larmor's frequency. The correlation times, τ_C, and the activation energies, ΔE_a, quantitatively characterize the molecular motions.

By analogy with chemical kinetics, magnitude τ_C can be formulated as the time required by a molecule to be reoriented (tumbling time) and the activation energy is defined in the relationship

$$\tau_C = \tau_0 \exp\left(\frac{E_a}{RT}\right) \tag{1.17}$$

where the sense of magnitude $(\tau_0)^{-1}$ can be formulated as a frequency of attempts to overcome the energy barrier. However, it should be noted that a more accurate definition of the τ_C time is connected with the autocorrelation function in the theory of nuclear relaxation, where the correlation time τ_C is taken as the average time needed for the molecule to progress through 1 rad [4].

Molecular motions can be high-amplitude or low-amplitude and isotropic or anisotropic, producing quite different effects on NMR spectra, line shapes, and relaxation times. By definition, if a *single* value of the correlation time, τ_C, at a given temperature (and a *single* value of the activation energy, E_a, respectively) describes a molecular motion, this motion will be *isotropic*. It is obvious that, for example, fully symmetrical rigid spheres should undergo isotropic rotational reorientations. At the same time, even a symmetrical ellipsoid characterized by two directions of the spatial extension will already have two different inertia moments leading to rotational reorientations with two correlation times, $\tau_C(1)$, $\tau_C(2)$, and two activation energies, respectively. The behavior of the ellipsoid is a simple illustration of *anisotropic* molecular motions that often occur in chemistry.

Molecules in liquids and solutions undergo fast thermal motions: rotational reorientations, translational motions or their combinations, and various intramolecular motions such as rotations around the simple chemical bonds, the polarized double bond, conformational motions, etc. The correlation times τ_C of these motions are generally between 10^{-11} and 10^{-9} s at the τ_0 values of $\sim 10^{-13}$–10^{-14} s. These magnitudes correspond to relatively low energy barriers, for example, between 0.8 and 5 kcal/mol for small nonaggregated molecules. In the spectroscopic sense, such correlation times result in line narrowing (see the following) and the NMR signals take Lorentz forms.

Solids are not absolutely rigid and often show various motions of whole molecules or separated groups within molecules or aggregates. It is obvious however that these motions, dependent on the nature and structures of molecular systems, are much slower. Generally in solids at a temperature between −150°C and +250°C the correlation times τ_C take values between 10^{-4} and 10^{-6} s and activation energies increase up to 18–25 kcal/mol or more. Such correlation times of molecular motions cause a slow T_1 relaxation (10, 20, 100 s or minutes). From Figure 1.7, this relaxation will strongly elongate the duration of NMR experiments necessary for collection of NMR data with a good signal-to-noise ratio. In fact, it is well known that the single pulse NMR experiments (Figure 1.7) should be performed at relaxation delays RD that must be longer than the T_1 times by a factor of 5 providing full nuclear relaxation, if the RF pulses applied in the pulse sequence were of 90°. At a small sensitivity, these experiments will require large numbers of scans to reach good signal-to-noise ratios in the resulting NMR spectra. The Ernst rule [12] optimizes the acting angle of RF pulses and relaxation delays to reach the best sensitivity in the NMR experiments: at the total relaxation delay of 1 s, the angle should be of 8°, 25°, 33°, 68°, 86°, and 90° if the T_1 time is 100, 10, 4, 1, 0.4, and 0.1 s, respectively.

1.4.2 Contribution of Molecular Mobility to Spin–Spin and Spin–Lattice Nuclear Relaxation

The mechanisms of nuclear relaxation are different. Nuclei can relax via nuclear dipole–dipole, quadrupole, spin rotation, scalar, and chemical shift anisotropy interactions or by spin diffusion. Some of these processes important for solids are discussed later. This section considers effects of molecular mobility on spin–spin and spin–lattice relaxation times to show the difference between NMR in solutions and the solid state.

Independently of mechanisms and the nature of condensed phase, the relaxation rates of nuclei, $1/T_1$ and $1/T_2$, can be generally written as

$$\frac{1}{T_{1,2}} = C\,J(\omega_0, \tau_C) \tag{1.18}$$

where

C is the force constant (dipolar or quadrupolar coupling, etc.)
$J(\omega_0, \tau_C)$ is a function of spectral density

This function shows how a combination of resonance frequency ω_0 and molecular motion correlation time τ_C weakens the influence of C. According to the BPP theory, functions $J(\omega_0, \tau_C)$ corresponding to the dipolar spin–lattice and spin–spin relaxation are

$$J(\omega_0, \tau_C)(T_1) = \left\{\frac{\tau}{(1+\omega_0^2\tau^2)} + \frac{4\tau}{(1+4\omega_0^2\tau^2)}\right\} \tag{1.19}$$

$$J(\omega_0, \tau_C)(T_2) = \left\{3\tau + \frac{5\tau}{(1+\omega_0^2\tau^2)} + \frac{2\tau}{(1+4\omega_0^2\tau^2)}\right\} \tag{1.20}$$

The expressions show clearly the quantitative difference between T_1 and T_2 relaxation times through the different influence of the τ_C. Because the molecular motion correlation times τ_C depend on the temperature as $\tau_C = \tau_0 \exp(E_a/RT)$, the T_1 plots in semilogarithmic coordinates in Figure 1.11 will be symmetrical, V-shaped, and go through minima at

$$\tau_C = \frac{0.62}{\omega_0} \tag{1.21}$$

when the temperature decreases. In contrast, on cooling the T_2 time constantly reduces particularly strongly after the T_1 minimum. Thus the pattern shows that in spite of T_1 elongations expected after the T_1 minima, line widths of the signals

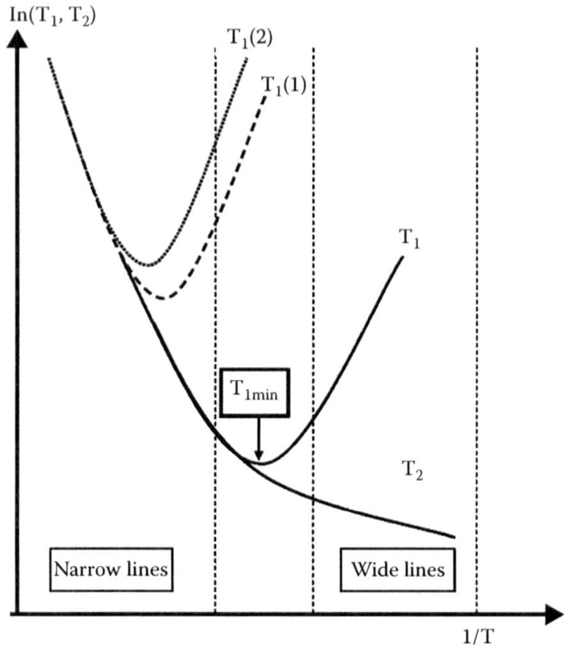

FIGURE 1.11 The typical T_1, T_2 temperature dependences. The dashed T_1 curves show the influence of strength of the external magnetic field increasing from curve T_1 to $T_1(1)$ and $T_1(2)$.

will strongly increase due to the T_2 shortening. This effect will be particularly significant in solids. In fact, molecular reorientations in liquids and solutions are fast and T_1 minima for relatively small molecules can generally be reached only at the lowest temperatures. In solids, however, due to slow molecular motions, the T_1 minima should be observed at high temperature zones, that is, at temperatures that are remarkably higher than room temperature. For this reason, the resonances in solids will be *strongly broadened* due to short T_2 times. The results of relaxation measurements performed in the silicate octadecasil, containing ions F^- [13], clearly illustrate the difference between T_1 and T_2 times in solids: $T_1(^{19}F) = 2.6$ s versus $T_2(^{19}F) = 3.5$ ms or $T_1(^{29}Si) = 22$ s versus $T_2(^{29}Si) = 18$ ms.

Following the BPP theory, it is easy to show that the relative dispositions of the T_1 minima on the curves in Figure 1.11 will depend on the strength of the external magnetic fields: the T_1 minima will shift toward high temperature zones on increasing the magnetic field strengths. In some sense this illustrates the fact that applications of NMR spectrometers working at highest magnetic fields for solids are not always productive. On the other hand, as the curves in Figure 1.11 show, the wide-line zone (where T_1 and T_2 times are not equal) provides determination of correlation times (τ_C) for molecular (or group) motions by the T_1 and T_2 measurements via T_1/T_2 ratios. This is typical of solids.

To conclude the discussion of molecular mobility, it should be emphasized that some molecular (or group) reorientations can be very fast even in solids.

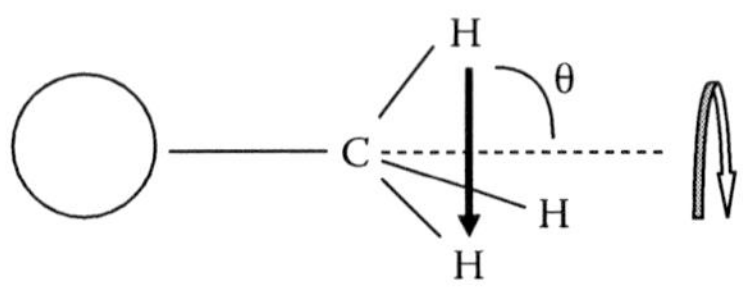

FIGURE 1.12 The CH_3 group spinning around the axis perpendicular to the dipolar vector H–H ($\theta = 90°$). Independent of rotational mechanism, the motion causes space reorientations of the vector affecting, therefore, relaxation times.

Generally, such motions have an intramolecular character as in the case of intramolecular rotations around single bonds (or partially double bonds), which have been the subject of numerous solid-state magnetic resonance studies. For example, the high-temperature tetragonal phase of solid BD_3ND_3 shows above 225 K fast reorientations for the BD_3 groups with the low activation energy of 1.4 kcal/mol at the τ_0 parameter of 1.1×10^{-13} s and for the ND_3 groups with $E_a = 1.7$ kcal/mol at $\tau_0 = 4.4 \times 10^{-14}$ s. In the low-temperature orthorhombic phase, these motions require higher energies: $E_a = 3.2$ and 6.2 kcal/mol, remaining, nevertheless, still fast [14].

Figure 1.12 demonstrates this type of rotation motions for methyl groups. These groups in a molecular crystal can reorient either by thermally activated jump rotation over the low energy barrier or even by tunneling through the barrier [15]. Solid toluene d_5, which was diluted in toluene d_8 to remove dipole interactions of methyl protons with protons outside the CH_3 group, shows the two types of activation process: random jump reorientation ($\Delta E_a = 1.3$ kJ/mol) over the barrier and torsional excitation ($\Delta E_a = 0.55$ kJ/mol) connected with tunneling.

However, in spite of the rapidity of these intramolecular rotations, they cannot result in the conditions of line narrowing, as Figure 1.11 shows, because of their anisotropic character relative to the direction of the external magnetic field (see the proton–proton vector).

When correlation times τ_C of molecular (or group) motions reach magnitudes of $\sim 10^{-4}$ s and slightly longer, they do not affect the T_1, T_2 times, but shapes of NMR signals in solids can become temperature dependent, which is often observed in solutions. These effects of the temperature evolution, which are discussed in Chapter 5 for different nuclei, are illustrated by an experiment performed for solid complexes $[M_2(\eta\text{-HC}^1C^2C^3Me_2)(CO)_4(\eta C_5H_5)_2]^+ [BF_4]^-$ (M = Mo, W) in Figure 1.13 [16]. The ^{13}C CP MAS NMR spectrum exhibits two $^{13}CH_3$ signals with chemical shifts of 36.6 and 27.2 ppm at the temperature of 223 K.

However, on heating, the resonances broaden strongly to show at 343 K only a single line characterized by an averaged chemical shift of 32.3 ppm. These spectra unambiguously reveal a fast chemical exchange (generally typical of solutions), which corresponds to a restricted rotation about the $C–C^+$ bond on the NMR time scale. The rotational free energy $\Delta G^{\neq}$ has been roughly estimated as 13–14 kcal/mol, again typical of solutions. Finally it should be noted that among the various intramolecular motions shown in Figure 1.13, only the rotation around $C–C^+$ bond is detected in the solid state.

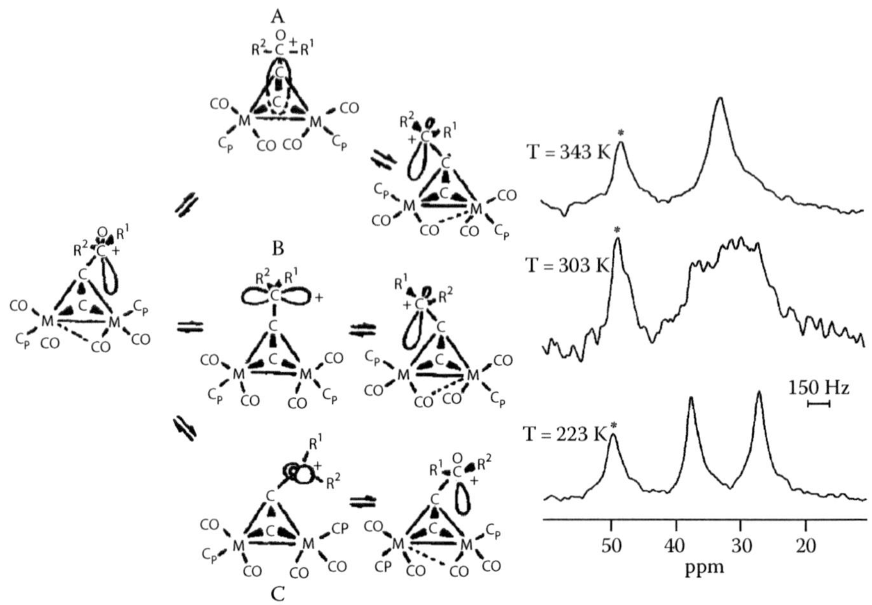

FIGURE 1.13 The temperature evolution of the $^{13}CH_3$ resonances in solid complex $[M_2(\eta\text{-}HC^1C^2C^3Me_2)(CO)_4(\eta C_5H_5)_2]^+ [BF_4]^-$ (M = Mo, W). The resonance noted with asterisk is the sideband. The CP $^{13}C\{^1H\}$ MAS NMR spectra were recorded at a mixing time of 1500 μs and at a spinning rate of 5 kHz. (Reproduced with permission from Galakhov, M.V. et al., *J. Magn. Reson. Chem.* 29, 506, 1991.)

1.5 PARAMETERS CHARACTERIZING NUCLEI IN THE EXTERNAL MAGNETIC FIELD

The nuclear spins, being the objects in NMR studies of different molecular systems as well as materials, are not isolated. Their closest environments give rise to the phenomena of chemical shift, spin–spin coupling, and quadrupolar coupling. First, these are main parameters in NMR; second, they affect the NMR spectra and shapes of the resonances; and third, they are related to chemical structures of investigated systems.

1.5.1 Chemical Shift

When electrons surrounding nuclei circulate in the external magnetic field, B_0, they create a new magnetic field. This new local magnetic field, B_{loc}, will affect the nuclei and shift a resonance frequency ν_0 characterizing isolated nuclei, leading to isotropic magnetic shielding constant σ

$$B_{loc} = (1 - \sigma)B_0 \tag{1.22}$$

or isotropic chemical shift δ

$$\delta = -\sigma \tag{1.23}$$

In practice, chemical shifts δ are measured in ppm and generally refer to internal or external (rather typical of solids) standards via

$$\delta = \frac{\{\nu(\text{sample}) - \nu(\text{reference})\}10^6}{\nu(\text{reference})} \tag{1.24}$$

A number of compounds can be used as standards, for example, TMS in the case of ^{1}H and ^{13}C nuclei, a solution of H_3PO_4 for ^{31}P nuclei, a solution of CF_3COOH for ^{19}F nuclei, etc. Expression (1.24) shows that the isotropic chemical shift does not depend on the external magnetic field strength.

Following Abragam [4], the total shielding constant σ consists of three terms: diamagnetic term σ^{local}(dia), paramagnetic term σ^{local}(para), and magnetic anisotropy term σ*. The first diamagnetic term originates mainly from the circulation of unperturbed spherical electrons (s) with mass m_e and can be expressed in a simple form by Lamb's equation,

$$\sigma^{local}(\text{dia}) = \left(\frac{\mu_0 e^2}{8\pi m_e}\right)\int r\rho_e dr \tag{1.25}$$

where

μ_0 is the permeability constant
ρ_e is the charge density
r is the electron–nucleus distance

In other words, a larger electronic density produces a larger diamagnetic chemical shift and vice versa. The second (paramagnetic) term, recognized by McConnel and coworkers [17], is caused by perturbed nonspherical electrons, that is, p-, d- and f-electrons. This term can be approximately described in a simple form via the average energy of the electron excitation, ΔU, as

$$\sigma^{local}(\text{para}) = -\left(\frac{\mu_0 e^2 \hbar^2}{6\pi m^2 \Delta U}\right)\langle r^{-3}\rangle P_U \tag{1.26}$$

Function P_U is named the p-electron "imbalance," which is expressed via charge densities and bond orders.

The last relatively small term σ*, comes from electrons situated at neighboring groups, for example, bonds C–C, C=C or C≡C, typical of organic systems. Such groups create magnetically anisotropic fields and therefore the observed shifts will depend on orientations of target nuclei relative to these bonds. Finally, combinations of these terms lead to variations in chemical shifts δ, which are dependent on the nature of nuclei. For example, chemical shifts of ^1H nuclei are generally produced by diamagnetic terms. Therefore, the ^1H chemical shift range is relatively narrow: from +20 to −2 ppm for regular diamagnetic organic or inorganic molecules. At the same time the range increases to +10 to −50 ppm for organometallic molecular systems. It should be emphasized that this relatively narrow δ diapason complicates applications of ^1H NMR in solids. In fact, due to strong dipolar proton–proton coupling, a δ analysis is difficult, requiring specific methods to improve the spectral resolution.

The chemical shift ranges for other nuclei produced by significant paramagnetic terms are remarkably larger: from −250 to +400 ppm for ^{19}F nuclei, from −125 to +500 ppm for ^{31}P nuclei, from −500 to +850 ppm for ^{15}N nuclei, or from −1,370 to +12,000 ppm for ^{195}Pt nuclei. It is clear that such target nuclei are preferable in the context of solid-state NMR studies because they potentially provide more detailed and accurate information on the presence of various structural units.

In principle, the quantum mechanic theory of NMR and quantum-chemical calculations can be used to obtain and rationalize chemical shift values observed as a function of the chemical structure. However, these calculations are not simple (some of them are considered in Chapters 3 through 5). Therefore in practice structural studies of materials are rather based on numerous experimentally established spectrum/structure relationships showing, for example, the δ as a function of atomic coordination numbers or atomic oxidative states or the $\delta(^1H)$-changes in hydrogen bonding and/or increasing the acidity on the surface of materials, etc.

To conclude this section, it should be noted that the *isotopic* chemical shifts (e.g., for ^1H at ^{12}C versus ^1H at ^{13}C, or ^{19}F at ^{13}C versus ^{19}F an ^{12}C, or ^{19}F at ^{121}Sb and ^{123}Sb) are relatively small (no more than 0.5–0.6 ppm) and they are generally unresolved in solid-state MAS NMR spectra. Therefore, their role in the structural analysis of materials is not significant. Finally it is obvious that due to close mutual locations of molecules in the solid state and the absence of high-amplitude motions, chemical shifts of the same compound can differ in solutions and the solid state. In practice one can formulate the following statement: if the chemical shift difference is ≤10 ppm, then the structure of a compound does not change in going from a solution to the solid state.

1.5.2 Spin–Spin Coupling

According to the quantum mechanical theory [4,8], the spin Hamiltonian, describing NMR in isotropic media, takes the form

$$h^{-1}\hat{H} = -\sum_j (2\pi)^{-1}\gamma_j B_0(1-\sigma_j)\hat{I}_j^z + \sum_{j<k} J_{jk}\,\hat{I}_j\hat{I}_k \qquad (1.27)$$

The first part, valid for isolated nuclear spins in the external magnetic field, B_0, and known from the previous section, corresponds to the screening constant σ. The second term represents interaction "spin–spin," which is characterized by the spin–spin coupling constant J_{jk}. Because these J_{jk} constants are relatively small (typically they take values from 1–2 to 1200 Hz), spin–spin coupling is always considered as perturbation of the first term.

Effects of the perturbation of Zeeman's energy levels can be easily represented by attributing the symbols α and β to the parallel and anti-parallel mutual orientations for nuclear spins of 1/2. Then the spin states for a system consisting of two nuclei will be ββ ($m_I = -1$), αβ ($m_I = 0$), βα ($m_I = 0$), and αα ($m_I = 1$). Thus, the first nuclear spin can "see" the second one in two states and its energy levels undergo an additional splitting by energy J measured in Hz. As the theory shows, the J constant is a scalar magnitude, it has a sign, and it is independent of the external magnetic field B_0, and reflects the energy of coupling.

Magnetically nonequivalent nuclear spins are well known to form simple spin systems AX and strongly coupled spin systems AX_2, AXX′, ABX, AMX, AB_2, or ABC, showing very complex NMR patterns requiring a computer analysis due to the presence of combinational transitions [6,8]. In solids such spin systems do not appear due to small spin–spin coupling constants or due to high power decoupling at proton frequency for C–H, N–H, Si–H fragments, for example. However, because the simple systems AX with the large hetero and homo spin–spin coupling constants can be still observed in solids (see Figures 1.1 and 1.2), it is pertinent to show the mechanisms of spin–spin coupling.

Spin–spin coupling is transmitted via electrons surrounding the nuclei, and therefore the J constants depend on molecular structures: electron distributions, atomic hybridizations, and interbond angles. Generally, a total J energy consists of three contributions coming from the spin–orbital (SO), spin–dipolar (SD), and Fermi contact (FC) interactions [4] written as

$$H_{SO} = \left(\frac{\mu_0}{4\pi}\right) g_L \mu_B \gamma_N \hbar \sum 2L_e I_N r_{eN}^{-3} \tag{1.28}$$

$$H_{SD} = -\left(\frac{\mu_0}{4\pi}\right) g_S \mu_B \gamma_N \hbar \sum \{r_{eN}^{-3} I_N S_e - 3r_{eN}^{-5} (I_N r_{eN})(S_e r_{eN})\} \tag{1.29}$$

$$H_{FC} = \left(\frac{2}{3}\right) \mu_0 g_S \mu_B \gamma_N \hbar \sum \delta(r_{eN}) I_N S_e \tag{1.30}$$

where

e and N correspond to electron and nucleus, respectively
L_e is the orbital angular moment operator
r_{eN} is the distance between electron and nucleus
$\delta(r_{eN})$ is the Dirac delta function

This function takes a value of 1 at $r_{eN} = 0$ and 0 otherwise, requiring, thus, the presence of electron density directly on the nuclei. Calculations show that Fermi contact contribution usually strongly dominates and hence J constants can be interpreted in terms of electron densities.

Spatial proximity of some nuclei, for example, ^{19}F, can lead to the through-space spin–spin coupling constants, sometimes reaching 100 Hz and more. However, again, according to Contreras [18], Fermi contact interaction dominates but the coupling goes via nonbonding electron orbitals. The main feature of spin–spin coupling constants is their direct dependence on molecular structures and their independence of molecular orientations in the external magnetic field. This circumstance (in combination with their low values) explains why spin–spin coupling plays a minor role in the solid-state NMR spectra. Finally, in contrast to spin–spin coupling considered earlier, dipolar interaction between nuclei spins or dipolar coupling, which is one of the important factors affecting line shapes of NMR signals in the solid state, is transmitted directly through space but not via electrons surrounding the nuclei.

1.5.3 Quadrupolar Coupling

Distributions of charges at nuclei with spin numbers >1/2 [1,4] are nonspherical and responsible for the appearance of nuclear quadrupole moments Q. Spins of such nuclei (e.g., ^{2}H, ^{14}N, ^{27}Al, ^{81}Br, ^{127}I, ^{35}Cl, etc.) interact with not only the external and local magnetic fields but also with electric field gradients EFG at these nuclei. A number of such nuclei, for example, ^{81}Br and ^{79}Br, or ^{35}Cl and ^{37}Cl, have extremely large quadrupole moments, measured as 25.4 and 30.5 or −8.16 or -6.44×10^{-30} m^2, respectively (in contrast to 0.286×10^{-30} for deuterium). Therefore they become objects in the nuclear quadrupole resonance (NQR) studies, based by analogy with NMR on quantized energy levels corresponding to different orientations of quadrupole moments Q with respect to the electric field gradients, EFG.

The EFG characterizes a nonhomogeneous electric field, E, shown in Scheme 1.2, which changes, for example, along chemical bond A–B.

The field can be expressed mathematically as a tensor:

$$\begin{vmatrix} eq_{XX} & 0 & 0 \\ 0 & eq_{YY} & 0 \\ 0 & 0 & eq_{ZZ} \end{vmatrix} \tag{1.31}$$

containing the zero off-diagonal elements. By convention, the largest element of the EFG tensor, eq_{ZZ}, is always oriented along the A–B bond (the Z axis).

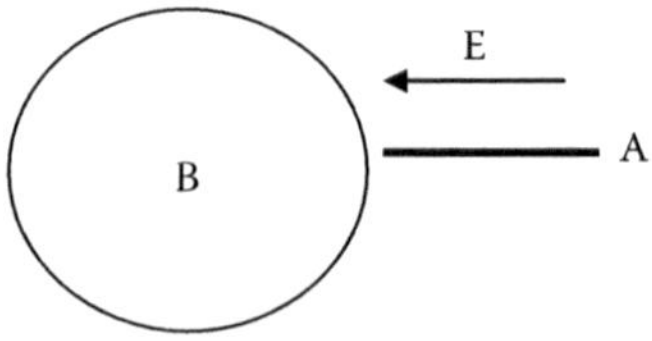

SCHEME 1.2 The electric field gradient characterizes a nonhomogeneous electric field, E, changing along chemical bond A–B.

Interactions between the nuclear quadrupole moments, Q, and the EFG are energetically characterized via the nuclear quadrupole coupling constants, NQCC,

$$\mathrm{NQCC} = \frac{e^2 q_{ZZ} Q}{h} \tag{1.32}$$

where

eq_{ZZ} term is the principal component of the EFG tensor
e is the elementary charge (measured as $1.6021773 \times 10^{-19}$ C)
NQCC is expressed in frequency units of an order of 10^6–10^9 Hz

As seen, the NQCC measures the "size" of the EFG, while a spatial extension or the "shape" of the EFG is defined via the asymmetry parameter, η,

$$\eta = \frac{|eq_{XX} - eq_{YY}|}{eq_{ZZ}} \tag{1.33}$$

If $eq_{XX} = eq_{YY}$ and hence $\eta = 0$, then the electric field gradient is axially symmetric. In spite of the presence of a large quadrupolar moment at a nucleus, shapes and sizes of the electric field gradients strongly depend on the symmetry of charge distributions around this nucleus. For example, a symmetrical charge distribution at ^{14}N nuclei in $NH_4^+Cl^-$ corresponds to a small NQCC value of 0.016 MHz versus 0.9 or even 3.98 MHz in compounds $EtONO_2$ or $MeNH_2$, respectively, where charge distributions at ^{14}N nuclei are not symmetrical. Following a semiquantitative level, the electric field gradient at any quadrupolar nucleus can be expressed as the sum of nuclear and electronic terms:

$$eq_{ZZ} = +\frac{\sum_n K_n\left(3z_n^2 - r_n^2\right)}{r_n^5} - e\left\langle \psi^* \left| \frac{\sum_i \left(3z_i^2 - r_i^2\right)}{r_i^5} \right| \psi \right\rangle \tag{1.34}$$

where

K and e are charges of the neighboring nuclei and electrons, respectively
r_n and r_i are the corresponding distances

Thus, the electric field gradient is an important parameter of bonding modes and can be useful for characterizations of investigated molecular systems.

It should be emphasized that the quadrupolar coupling can reach exclusively large magnitudes that can be comparable with or surpass significantly the strongest dipolar interactions. Therefore they play a major role in shapes of resonances observed in solids. This is particularly important for rigid solids with limited molecular motions, where shapes of NMR resonances observed for quadrupolar nuclei are completely governed by quadrupolar interactions. Figure 1.14 illustrates schematically a ^{2}H resonance in a static powder sample that shows the quadrupolar splitting, marked as $\Delta\nu$. This shape is typical of systems with axially symmetric electric field gradients.

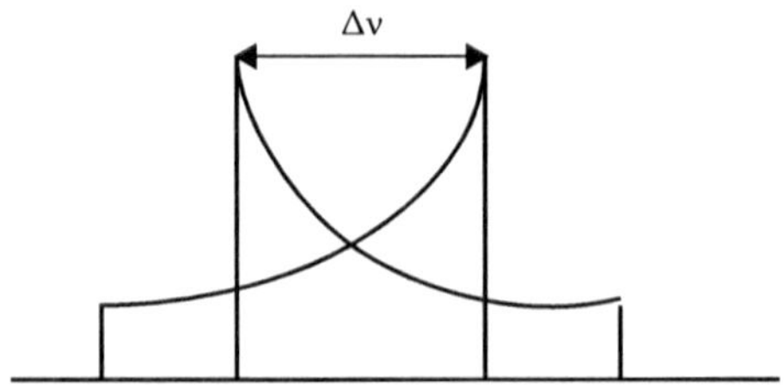

FIGURE 1.14 Schematic representation of a resonance observed for ^2H nuclei in a static powder sample, where the electric field gradient is axially symmetric and $\Delta\nu$ is the observed quadrupolar splitting with a magnitude dependent on the type and rates of molecular (or group) reorientations.

When solids are rigid, the splitting corresponds to the NQCC via the relationship, $\Delta\nu = 3/4(e^2q_{ZZ}Q/h)$. Thus, if the experimentally observed splitting corresponds to a NQQC value similar to a static NQQC, then this fact is a good criterion for the absence of molecular (or group) reorientations.

1.6 INTRODUCING NMR EQUIPMENT AND NMR EXPERIMENTS

As noted in preface, rapid successes in radio frequency technology resulted in the creation of the first commercial NMR spectrometers in 1953. Subsequently, however, chemical needs rapidly required more powerful NMR spectrometers and new spectroscopic pulse approaches aimed at investigation of increasingly complex objects. The complexity of NMR spectrometers increased and this process is constantly in progress.

In spite of the complexity, the principal scheme can be represented by spectrometers applied in students' laboratories in Figure 1.15.

As seen, the spectrometers should include the following basic units needed in NMR experiments: a radio frequency reference generator (or generators for multichannel devices), a transmitter, an amplifier, a receiver, a pulse generator, a powerful computer, a magnet, and an NMR probe.

The quality of all these units can strongly affect the signal-to-noise ratio in NMR spectra, which is a main practical measure of sensitivity in NMR. Note that this ratio can be easily obtained by conventional software in NMR programs.

One can formulate that when all of the aforementioned units are "ideal," the NMR sensitivity is actually dictated by constants γ, the natural abundance of target nuclei, and strength of external magnetic fields. It is obvious that poor signal-to-noise ratios do not provide reliable conclusions even on the level of chemical shift determinations or determinations of structurally nonequivalent sites. In this connection, a simple elementary scheme applied in recording NMR signals is also important for spectroscopists.

The signal detected in a coil of an NMR probe takes the form expressed by function A $\exp(i(\omega_0 + \Delta\omega)t)$, where A is the signal amplitude, ω_0 is a carrier frequency, and $\Delta\omega$ is the desired offset. Since the signal in the coil is weak, it should be amplified with the pre-amplifier. It should be emphasized that the NMR probe should be

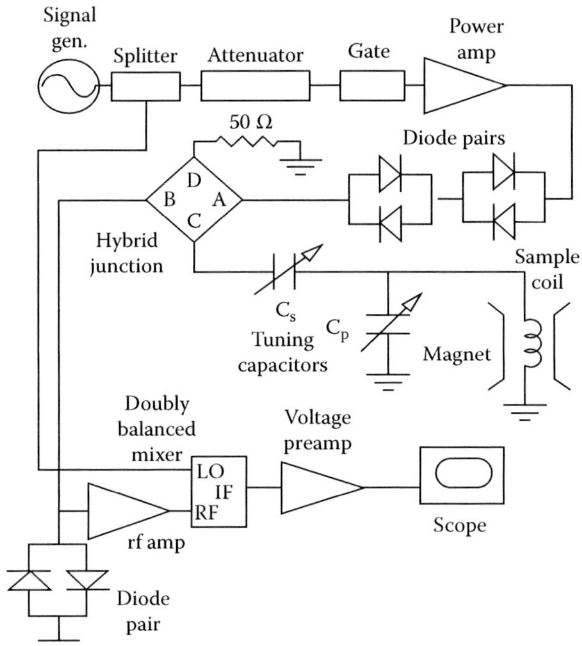

FIGURE 1.15 The principal scheme of an NMR spectrometer used by students in the Department of Physics, University of Florida. http://www.phys.ufl.edu/courses/phy4803L/group_II/nmr/nmr.pdf

carefully tuned to the carrier frequency because the tuning plays a dramatic role in the resulting spectra. The signal, then, has to be filtered and mixed with an intermediate frequency.

The next step should separate the real and imaginary parts of the signal components. After the separation, the analogue to digital converter gives a digitized signal, which can now be processed by a computer.

Solid-state NMR spectra are generally characterized by the large spectral widths needed for registration of NMR signals. For this reason, the intense radio frequency (rf) pulses operating in channels of the spectrometer are required to excite uniformly the corresponding spectral range. This condition is important and necessary to produce the undistorted NMR spectra.

In the case of the nuclei "attached" to ^{1}H (or ^{19}F) nuclei, direct dipolar couplings would be suppressed. The decoupling requires additional radio frequency energy. In solution NMR, a decoupler power of around a few watts is quite sufficient to decouple completely the detected nuclei, while in the solid state, there are other requirements for decoupling powers that should be increased by two orders of magnitude. For these reasons, all the channels in modern solid-state NMR spectrometers to observe the signal and to carry out the decoupling should be equipped with high-power amplifiers capable of producing a kilowatt of power. The technical details can be found in the work of Wasylishen and coworkers [19]. The authors note that the amplifiers can be linear class AB or nonlinear class C.

After the action of RFP, the FID collected in NMR experiments are digitized by the use of an analogue to digital converter before FT. One of the very important parameters of the analogue to digital converters is the maximum resonance frequency range that can be properly digitized. Again, in contrast to solutions, the solid-state NMR spectrum generally covers a very large frequency range. For this reason, the solid-state NMR spectrometers should be equipped with high-speed digitizers with much faster speeds.

In practice, Wasylishen [19] offers to compare the spectrometers working in solutions and the solid state. Spectral windows of 1 MHz or more are generally applied for solid-state NMR experiments. Under this condition the NMR spectrometer should have a 12-bit analogue to digital converter capable of digitizing spectral windows of up to 5 MHz. At the same time solution NMR spectrometers are equipped commonly with 16-bit converters. They can observe a maximum spectral window of only 125 kHz.

Finally, intensity of the radio frequency fields B_1 can be defined as

$$2\pi\nu_1 = \gamma B_1 \tag{1.35}$$

In turn, at the RFP operating with the 90° acting angle and duration t_p, the radio frequency field ν_1 is expressed by

$$\nu_1 = \frac{1}{4t_p}(90°) \tag{1.36}$$

Thus, with a duration of the 90° pulse of 5 μs, the corresponding radio frequency field will be of 50 kHz only. This important rule should be taken into account for running solid-state NMR experiments.

1.6.1 Magnets and NMR Probes

A magnet is a source of the strong, stable, and homogeneous magnetic field needed for initial nuclear polarization and observation of the high-resolution NMR spectra. Increasing the magnetic field strength, B_0, as expected for the theory, leads to the larger energy gap, ΔE, a larger population difference $\Delta N/N$, and hence to larger sensitivity of an NMR spectrometer.

Andrew and Szczesniak [2] note that the earliest NMR experiments have been performed with iron permanent/electromagnets that generated magnetic fields of 1–2 T and the NMR signals have been detected by the continuous-wave one coil methods or later by a two-coil method. However, the iron magnets were relatively unstable, very sensitive to temperature, and capable of creating only moderate field strengths, corresponding to the working ^{1}H frequency of ≤100 MHz. Nevertheless, even now such magnets are acceptable to provide, for example, the routine NMR relaxation or spectral measurements in industry.

Modern superconducting electromagnets show excellent stability, good homogeneity, and their fields are capable of routine working at a proton frequency up to 900 MHz. In addition, stability of the magnetic field, necessary in the high-resolution

NMR experiments in solutions and the solid state, can be improved by field-locking. It should be noted that the NMR experiments in the solid state generally require wide-bore magnets that can provide various manipulations with samples.

Normally, an NMR probe represents a single tank circuit needed for irradiation and detection of an NMR signal. The design of a solid-state NMR probe depends significantly on NMR experiments performed. For example, variable-temperature measurements, usually between −120°C and +120°C, require, on the one hand, the corresponding heating/cooling equipment at accurate temperature settings and on the other hand, the absence of remarkable temperature gradients through a sample. In addition, a special design of solid-state NMR probes provides NMR measurements even at high temperatures up to 300°C. It should be emphasized that all the temperature experiments obviously require the corresponding procedures for reliable calibration of temperature, particularly at the lowest and highest temperatures. Some aspects of temperature changes in spinning samples are considered later. Double and triple resonance or gradient-field experiments are also available, when NMR probes have a special design.

Experiments with static samples are usually performed with wide-line NMR probes where a sample is placed into a glass NMR tube located perpendicular to the external magnetic field. Such wide-line probes can be equipped with flat coils if necessary for biomolecular investigations. NMR probes can be equipped with tools for a mechanical rotation of samples especially at high spinning rates of tens of kHz. Such rates are needed to suppress strong dipole–dipole interactions in solids, when the MAS NMR rotors are oriented at the magic angle relative to the external magnetic field (Figure 1.16). As we will consider later, this angle transforms term ($3\cos^2\theta - 1$) to zero thus creating the conditions of dipolar decoupling in the solid state.

It should be emphasized that a nonaccurate magic angle setting requiring specific procedures can lead to unpredictable shapes of NMR signals particularly for nuclei

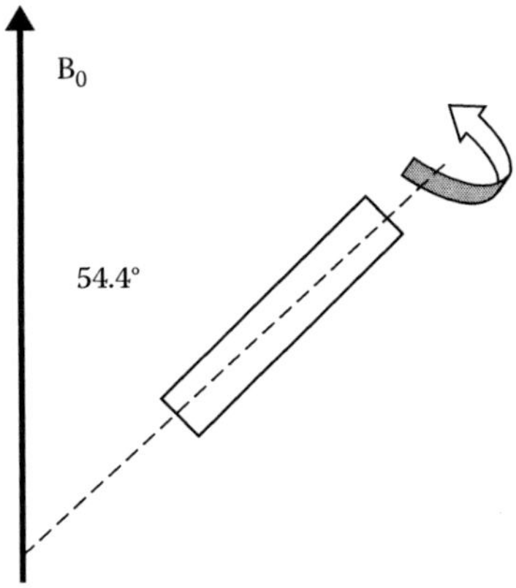

FIGURE 1.16 Angle θ of 54.4°, named "the magic angle" by Andrew. This angle can potentially reduce dipolar term ($3\cos^2\theta - 1$) to zero, leading to the high resolution MAS NMR spectra in the solid state. The best narrowing line result can be reached when the spinning frequency compares well with line widths observed in static samples.

with large magnetic shift anisotropy. Generally for MAS experiments standard 7-, 4-, or 2.5 mm zirconium oxide rotors can be used to routinely reach spinning rates of 7, 18, or 25 kHz. Higher spinning rates are also available in modern NMR spectrometers. Variable-temperature NMR experiments, performed in spinning samples below −50°C and above 50°C, generally require very fragile boron nitride caps for rotors. Finally, we stress again that the ultra-fast mechanical rotations can cause the samples to heat up.

1.6.2 Homogeneity of the External Magnetic Field and Digital Resolution in NMR Spectra

Since one of the major tasks in an NMR analysis is determination of a number of chemically or structurally nonequivalent nuclei in investigated systems, the final result will be dependent on a spectral resolution that can be formulated in practice as a minimal distinguished distance between two peaks measured in Hz. On one hand, the spectral resolution is dictated by homogeneity of the external magnetic field, which can be improved by standard manipulations with shimming coils. On the other hand, since the signals detected in NMR experiments transform to digitized forms by the analogue to digital converters and are treated with a computer, the resolution in an NMR spectrum obtained after FT will be dependent on memory sizes used for this treatment. It is obvious that the digital resolution in the resulting spectrum will be a function of sweep widths applied for the data collection and numbers of points [20]. Under these conditions, even careful manipulations with shimming coils will be limited by a relatively small number of the points leading to a poor resulting resolution. On the other hand, a poor homogeneity of the magnetic field does not require a good digital resolution.

In solutions where nuclear relaxation times T_2 are generally long and therefore lines in NMR spectra are sharp ($\Delta\nu = 1/\pi T_2$), the shimming coil manipulations are critical for each sample and collection of NMR data requires a relatively large computer memory. In solids, however, resonance lines are broad due to strong dipolar internuclear interactions and relatively short T_2 times. Therefore, the aforementioned shimming coil manipulations are not critical and a reasonable compromise between sweep widths and computer memory sizes is possible. The situation changes if, for example, a solid-state NMR spectrum exhibits overlapped resonances after FT, which should be resolved by Gauss multiplication procedures (Figure 1.2).

1.6.3 Dead Time of NMR Spectrometers: Intensity of Signals and Distortion of Baselines in NMR Spectra

A radio frequency coil located in NMR probes is used for irradiation of a sample and detection of NMR signals. According to the principles of Kisman and Armstrong [21], the radio frequency coming from the transmitter must be distinguished from the signal from the probe. Under these circumstances, a relatively weak signal should be measured immediately after the action of a high-power RFP. This switching problem gives rise to the dead time of an NMR spectrometer. It should be noted that this problem is technically minimized in modern NMR spectrometers, and their standard components significantly reduce the dead times to a few microseconds, comparable with RFP lengths.

In NMR experiments with liquids and solutions, the dead time is not an important problem because relaxation times and the FIDs are long and therefore the dead time does not affect signal intensities. In contrast, solids generally show strongly much shorter FIDs. For this reason, if the dead time of an NMR spectrometer is relatively long, the NMR signal can lose its intensity or under certain conditions the signal can become "invisible." Figure 1.17 illustrates this situation.

If a magnetization created along the OX (or OY) axis by a powerful RFP decays very rapidly and the dead time of a spectrometer is long, FT converts the FIDs into NMR spectra, where baselines are strongly distorted. This effect, particularly important for solid-state NMR, is not easy to eliminate in spite of software procedures available in spectrometers. Moreover, in the presence of the baseline distortions, it is difficult to distinguish the distorted baselines and broad resonances that can appear in the spectrum. A priori, to avoid these problems, such solid-state NMR experiments generally requiring large sweep widths, should be performed at "dead time parameter," which is longer than "dwell time parameter." However, even in such a situation, the baseline problem is still present. Applications of pulse sequences Hahn-echo, spin-echo, or solid-echo, where the NMR data collection starts immediately after the second registering RFP, can minimize this problem. It should be added that a procedure available in NMR software for cutting first FID points is very successful in correcting baselines (Figure 1.18).

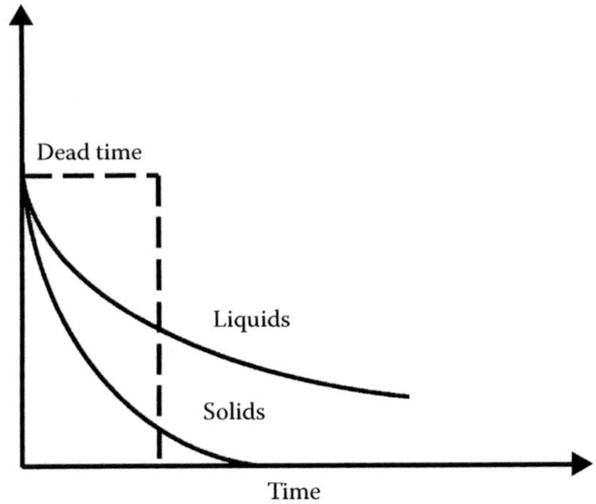

FIGURE 1.17 Schematic representation of how the dead time of an NMR spectrometer affects the observed intensity of NMR signals in liquids and solids with the long and short FIDs respectively.

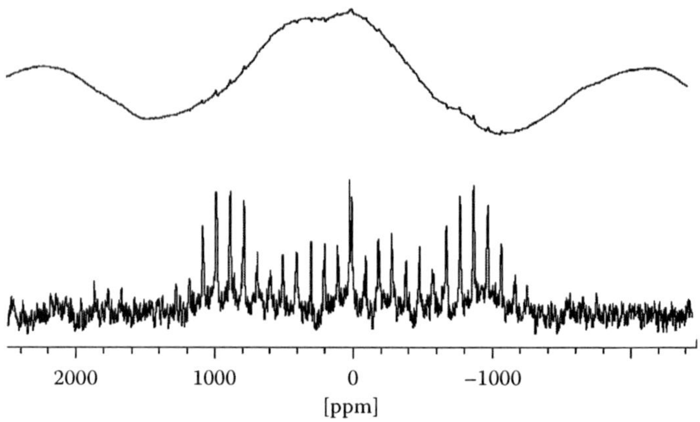

FIGURE 1.18 The ^{2}H MAS NMR spectra recorded at a large sweep for a material containing phenyl phosphine groups, which have been deuterated in para-position of the aromatic rings. The spectra are obtained in a sample spinning at a rate of 5 kHz (a standard 7 mm NMR rotor). The top spectrum, obtained with a single-pulse sequence, shows the very strong baseline distortions. The bottom spectrum, recorded with a Hahn-echo pulse sequence, synchronized with the spinning rate, and treated with a cutting procedure, demonstrates a good baseline.

1.7 MAIN PRINCIPLES OF TWO-DIMENSIONAL NMR

As shown earlier, the simplest 1D NMR experiment entailing direct (or indirect) excitation of target nuclei is based on the single-pulse sequence shown in Figure 1.7. Here AT to collect the FID is a single time for spin evolution. As a result, FT leads to a pattern of "intensity versus frequency."

Any 2D NMR experiment adds the second frequency axis. The pulse sequences currently applied for the 2D NMR experiments are varied and numerous. In some sense, they even depend on the nature of target nuclei and objects investigated. Nevertheless, the main principles of 2D NMR can be easily illustrated in Figure 1.19. As can be seen, after the first 90° pulse, a nuclear spin system develops during time, noted as t_1, for example, due to *nucleus–nucleus spin–spin coupling*. Then, the first set of time-dependent data, noted as $f(t_2)$, can be collected after the action of the second RFP during AT t_2. It should emphasized that the $f(t_1)$ datasets can be collected at varied t_1 values. Finally, a double FT performed for domains t_1 and t_2 leads to two frequency axes. This is a simple type of correlated spectroscopy. The resulting NMR spectrum can be represented as a square plot with the diagonal corresponding to the 1D NMR spectrum, and cross-peaks that appear due to scalar spin–spin coupling. Ideologically similar pulse sequences can be used to collect the 2D NMR data corresponding to heteronuclear chemical shift correlations. Figure 1.20 shows an example of the ^{13}C-^{1}H correlation spectrum. Here the F_2 and F_1 coordinates represent ^{13}C and ^{1}H frequencies, respectively, and the cross-peaks show their spin–spin couplings. Finally standard software of modern NMR spectrometers can convert Figure 1.20 into Figure 1.21, where the

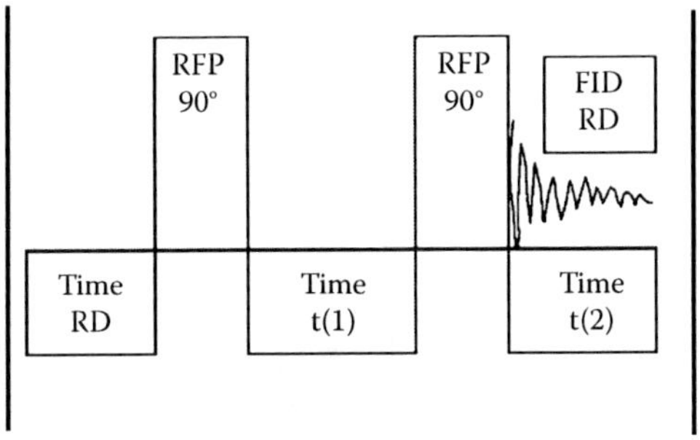

FIGURE 1.19 Schematic representation of a pulse sequence and time sections operating in the simple 2D NMR experiments, COSY. Typical values of pulse lengths and RD and t(2) are similar to Figure 1.7.

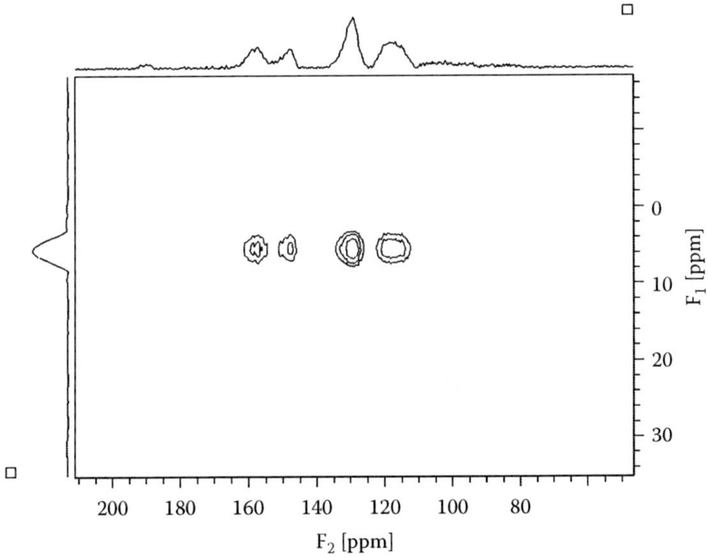

FIGURE 1.20 The heterocorrelation ^{13}C-^{1}H solid-state MAS NMR spectrum recorded at a spinning rate of 10 kHz for a polymer system, containing aromatic rings and carbonyl groups. Chemical shift values refer to an external TMS.

intensity coordinate is added. Both of the NMR spectra show clearly the presence of the distinguished resonances of aromatic rings in the ^{13}C NMR part. At the same time, all of the signals correlate with a single broad resonance detected in the ^{1}H NMR part. This is normal because due to strong proton–proton dipolar coupling, protons of the rings remain unresolved. As shown in the following, wide-line separation NMR spectroscopy reported by Schmidt-Rohr et al. [22] can "resolve" ^{1}H resonances in terms of different molecular mobility in solid polymers.

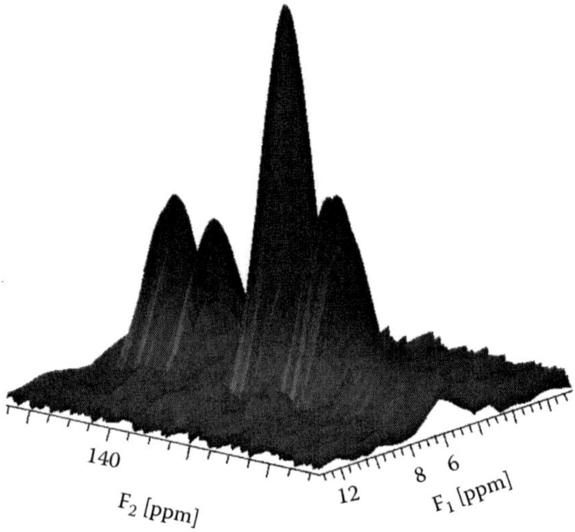

FIGURE 1.21 A heterocorrelation $^{13}C(F_2)$-$^{1}H(F_1)$ MAS NMR spectrum obtained under conditions as in Figure 1.19 and represented as the pattern where the new axis corresponding to the signal intensity is added.

1.8 ENHANCEMENT OF SENSITIVITY IN NMR

This section ends the brief description of the main principles in pulsed NMR by introducing the methods that can potentially increase the sensitivity in NMR experiments. In fact, applications of solid-state NMR spectroscopy in many fields of science, primarily in biology, biochemistry, and materials chemistry are significantly limited by the relatively low sensitivity of NMR.

Besides increasing the NMR sensitivity by stronger initial nuclear polarization in the highest external magnetic fields, it can be achieved with the help of cryogenic NMR probes, where the radio frequency coil and associated electronics are cooled to a low temperature, for example, to 25 K and lower. The high sensitivity, increased by a factor of 12–16 in cryogenic NMR probes, is exploited for studies of natural products containing rare nuclei such as ^{13}C, ^{17}O, ^{15}N, etc.

NMR sensitivity can be dramatically enhanced by the specially developed approaches based on optical pumping [23], para-hydrogen-induced polarization [24], and dynamic nuclear polarization, DNP [25]. In the gaseous phase, according to Bifone et al. [26], optical pumping with laser light can enhance the nuclear polarization by five orders of magnitude for gases like xenon. For solids, however, DNP seems to be most important.

The DNP effect is based on the presence of species with unpaired electrons (paramagnetic metal ions, organic groups, etc.). Such electron–nucleus systems are described via the Hamiltonian,

$$H = -\omega_e S_Z - \omega_I I_Z + H_{ee} + H_{en} + H_{nn} \tag{1.37}$$

where the first two terms correspond to Zeeman's interactions of electron and nuclear spins, respectively, H_{ee} and H_{nn} describe electron–electron and nucleus–nucleus coupling, and the H_{en} term is the hyperfine electron–nucleus interaction. The latter term plays a major role in experiments with DNP generation.

Irradiating the samples at (or near) the electron Larmor frequency, ω_e, can lead to simultaneous electron–nucleus spin flips due to electron–nuclear coupling H_{en}. In turn, these flips enhance the absolute value of the nuclear polarization via three possible mechanisms: an Overhauser effect, a solid effect, and a thermal mixing effect, described by Wind and coworkers [25]. In this context, the enhancement factors expected for the nuclear polarization will depend on the number of unpaired electrons, the ESR line-widths, the nuclear relaxation rates in the absence of electrons, and the amplitude of the microwave field that should be applied to the sample. Generally, in solids the enhancement factors can reach 660 and 2600 for ^{1}H and ^{13}C nuclei, respectively.

Technically the DNP experiments require specially designed NMR spectrometers and particularly NMR probes: the probes should be equipped with a coil that can be double-tuned to ^{13}C and ^{1}H frequencies, for example, and a horn antenna to transmit the microwaves. The simple enhancement experiments with ^{13}C nuclei entail their direct polarization. A proton polarization via DNP and its transfer to the ^{13}C spins via ^{1}H-^{13}C cross-polarization is also possible.

It is noteworthy that Addenkjaer-Larsen [27] applied a similar principle for the NMR spectra in liquids to increase the signal-to-noise ratio by a factor of $\gg 10^3$. Finally, the DNP is not popular among chemists working with materials. First, the samples should be initially treated with a paramagnetic compound providing the DNP and distributed homogeneously through the sample volume. Second, the electrons in paramagnetic metal ions generally applied to modify properties of materials show very short relaxation times.

REFERENCES

1. Bakhmutov, V. I. 2009. Magnetic resonance spectrometry. In *Encyclopedia of Applied Spectroscopy*, D. L. Andrew (ed.), pp. 933–963. Weinheim, Germany: Wiley.
2. Andrew, E. R., Szczesniak, E. 1995. *Prog. Nucl. Magn. Reson. Spectrosc.* 28: 11–36.
3. Bloembergen, N. 1949. *Physica* 15: 386.
4. Abragam, A. 1985. *Principles of Nuclear Magnetism.* Oxford, NY: Clarendon Press.
5. Farrar, C. T. 1987. *An Introduction to Pulse NMR Spectroscopy.* Chicago, IL: Farragut Press.
6. Friebolin, H. 1991. *Basic One- and Two-Dimensional NMR Spectroscopy.* Weinheim, Germany: VCH.
7. Lambert, J. B., Riddel, F. G. 1982. *The Multinuclear Approach to NMR Spectroscopy*, D. Reidel (ed.). Boston, MA: Boston Publishing Company.
8. Harris, R. K. 1983. *Nuclear Magnetic Resonance Spectroscopy.* Avon, U.K.: Bath Press.
9. Taylor, T. J., Bakhmutov, V. I., Gabbai, F. P. 2006. *Angew. Chem. Int. Ed.* 45: 7030.
10. Campos-Fernández, C. S., Schottel, B. L., Chifotides, H. T. et al. 2005. *J. Am. Chem. Soc.* 127: 12909.

11. Duer, M. J. 2002. *Solid-State NMR Spectroscopy: Principles and Applications*. Malden, MA: Blackwell Sciences.
12. Ernst, R. R., Anderson, W. A. 1966. *Rev. Sci. Instrum.* 37: 93–102.
13. Fyfe, C. A., Lewis, A. R., Che'zeau, J. M., Grondey, H. 1997. *J. Am. Chem. Soc.* 119: 12210–12222.
14. Penner, G. H., Chang, Y. C., Hutzal, J. 1999. *Inorg. Chem.* 38: 2868.
15. Mcjller-Warmuth, W., Schujler, R., Prager, M., Kollmar, A. 1979. *J. Magn. Reson.* 34: 83–95.
16. Galakhov, M. V., Bakhmutov, V. I., Barinov, I. V. 1991. *J. Magn. Reson. Chem.* 29: 506–508.
17. McConnel, H. M., Chesnut, D. B. 1958. *J. Chem. Phys.* 28: 107–117.
18. Contreras, R. 2001. *J. Am. Chem. Soc.* 123: 9144–9152.
19. Bryce, D. L., Bernard, G. M., Gee, M. et al. 2001. *J. Anal. Sci. Spectrosc.* 46: 46–82.
20. Hoch, J. C., Stern, A. S. 1996. *NMR Data Processing*. Chichester, U.K.: Wiley.
21. Kisman, K. E., Armstrong, R. L. 1974. *Rev. Sci. Instrum.* 45: 1159–1163.
22. Schmidt-Rohr, K., Clauss, J., Speiss, H. W. 1992. *Macromolecules* 25: 3273.
23. Walker, T. G., Happer, W. 1997. *Rev. Mod. Phys.* 69: 629–642.
24. Bargon, J., Natterer, J. 1977. *Prog. Nucl. Magn. Reson. Spectrosc.* 31: 293–315.
25. Wind, R. A., Duijvestun, M. J., Van Der Lugt, C., Manenschijn, A., Vriend, J. 1985. *Prog. NMR Spectrosc.* 17: 33–67.
26. Bifone, A., Song, Y. O., Seydoux, R. et al. 1996. *Proc. Natl. Acad. Sci. U.S.A., Biophys.* 93: 12932–12936.
27. Addenkjaer-Larsen, J. H., Fridlund, B., Gram, A. et al. 2003. *Proc. Natl. Acad. Sci. U.S.A.* 100: 10158–10163.

2 On Amorphous Materials as Potential Objects in Solid-State NMR Studies: Common and Key Problems

Due to rapid development of current new methodical approaches for solid-state NMR experiments particularly under conditions of ultra-fast magic angle spinning (MAS), NMR became a central tool in investigating and understanding the structural and dynamic properties of amorphous solids on the molecular and macroscopic levels. The world of modern materials is impressively broad and diverse due to unlimited possibilities in variations of their chemical compositions and morphology, on the one hand, and due to the constant progress in chemical approaches to the controlled design and creation of new three-dimensional (3D) networks with desired physical and chemical properties, on the other hand.

Today, much attention of chemists is focused on different classes of materials such as catalysts, polymers, glasses, zeolites, batteries, liquid crystals, natural products, pharmaceutical compounds, proteins, membranes, amyloid fibrils, and many others. The materials can be crystalline or noncrystalline. Generally speaking, noncrystalline materials have an advantage because their properties can vary continuously with variation in the composition.

This chapter represents a brief summary considering only some of them.

The choice of the materials is certainly subjective and the goal of the chapter is to show general and specific properties of the materials that can be important in the context of solid-state NMR. In addition, the chapter shows the physicochemical methods other than solid-state NMR that are generally applied in materials science. First, one can see their advantages and limitations. Second, the summary demonstrates that in many cases, valuable conclusions can be made only on the basis of solid-state NMR experiments.

Biological and biochemical systems (such as amino acids, peptides, proteins, membrane active peptides, etc.) are not the focus of this book. The reader is referred to numerous reviews and books available currently that describe NMR of biological macromolecules, determination of their conformation, alignment, and interaction with binding partners, and their response to changes in the environment. It is particularly difficult to characterize amorphous biological solids and immobilized systems,

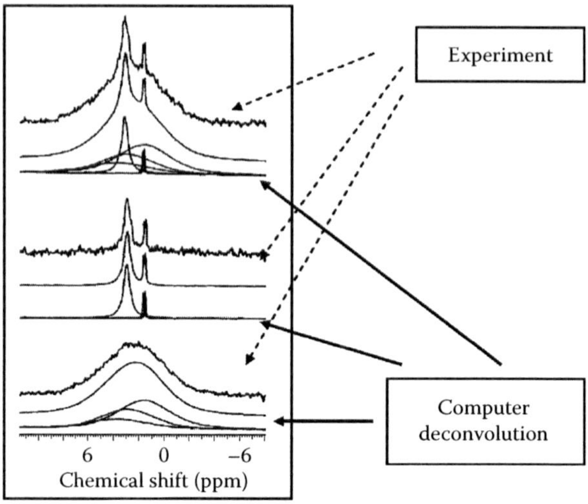

FIGURE 2.1 The ^{31}P MAS NMR spectra of native casein micelles recorded by a single pulse (top), a spin-echo (middle), and a cross-polarization experiment (bottom). (Reproduced with permission from Spyros, A. and Dais, P., *Prog. Nucl. Magn. Reson. Spectrosc.*, 54, 195, 2009.)

such as biological membranes or fibers. Again solid-state NMR plays a major role in these studies. For example, advantages of solid-state ^{19}F NMR in probing the biomembranes can be found in a recent review by Ulrich [1]. Solid-state NMR is widely used also in food chemistry, where a great role belongs to solid-state ^{31}P NMR [2].

Figure 2.1 illustrates these applications by characterization of native casein micelles with the help of the ^{31}P MAS NMR spectra, recorded by the single pulse, spin-echo, and cross-polarization experiments. As the figure shows, the two sharp lines observed at δ of 1.2–1.4 ppm can be assigned to SerP in *k*-casein, the more broadened resonance with δ of 3.1 ppm is colloidal calcium phosphate, while the very broad line with δ of 2–3 ppm belongs to immobile phosphorous compounds. In addition, computer simulations provide a more accurate NMR analysis. Finally, following the spectra it become obvious that even simple ^{31}P relaxation time measurements in such systems are capable of distinguishing immobile phosphorous fragments.

2.1 HYBRID MATERIALS

Following Ferey [3], hybrid materials, representing a new class of porous solids, have been developed in the last 15 years. Due to the numerous works, a new domain in chemistry of materials has been created. Hybrid systems can be synthesized by the various reactions between organic and inorganic species. The goal of these reactions is to build up 3D frameworks with skeleton including organic and inorganic moieties that mutually bind by strong covalent chemical bonds.

Ferey emphasizes that there is no conceptual difference between classical porous systems, containing only inorganic structural units, for example, tetrahedral species SiO_4, PO_4, AsO_4, or SO_4, associated with metallic cations, on the one hand, and hybrid materials on the other hand. However, even the simple structures

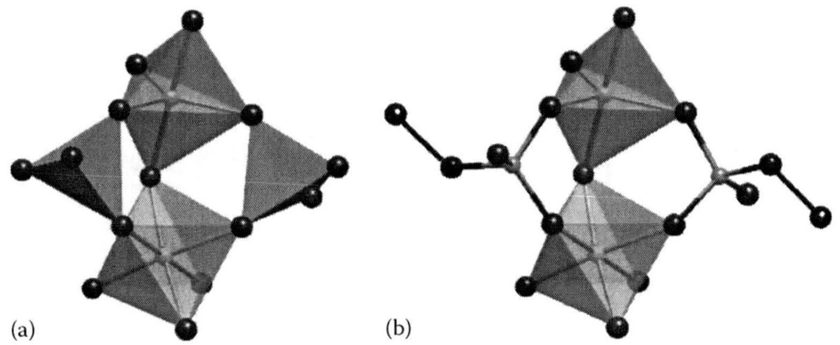

FIGURE 2.2 The schematic representation of a material forming the network on the basis of inorganic structural units (a) and a hybrid material (b), where the inorganic units are connected through organic moieties. (Reproduced with permission from Ferey, G., *Chem. Soc. Rev.*, 191, 191, 2008.)

in Figure 2.2 show that hybrid materials can potentially possess an inorganic matrix, having a larger dimensionality due to the presence of organic fragments. This matrix forms chains, layers, or 3D inorganic frameworks, often called metal–organic frameworks (MOF).

It has been found that even the reaction temperature strongly affects the coordination of the metallic species, the nuclearity, and the dimensionality of the inorganic subnetworks in the MOF synthesis. Generally, at the same chemical composition, higher temperatures result in an increased condensation of supplementary metallic polyhedra on the starting clusters, leading to increasing their nuclearity, and a change into chains and then, layers of inorganic polyhedra. In turn, the increased inorganic dimensionality can lead to appearance of long-range interactions discovering new fields of their applications. Thus, the advantage of these systems is obvious.

Among practical applications of hybrid molecular systems, two fields seem to be particularly important: catalysis and gas adsorption/separation. The catalysis by porous hybrid solids implies creation of metal–organic open framework materials with different active sites that will correspond to different types of catalytic systems. For example, recently chemists have focused on the design of homochiral MOF, that are important for enantioselective catalysis, and on metal ions or ligands in the MOF, metal–organic porous materials, where metal centers are coordinationally unsaturated. Finally great attention has been concentrated on creation of systems with highly dispersed metal or metal oxide nanoparticles.

It is obvious that the synthesis of these complex systems requires a detailed knowledge of their structures and control during chemical manipulations. Computer simulations [3] can be used to predict a list of potential hybrid candidates. However these simulations should be based on the already known structural parameters, such as the space groups, cell parameters, and atomic positions. X-ray diffraction (XRD) can definitely provide these data, when the products form suitable crystals. From Figure 2.1, one could think that solid-state NMR is useful if the systems are amorphous: inorganic parts contain ^{29}Si, ^{31}P, and ^{17}O nuclei chemical shifts that are sensitive to their environments and organic linkers can be well characterized by ^{1}H, ^{13}C, and H–C correlation NMR spectra. In addition different mobility of inorganic and organic components will affect the NMR spectra.

The important direction in practical applications of the MOF hybrids is "gas adsorption/separation/storage and energy" that is based on the well-known capacity of the MOF systems to adsorb large amounts of strategic gases such as H_2, CO_2, CH_4, CO, O_2, NO_x, and C_2H_2 within the cells. It is well known that the total capacity can be easily determined by the adsorption–desorption techniques [3]. However, localization of these gases is not simple and requires independent data obtained experimentally or theoretically by computer simulations. Again, NMR can help in this context, particularly relaxation time measurements performed for gaseous molecules.

2.2 POLYMER SOLIDS

The chemistry of polymers today represents one of the important fields in materials science due to the unique chemical and physical (macroscopic) properties of the polymers. In addition, relatively low costs of the polymer systems provide their wide usage even in high technology.

Since variety in polymerization reactions is high, they result in various polymer products. Depending on formulated goals, the reaction can result in homopolymer systems, copolymer systems or cross-linked polymers, the nature and structures of which will dictate their physical and chemical properties. For example, fluoropolymers are well known to show the unique combination of chemical and physical properties: high-temperature stability, specific dielectric properties, and low friction coefficients.

Generally the synthesis of final products with specific properties can be achieved when structure to property relationships are known. This requires a structural analysis of solid polymers by convenient methods.

Among different domains in the synthetic chemistry of polymeric materials, the controlled design of well-ordered superstructures is of greatest interest [4]. Such superstructures can be generated with self-assembling building blocks. It is obvious that these blocks should be carefully chosen to obtain systems with the desired physical and chemical properties. Following Spiess and coworkers [4], even non-covalent interactions, such as hydrogen bonds or aromatic π–π interactions, can be structurally organizing forces to create these self-assemblies. For example, linear polymers and reversible polymer networks can be formed by the self-assembly of monomer units by their dimerization in a self-complementary array, where cooperative hydrogen bonds play a major role. On the other hand, again following Spiess and coworkers [4], the self-assembly in the bulk of dendritic building blocks into spherical, cylindrical, or other supramolecular architectures can be reached based on the shapes and the demixing of aliphatic and aromatic segments.

One of the important problems in this design is related obviously to a detailed structural analysis of intermediate products and resulting assemblies. It has been established that in spite of the presence of a considerable order in the aforementioned polymer networks on different length scales, single crystals suitable for XRD studies are often unavailable. In other words, solution of a full crystal structure by the

XRD method is difficult or even impossible for many self-assembled supramolecular systems. That is why solid-state NMR leads in this field.

Another important aspect in the synthesis and application of polymeric materials is their dynamics [5]. The dynamics can be quite rich: polymeric systems can show various segmental motions, sometimes very intense, as well as molecular reorientations that strongly affect their mechanical, transport, and optical properties [6]. The dynamics plays a particularly important role in the activity of biological systems, for example, proteins [7] and the stability of pharmaceuticals [8]. In the case of amorphous polymeric materials the dynamics defines their behavior near the glass transition [9,10]. It is obvious that an accurate knowledge of distributions of the mobility in polymer materials is very important from different points of view. Generally this information can be obtained by classical dynamic methods like dielectric relaxation measurements at different temperatures.

Besides segmental dynamics and description of local conformations in solid polymer materials, knowledge of their phase transitions and determination of the conditions for these transitions is also necessary. These transitions can be observed and investigated by differential scanning calorimetry. However, as we will show in Chapter 5, solid-state NMR is a unique instrument uniting structural and dynamic aspects in studies of polymers.

2.3 SYSTEMS BASED ON NANOSTRUCTURES

During the past 10–15 years, the interest of materials chemists has focused on aggregated nanostructures of organic molecules due to their wide applications as light-emitting diodes, photovoltaic cells, sensors, lasers, field emitters, and field-effect transistors. In addition, the aggregated systems characterized by low-dimensionality structures, such as nanowires and nanotubes, are of great interest in the development of new optoelectronic and nanoelectronic devices. All these aspects in synthesis and studies of nanostructures can be found in a recent review [11]. According to Liu and coworkers, nanomaterials can be created on the basis of self-assembled aggregate nanostructures containing conjugated small molecules. Among them, porphyrins and organic π-conjugated polymers, such as polydiacetylenes, or organic charge-transfer complexes (AgTCNQ and CuTCNQ, where TCNQ is 7,7,8,8-tetracyanoquinodimethane) are most popular as building blocks. Great attention was also concentrated on organic/inorganic hybrid nanocomposites. They enable control of many useful physical properties. It is important that the size and shape of nanostructures can be tuned in these composites. Finally, the assemblies can be organized by metal–metal junctions, metal–polymer junctions, inorganic semiconductor junctions, inorganic semiconductor–metal junctions, and semiconductor–carbon nanotube junctions, leading to great variety in their properties.

The design of new systems is aimed at unusual optical, electrical, and electronic properties that in turn strongly depend on the size or morphology of the nanostructures. Generally, the morphology of the systems can be successfully probed by scanning electron microscopy (SEM), transmission electron microscopy (TEM), and high-resolution TEM (HRTEM) images, while the fine structural details can be obtained by the XRD patterns or the selective area electron diffraction patterns.

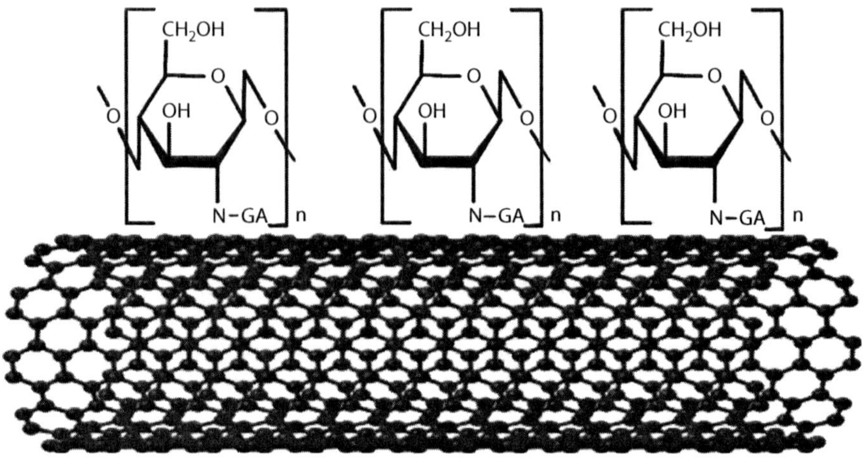

FIGURE 2.3 Carbon nanotubes modified with chitosan. (Reproduced with permission from Salam, M.A. et al., *J. Alloys Compd.*, 509, 2582, 2011.)

When the systems contain spectroscopically convenient nuclei, solid-state NMR spectroscopy can help to solve a number of important problems. For example, carbon nanotubes, modified by chitosan with a general structure shown in Figure 2.3, possess a number of unique structural, mechanical, electrical, electromechanical, and chemical properties [12]. They can be characterized by Fourier transformation infra red spectroscopy (FTIR), Thermo gravimetric analysis (TGA), SEM, and TEM techniques. Nevertheless, only solid-state ^{13}C MAS NMR spectroscopy is capable of providing the direct evidence for binding the chitosan molecules. It has been shown [13] that in spite of paramagnetic behavior of the carbon nanotubes, the ^{13}C MAS NMR spectra can be collected at short relaxation delays of 0.5 s in the single pulse sequence and successfully interpreted. It has been also noted that cross-polarization ^{1}H-^{13}C NMR does not work because of short proton relaxation times.

2.4 WOOD AND WOOD COMPONENTS

Wood has a complex chemical composition and represents a natural composite material. As with other objects in materials science, the scope of investigations of wood and wood products is very large [14]. Scheme 2.1 shows three major components of wood: cellulose (40%–45%), hemicelluloses (20%–30%), where the main component is xylose, and lignin (20%–30%).

Cellulose is well known to be the crystalline part of wood and forms fibrils. In turn, the biological functions of hemicelluloses and lignin are to buttress the fibrils. When a plant is growing, the local molecular organization of hemicelluloses and lignin in the cell wall (and also around the vegetable cell) results in quite different 3D morphological structures [15,16]. All these functions contribute largely to the strength of wood.

It is obvious that isolated wood components can be characterized with a very good accuracy by using the modern arsenal of structural techniques. In spite of these

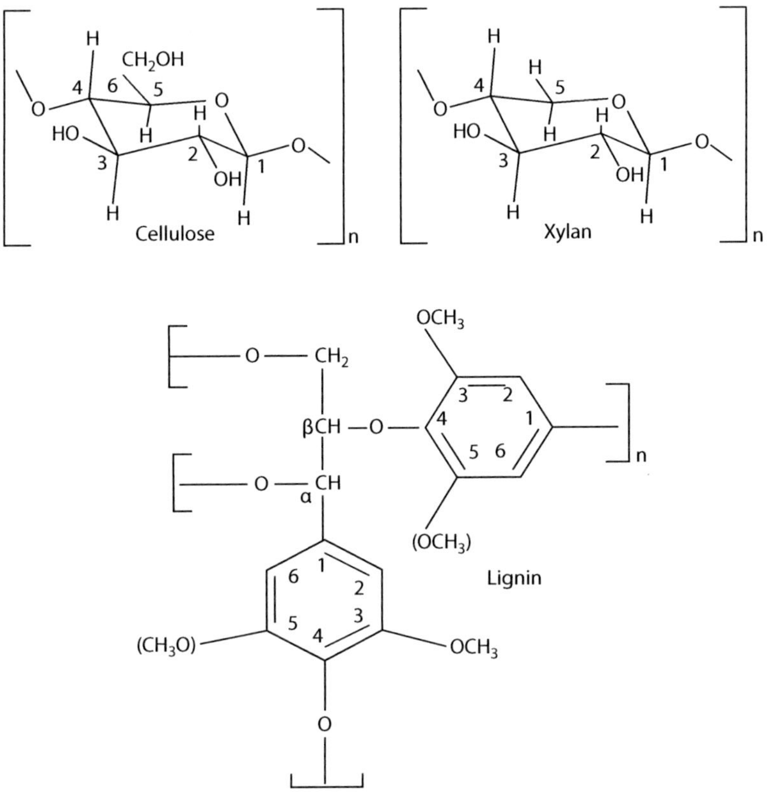

SCHEME 2.1 The major components of wood: Cellulose (40–45%), hemicelluloses (20–30%), where the main component is xylose, and lignin (20–30%).

characterizations, discovering to which extent the separated fractions represent its original properties and interactions in the object as a whole remains very difficult. In fact, due to the structural complexity of wood, regular structural approaches often do not work; the morphology of wood and particularly its degradation expressed as a decrease in its mechanical resistance are of great practical interest and require detailed investigation.

The next important aspect in studies of wood is its biodegradation. In fact, wood biodegradation routes contribute to the mineralization of organic, carbon containing moieties in the wood. This is part of the well-established carbon cycle in nature. Any dead wood, without human intervention, can be degraded within a few years under natural aerobic conditions. This process is particularly important for archaeologists [17] because these intrinsic properties of wood can be challenging for the conservation of wooden objects. Finally, wood porosity and behavior of small molecules within the pore space of the wood matrix are also very interesting theoretically and practically, and require detailed knowledge. The chemical composition in Scheme 2.1 shows that ^{13}C MAS NMR is the most powerful alternative solid-state method that can be applied for an analysis of such systems, providing structural information at the molecular level and determination and study of their crystal and amorphous phases.

2.5 GLASSES

Research in the field of glasses and glassy materials is traditional. At the same time, it is still very active and is splitting now into different scientific disciplines. Among them, common and specific properties of glassy amorphous systems (such as vitreous polymers, noncrystalline semiconductors, frozen organic liquids, glassy ionics, etc.) are particularly interesting [18].

Generally chemists define the structure of any crystalline solid as the space group and set of coordinates for a relatively small number of atoms in the unit cell. It is well known that this information is sufficient to describe all the atoms and their locations in the investigated crystal object. In contrast, glasses, prepared usually by *nonequilibrium* cooling procedures from liquids, are not crystalline objects. They are even often termed *undercooled liquids*. For example, Warren emphasizes that in spite of mechanical properties of glasses as true solids, they are rather similar to liquids [19]. However, there is one important difference: in glasses, each atom is surrounded by permanent neighbors at fairly definite distances, while the neighbors in liquids are constantly changing. On the other hand, following Stebbins, glass-forming liquids are often considered as highly structured objects, where an energetically significant order extends beyond the first coordination sphere [20]. Due to these features, glasses and the glass-forming liquids can be defined as amorphous states of matter.

Structurally, silicate glasses can be described, for example, as systems with a short-range order dominated by the SiO_4 and AlO_4 tetrahedral units, while the intermediate range order is dictated by the coordination type for each oxygen atom. In addition, the oxygen atoms can be chemically connected with different cations, for example, Na^+. Thus, the structure of glasses can be characterized semiquantitatively in terms of the tetrahedral silicon sites, Q^n. Then, distributions of these sites are key parameters for descriptions of intermediate range orders and explain the large variations in physical and chemical properties of glasses.

In practice, the Q^n characterizations of glasses are not simple, and often controversial because they are remarkably dependent on the physical method applied. There are situations when, for example, Raman spectroscopy reveals the presence of many different SiO_4 species while the x-ray scattering experiments show only one or two species.

The so-called bioactive glasses, discovered 30 years ago, are still subject to very extensive investigations [21]. The bioactivity of the glasses is connected with the presence of surface layers, containing the Ca-deficient hydroxycarbonate apatite. Therefore, developing new synthetic approaches and procedures that result in new glass systems having enhanced bio-effects, is constantly the focus of researchers. One such approach is based on the mesoporous bioactive glasses of the formula $CaO–SiO_2–P_2O_5$, where the calcium content can be variable in large limits. The main problem in probing such systems is their structure, which cannot be solved by regular diffraction methods due to their strongly amorphous nature. A simple model describing their structures is shown in Figure 2.4, where SiO_2 units are the main pore-wall components while the phosphorous sites form a separate amorphous calcium orthophosphate phase. This phase is dispersed over the pores as clusters.

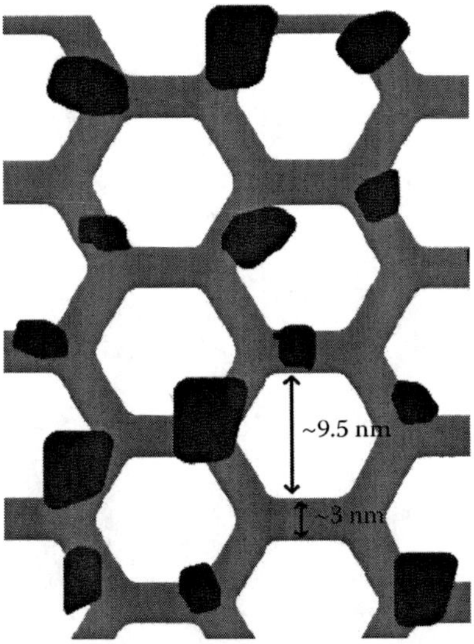

FIGURE 2.4 A structural model proposed for bioactive glasses: the $CaO-SiO_2$–amorphous pore-wall component is noted in gray while the inclusions into this network, representing the calcium orthophosphate phase, are shown in black. (Reproduced with permission from Leonova, E. et al., *J. Phys. Chem. C*, 112, 5552, 2008.)

Due to the amorphous nature, such structural models can be obtained only on the basis of solid-state multinuclear NMR experiments.

Finally, besides complex structural organizations, glasses and glassy materials exhibit a complicated dynamics [20]. Since the dynamic behavior is one of the fundamental characteristics of glasses and glassy compounds, it is constantly of great interest. Currently a few experimental approaches are capable of detecting heterogeneity in dynamic behavior of glassy systems: optical techniques with dye probes, nonconventional dielectric methods, the nonresonant spectral hole-burning, or atomic force microscopy. Again solid-state NMR can be an independent tool for these tasks.

2.6 ZEOLITES

Zeolites, physical and chemical properties of which are well described in reviews [22–24], represent the most important class of known materials playing a very important role as catalysts, sorbents, and ion exchangers. Zeolites are built as systems with the corner-sharing SiO_4^{4-} and AlO_4^{5-} tetrahedral units. Generally they are formulated as open framework aluminosilicates. Since the commercially available zeolites, widely used in industrial processes, are more crystalline, their structures are well established by the XRD methods. It should be added that the lattice frameworks of the zeolites can be assembled from single building units which form, however, many different structural motives. One of the frameworks is illustrated in Figure 2.5.

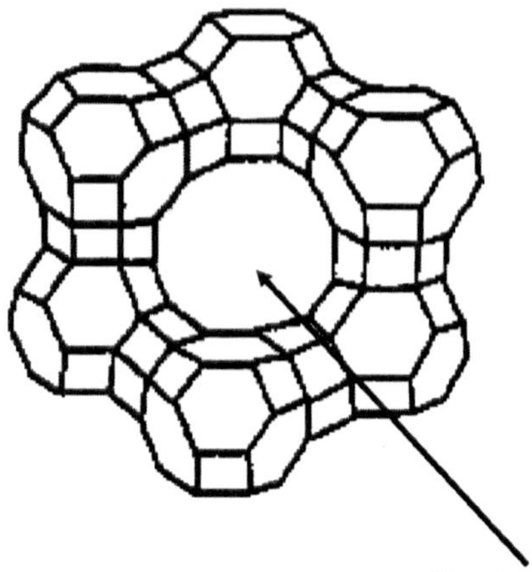

FIGURE 2.5 A lattice structure established for typical zeolites formed from the sodalite cage building blocks. These zeolites also known as zeolite-X or zeolite-Y, obtained synthetically at different silicon/aluminum ratios, and widely used as a "cracking" catalyst in petroleum refining.

The main feature of the frameworks is the presence of a large central cavity. The size of the cations situated in these cavities can be different and this circumstance dictates, therefore, the sizes of guest molecules that can be potentially absorbed. The latter factor plays an important role in practical applications of zeolites. For example, the lattice of the well-known zeolite-X or zeolite-Y, similar to Figure 2.5 and obtained at different Si/Al ratios, provides their wide applications in petroleum refining [25].

Klinowski [26] formulates the following important factors that strongly affect the catalytic activity of zeolite systems:

1. The chemical composition and the framework ordering
2. The nature and quantity of aluminum in extra-framework
3. Properties of the Bronsted acid sites on the matrix surface
4. The nature of guest organic molecules, situated in the intracrystalline space
5. Shape selectivity

The conclusion is obvious: regulating these factors, based on modifications of the zeolite frameworks and adjustments of their pore sizes and also their surface properties, opens the way to creation of new catalytic systems. However, new procedures involved in the synthesis more often result in zeolite systems with modified structures that are difficult to describe by conventional techniques for structural elucidations. Again solid-state NMR on ^{29}Si, ^{27}Al, and other nuclei can help in this context.

As seen from the previous sections, the materials considered differ remarkably due to different chemical compositions and structures. In spite of this variety, one can formulate two common features: first, the systems are generally strongly amorphous and second, they have a porous nature. Both these circumstances create a number of common problems in structural characterizations that can affect the progress in design and practical applications of these materials. These common problems and methods of their solution are discussed in the next section.

2.7 DESIGN OF POROUS AND LAYERED MOLECULAR SYSTEMS

Since the controlled design of porous and layered molecular systems with 3D porous network became the most exciting direction of modern materials science, numerous research papers and reviews have been published on the synthesis and applications of these materials. The properties of these materials can be found, for example, in recent reviews [3,27]. The properties and structures of porous metal phosphates and phosphonates, which change with the nature of the metals, can be found in Refs. [28–33]. It should be noted that these materials are widely used as ion exchangers for processing of radioactive waste streams, sorption, catalysts, sensors, and in nonlinear optics. Porous supramolecular assemblies, based on metal cyanides, are described, for example, in articles by Bignozzi et al. [34] and Chesnut et al. [35]. These systems are very perspective in the context of molecular magnetism. Finally, considerable progress has been made in the synthesis of porous molecular sieves [36]. They can be applied as contrast agents for diagnostic magnetic resonance imaging, semiconductor nanowires, and even as hosts for laser dyes.

As we have seen, the area of applications of the porous systems depends on their physical and chemical properties and structure, especially on the morphology of the porous medium. In addition, it is critical to know the pore morphology in sufficient detail because it determines dynamics of the pore-confined molecules. This parameter can be probed by different physical methods such as mercury intrusion, N_2-adsorption, neutron scattering, thermoporometry, NMR spectroscopy, and NMR relaxation (see the following), and a combination thereof [37]. At the same time, the nature, concentration, and strength of the active sites and also the size of the channels, important for catalysis, can be studied by x-ray and neutron scattering [38], FTIR spectroscopy [39], TEM [40], calorimetry [41], and solid-state NMR. Computational methods based on the *ab initio* or Monte Carlo calculations can be successfully applied for structural studies and establishment of the driving forces for the framework assemblies [42].

The scenario in the design of porous systems can be represented by zeolites. In commercially available zeolites, the pore size is relatively small and pore entrances are only ca. 10 Å in diameter, resulting in strong binding of the sorbed substrates to the cavities and tunnels. To avoid this problem, important in the context of catalysis, significant synthetic efforts have led to the development of systems with larger pore sizes [43]. However, such materials have not found broad commercial utilization because of their lattice instability and the pore-blockage phenomenon. The development of new mesoporous phases such as FSM-16 [44] and MCM-41 [45] and their intense studies have led to silica-based materials of high surface areas where the pore size ranges from 20 to 100 Å [46]. Finally the gap in pore sizes of 10–25 Å has been filled by the amorphous mixed oxides [47] obtained by sol–gel techniques with various amines used as the templates. These "sol–gel" materials represent systems where a matrix is created by *structure-directing reagents* (SiO-moieties or transition metal oxides in variable oxidation states). This matrix has pores, channels, and cavities with the pore sizes and shapes of the cavities dependent on both the synthetic method employed and postsynthesis treatments. For example, calcination at high temperatures strongly affects the matrix structure.

Finally, besides the architecture considered earlier, the nature, concentration, and strength of the active sites important for catalysis can be modified by different transition and main-group metal ions that can be added in high or low concentrations. It is clear that structural control of such systems requires reliable characterization methods.

2.7.1 General Problems in Structural Characterizations of Amorphous Porous Materials

Characterization of porous materials can be performed by various physical methods and/or theoretical approaches. It is obvious that the nature of the material prompts the choice of the corresponding and adequate methods or combinations thereof. For example, *tin*-containing materials can be generally characterized by UV and Mossbauer spectra [48], while their porosity can be measured by N_2-absorption-desorption isotherms. When targets in materials are active in the Mossbauer spectra, then this spectroscopy is the best method for determination of oxidative states of metal ions. Figure 2.6 shows the Mossbauer spectra recorded

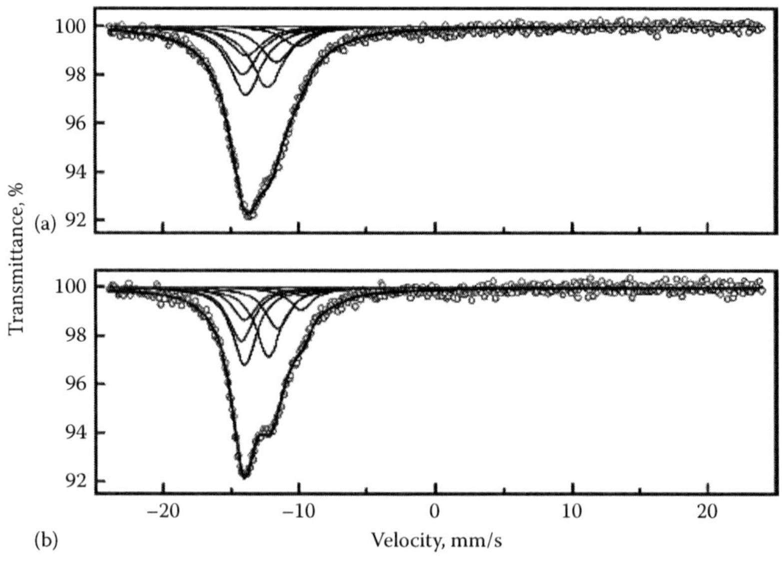

FIGURE 2.6 The Mossbauer spectra recorded for $Eu_2MgSi_2O_7$ (a) and $Eu_2MnSi_2O_7$ (b). The observed spectra are shown as the circles. The experimental spectra can be rationalized in terms of the conventional theory. The fitting curves are shown as solid lines. The fittings lead to determination of Mossbauer's parameters, which can be even treated by deconvolution procedures. (Reproduced with permission from Endo, T. et al., *Inorg. Chem.*, 49, 10809, 2010.)

for two inorganic solids, $Eu_2MgSi_2O_7$ and $Eu_2MnSi_2O_7$, which are perspective as materials with unusual magnetic properties [49]. Following from the Mossbauer spectra, an absorption peak is observed at 13 mm/s, well corresponding to the divalent Eu ions.

The microstructures and/or the morphology of porous systems, for example, microporous Mg/Al/Si materials or the hybrid mesoporous organo-Al-silicates can be successfully probed by TEM and SEM [50]. As noted earlier, sometimes, FTIR and UV spectroscopy can be also applied. Clearfield and coworkers [51] have demonstrated applications of x-ray photoelectron spectroscopy (XPS) to characterize mixed oxide catalysts $NiO/SiO_2/Al_2O_3$. Applications of the Raman and extended x-ray absorption fine structure (EXAFS) methods for studies of hybrid aluminum and gallium systems can be found in the articles [52,53]. Electron nuclear double resonance (ENDOR) [54] and the temperature programming reduction method [51] are also available and can be useful in studies of porous molecular systems.

In spite of the large arsenal of modern physical methods, single crystal XRD still remains the most important and direct method for accurate molecular structural characterizations of solids. However, large and high quality single crystals required for this method are extremely difficult to obtain for many porous composite materials. Moreover this is a problem even for the silica-zeolites and their numerous derivatives widely used in catalysis. Therefore, structural investigations of such materials can be based only on powder XRD, involving three main steps [55]:

1. First, unit cell parameters and the space groups are determined from the XRD patterns, interpretations of which often require independent supporting data from another method, for example, solid-state NMR.
2. Second, an initial structural model is proposed, which should be consistent with general chemical concepts while being in agreement with available experimental data. The authors [53] emphasize that this step is most critical because of the well-known phase problems.
3. Third, the structural analysis can be completed by localization of the missing atoms in structural model refinement against the experimental diffraction data.

However, in spite of the impressive developments of this method and particularly data treatments, determination of the structure directly from the powder XRD pattern is often difficult or even impossible, particularly for complex heterogeneous materials. For example, the sol–gel synthetic procedures with appropriate organic template molecules, providing successful regulations of the pore sizes and entrances and even lengths and shapes of the pore channels, nonetheless lead to strongly amorphous materials. While zeolites (such as materials of the MCM-41 family) show informative XRD patterns, generally with an intense diffraction peak, three higher order peaks, and a diffuse peak, XRD data obtained for silica-based materials prepared by the sol–gel method are of insufficient quality for interpretation. Figure 2.7 illustrates a typical pattern where there are no additional peaks, $2\Theta < 2°$ at measurements with small angles [56] and thus no valuable structural conclusions can be

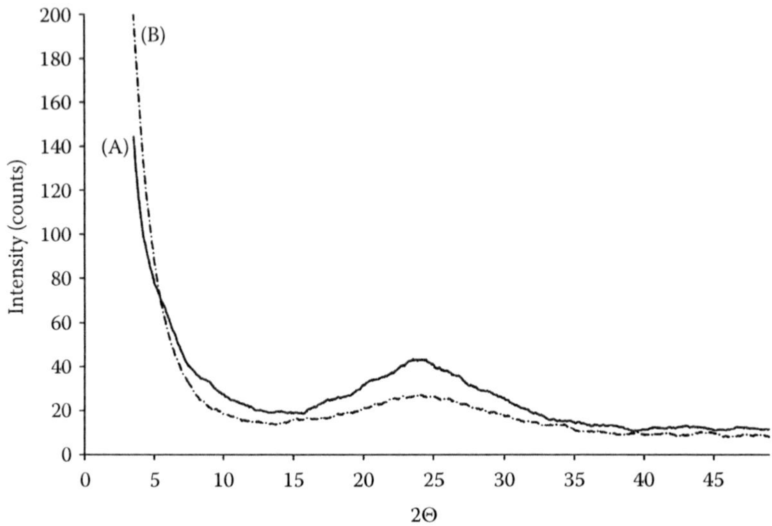

FIGURE 2.7 X-ray powder diffraction patterns typical of manganese–alumina–silica composites obtained by the sol–gel method: (A) the material as made and (B) the material calcined in air at 450°C. (Reproduced with permission from Bakhmutov, V.I., *Chem. Rev.*, 111, 530, 2011.)

drawn. Such XRD patterns only confirm the amorphous nature of the systems and point toward applications of other, more adequate physical methods. It is seen also that the structural information does not become richer after calcination procedures when the matrix changes.

2.7.2 Some Key Structural Issues in Characterizations of Amorphous Materials Modified by Metal Ions and Methods for Structure Solution

Insertion of the transition or main-group metal ions into the zeolite or other silica-based frameworks to tune their heterogeneous catalytic properties can be performed directly by synthesis in the presence of the metal ions, for example, by the sol–gel method, including applications of surfactants, by ion exchange [57], or even by simple impregnation [58] of the pre-made materials in the solutions. It is clear that the catalytic behavior of these systems is dependent on the metal–ion distribution (homogeneous or nonhomogeneous), interactions between the metal ions and the matrix of the materials, and finally on their interactions with the absorbed molecules. Since *a priori*, the metal atoms can be uniformly incorporated into the silica matrix and/or they can be situated within the cavities, or/and form separate phases, this key structural issue in studies of these materials is not simple. Add also that the final products are generally calcined at high temperatures in the air atmosphere. Under these procedures, the metal ions can change oxidation states.

The final products are often too amorphous and application of x-ray powder diffraction techniques is useless (see Figure 2.7), especially at low concentrations of the doped ions. Even the formation of a second phase at their low concentrations can be below the level of x-ray detection. For example, separate phases of supermicroporous mixed oxides of the $NiO/SiO_2/Al_2O_3$ family remain "invisible" in most of their powder XRD patterns even at relatively high nickel loadings [59]. It is interesting however that in some cases, the Ni^0 and NiO clusters can be observed experimentally in the powder XRD patterns.

As already mentioned, distribution of metal-ions throughout the modified porous materials is one of their very important characteristics. In principle, such information can be obtained by the TEM technique. However, the TEM patterns, as illustrated in Figure 2.8 for a typical porous material of the formula $MnO/Al_2O_3/SiO_2$, are not valuable in the aforementioned structural context.

If again, the doping metal ions in amorphous systems are active in the Mossbauer spectra, for example, ^{119}Sn and ^{57}Fe [60], chemical isomer shifts and quadrupole splitting, obtained for the nuclei of such atoms in the Mossbauer spectra, determine accurately their *oxidation states*, *coordination numbers*, and *geometries* (tetrahedral, octahedral, etc.). Moreover, probing porous materials of the Sn-SBA-15 family [49,61] has demonstrated that the Mossbauer spectra are capable of discrimination between the framework and extra-framework tin species.

At the present time, the x-ray absorption techniques, EXAFS and x-ray absorption near edge structure (XANES), are viewed by researchers as one of the most promising methods for structural characterizations of porous materials [62]. In fact these techniques can potentially provide detailed information on coordination numbers of metal ions and even their local environments, including the bond lengths. For example, these techniques, applied to structural analysis of the mesoporous materials Ni-MCM-41, have revealed that first, the Ni ions are tetrahedral, and second, they have the surrounding oxygen atoms, incorporated into the material matrix. It has been established, however, in characterization of the Ni-MCM-41 systems used in the synthesis of carbon nanotubes that the size of the Ni–O clusters can be measured from the coordination numbers determined by EXAFS only for particles smaller than 3–5 nm [63].

Among various physicochemical methods widely applied in chemistry, XPS, can be a successful alternative approach to evaluate different sites and oxidation states

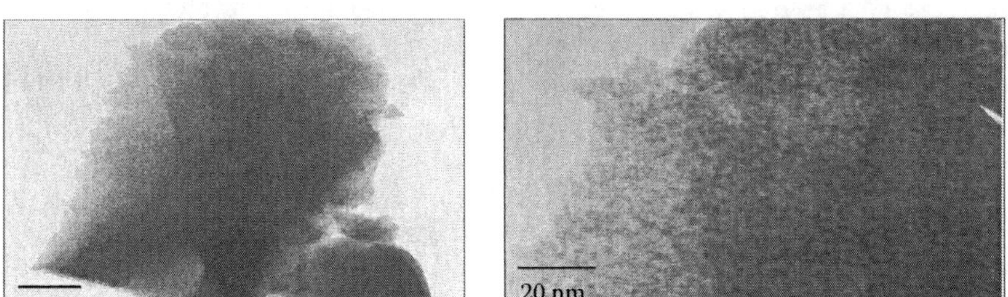

(a) (b)

FIGURE 2.8 Electron micrographs obtained for the porous material of the formula $MnO–Al_2O_3–SiO_2$, prepared at 19.8% Mn, 2.2% Al, and 26.9% Si. (a) The material preheated to 540°C and (b) The material preheated to 650°C. (Reproduced with permission from Bakhmutov, V.I., *Chem. Rev.*, 111, 530, 2011.)

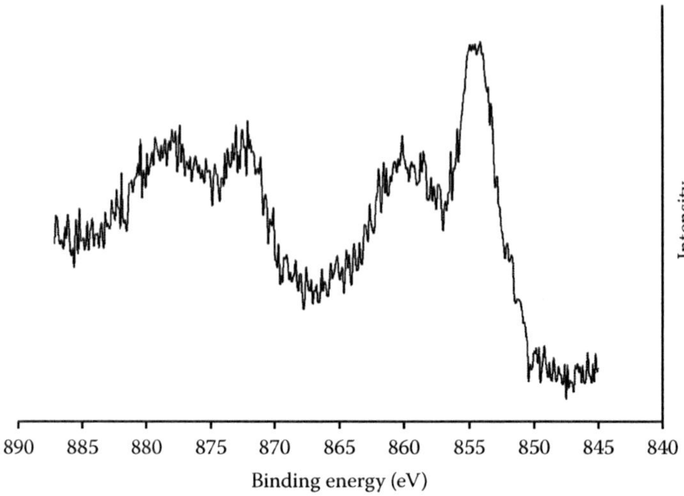

FIGURE 2.9 The XPS Ni 2p3/2 spectrum of material SiO_2–Al_2O_3–NiO with 13.4 wt% of Ni calcined at 540°C. (Reproduced with permission from Bakhmutov, V.I., *Chem. Rev.*, 111, 530, 2011.)

of metal ions in heterogeneous systems. It is important that this method does not require high crystallinity (unlike XRD) or a minimal particle size (unlike TEM). At the same time, an XPS analysis of some alumina-supported nickel catalysts [64] has shown that the final results can depend on the repartition of the catalyst components and therefore independent data and measurements are often required for correct interpretation.

Figure 2.9 illustrates an additional, rather intrinsic problem that often emerges in the studies of very amorphous porous systems such as supermicroporous materials of the SiO_2–Al_2O_3–NiO family [59]. As seen, the typical XPS pattern obtained for the material SiO_2–Al_2O_3–NiO containing 13.4% Ni (by weight) shows a peak centered at 854 eV and a satellite at 860 eV, corresponding to the NiO particles. However, the observed lines are very broad and the presence of more than one type of nickel species cannot be ruled out. Moreover even the standard deconvolution procedures for such XPS patterns are doubtful because the two broad bands might fit two peaks at 854.2 and 859.8 eV from "free" NiO, or three peaks at 854.3, 857.8, and 859.9 eV as a combination of free and silica-bound NiO, or even four peaks at 854.3, 855.9, 858.3, and 860.6 eV. Clearly, such XPS data are not particularly insightful.

The UV, IR, and Raman methods can be also applied to monitoring transformations of the porous systems during the synthesis and/or their specific treatments. A good example is functionalization of the MCM-41 surface to create active titanium sites [65]. The IR spectra show bands characteristic of silica lattice vibrations. It has been demonstrated that the intensity of the S–O band at 960 cm^{-1} increases considerably upon replacement of the proton in the Si–O–H moieties with Ti atoms. Generally, the presence of bands around 960 cm^{-1} can be considered as evidence for the presence of Ti, V, or Fe atoms in the zeolite frameworks [66,67].

EPR spectroscopy is known to be a direct method for determination of the nature and location of paramagnetic metal ions in porous materials. However, use of EPR (strongly limited by short electron relaxation times, leading to very broad EPR lines) also has a number of intrinsic problems, mainly dealing with interpretation and assignment of the EPR patterns. For example, Kovalchuk et al. [65] carefully analyze EPR spectra recorded for materials of the Fe(III)MCM-58 family that exhibit signals at g = 4.49 with shoulders at g = 9.05 and 2.30 along with intense broad signals at g = 2.00, typical of Fe(III) in zeolites. The analysis shows that assignments of the Fe^{3+} EPR signals "to various iron species in zeolites are controversially discussed in numerous papers and are still a matter of debate." In general, EPR data do not provide an opportunity to distinguish accurately between the Fe(III) ions in the framework and in the ion exchange sites. Similar problems can appear in the EPR spectra of porous systems containing manganese or cobalt ions, obviously requiring additional independent information, obtained, for example, by the XANES and EXAFS methods.

Some of the aforementioned "intrinsic" EPR problems can be avoided by the pulsed ENDOR technique, described in works [68,69]. Goldfarb and coworkers, applying the ENDOR techniques for aluminophosphate materials $AlPO_4$-20 and analogs, have revealed the presence of a single type of Mn(II) ions with a ^{55}Mn hyperfine coupling constant of 8.7 mT the field sweep echo-detected spectrum. Then, the ENDOR spectra, obtained with Mims and Davies sequences, displayed a ^{27}Al signal at the Larmor frequency and *a* ^{31}P *doublet corresponding to a hyperfine coupling of* 8 MHz. The large ^{31}P hyperfine coupling and the weak ^{27}Al coupling provide unique and direct evidence for Mn(II) substitution of the Al framework.

Finally the nature of paramagnetic centers in amorphous porous systems may be studied by variable-temperature magnetic susceptibility experiments represented, for example, in Ref. [59]. Data treated in terms of the Curie and the Curie–Weiss law provide the determination of the spin states and also for observation of both magnetically isolated and coupled centers. However, the method cannot distinguish between the incorporated and extra framework metal ions.

As the early brief summary shows, in many cases the large arsenal of physical methods still does not solve the key problem of structural determinations for amorphous, porous, metal ion-containing materials. It is obvious also that elucidating structural information is particularly problematic at low concentrations of the metal-ion dopant. We show in the following chapters that the solid-state NMR spectral and relaxation techniques can be most useful for these purposes, providing detailed structural information even for strongly amorphous porous materials including localizations of the doped metal atoms and also their oxidation states and coordination numbers. NMR is particularly powerful when the materials are mostly diamagnetic or if they contain paramagnetic centers as impurities that do not "distort" the NMR parameters. However the design of new catalytically active porous systems leading to the paramagnetic materials such as Ni-MCM-41, MnSBA-15, Co-MCM-41, Cr-MCM-41, Fe-MCM-58, and Fe-ZSM-5, etc. required development of new NMR approaches, strategies of which are considered after a theoretical introduction to the paramagnetic effects in solid-state NMR.

REFERENCES

1. Ulrich, A. S. 2005. *Prog. Nucl. Magn. Reson. Spectrosc.* 46: 1–21.
2. Spyros, A., Dais, P. 2009. *Prog. Nucl. Magn. Reson. Spectrosc.* 54: 195–207.
3. Ferey, G. 2008. *Chem. Soc. Rev.* 191: 191–214.
4. Brown, S. P., Spiess, H. W. 2001. *Chem. Rev.* 101: 4125–4155.
5. Azevedoa, E. R., Bonagambaa, T., Reichertb. J. D. 2005. *Prog. Nucl. Magn. Reson. Spectrosc.* 47: 137–164.
6. Inigo, A. R., Chiu, H. C., Fann, W. et al. 2004. *Phys. Rev. B.* 69: 075201.
7. Palmer, A. G., Williams, J., McDermott, A. 1996. *J. Phys. Chem.* 100: 13293.
8. Andronis, V., Zografi, G. 1998. *Pharmaceut. Res.* 15: 835.
9. Fujara, F., Petry, W., Diehl, R. M., Schnauss, W., Sillescu, H. 1991. *Europhys. Lett.* 14: 563.
10. Sillescu, H. 1999. *J. Non-Cryst. Solids* 243: 81.
11. Liu, H., Xu, J., Li, Y., Li, Y. 2010. *Acc. Chem. Res.* 43: 1496–1508.
12. Salam, M. A., Makki, M. S. I., Abdelaal, M. Y. A. 2011. *J. Alloys Compd.* 509: 2582–2587.
13. Carson, L., Kelly-Brown, C., Stewart, M. et al. 2009. *Mater. Lett.* 63: 617–620.
14. Bardet, M., Gerbaut, G., Giffard, M. C. et al. 2009. *Prog. Nucl. Magn. Reson. Spectrosc.* 55: 199.
15. Fengel, D., Wegener, G. 1991. *Wood: Chemistry, Ultrastructure, Reactions*. Berlin, Germany/New York: De Gruyter.
16. Knox, J. P. 2008. *Curr. Opin. Plant Biol.* 11: 308.
17. Maunu, S. L. 2002. *Prog. Nucl. Magn. Reson. Spectrosc.* 40: 151.
18. Bohmer, R., Diezemann, G., Hinze, G., Rosser, E. 2001. *Prog. Nucl. Magn. Reson. Spectrosc.* 39: 191.
19. Warren, B. E. 1940. *Chem. Rev.* 26: 237–255.
20. Stebbins, J. F. 1988. *J. Non-Cryst. Solids* 106: 359.
21. Leonova, E., Barba, I., Arcos, D., Hedin, A., Regi, M., Eden, M. 2008. *J. Phys. Chem. C* 112: 5552–5562.
22. Auerbach, S. M., Carrado, K. A., Dutta, P. K. 2003. *Handbook of Zeolite Science and Technology*. New York: Marcel Dekker, Inc.
23. Corma, A., Diaz-Cabanas, M., Martinez-Trigiero, J., Rey, F., Rius, J. 2002. *Nature* 418: 514.
24. Shpeizer, B. G., Bakhmutov, V. I., Clearfield, A. 2006. *Micropor. Mesopor. Mater.* 90: 81.
25. Fyfe, C. A., Feng, Y., Grondey, H., Kokotailo, G. T., Gies, H. 1991. *Chem. Rev.* 91: 1525–1543.
26. Klinowski, J. 1991. *Chem. Rev.* 91: 1459–1479.
27. Bakhmutov, V. I. 2011. *Chem. Rev.* 111: 530–562.
28. Kim, H. C., Park, S. M., Hinsberg, W. D. 2010. *Chem. Rev.* 110: 146.
29. Clearfield, A. 1998. *Progress in Inorganic Chemistry*, Karlin, K. D. (ed.), Vol. 47, pp. 371–510. Oxford, U.K.: Wiley & Sons.
30. Walawalkar, M. G. 1997. *Synthesis and Structural Studies of Model Compounds for Zeolite and Phosphate Materials*. Göttingen, Germany: Cuvillier.
31. Alberti, G. 1996. *Comprehensive Supramolecular Chemistry*, Lehn, J. M. (ed.), Vol. 7, pp. 151–187. New York: Elsevier.
32. Thompson, M. E. 1994. *Chem. Mater.* 6: 1168.
33. Cao, G., Hong, H.-G., Mallouk, T. E. 1992. *Acc. Chem. Res.* 25: 420.
34. Bignozzi, C. A., Argazzi, R., Bortolini, O., Scandola, F., Harriman, A. 1996. *New J. Chem.* 20: 731–738.
35. Chesnut, D. J., Plewak, D., Zubieta, J. 2001. *J. Chem. Soc., Dalton Trans.* 2567.

36. Trofymluk, O., Levchenko, O., Sarah A. A., Tolbert, H., Navrotsky, A. 2005. *Chem. Mater*. 17: 3772.
37. Hansen, E. W., Fonnum, G., Weng, E. 2005. *J. Phys. Chem. B* 109: 24295.
38. Flodstrom, K., Wennerstrom, H., Alfredsson, V. 2004. *Langmuir* 20: 680.
39. Haddad, E., Nossov, A., Guenneau, F., Gedeon, A. 2004. *C.R. Chim*. 7: 305.
40. Janssen, A. H., Van Der Voort, P., Koster, A. J., De Jong, K. P. 2002. *Chem. Commun*. 1632.
41. Petrovic, I., Navrotsky, A., Davis, M. E., Zones, S. I. 1993. *Chem. Mater*. 5: 1805.
42. Zwijnenburg, M. A., Bromley, S. T., Foster, M. D. et al. 2004. *Chem. Mater*. 16: 3809.
43. Davis, M. E. 1993. *Acc. Chem. Res*. 26: 111.
44. Yanagiwawa, T., Shimizu, T., Kuroda, K., Kato, C. 1990. *Bull. Chem. Soc. Jpn*. 63: 988.
45. Kresge, C. T., Leonowicz, M. E., Roth, W. J., Vartuli, J. C., Beck, J. S. 1992. *Nature* 359: 710.
46. Ciesla, U., Schuth, F. 1999. *Micropor. Mesopor. Mater*. 27: 131.
47. Speizer, B. G., Clearfield, A., Heising, J. M. 2005. *Chem. Commun*. 2396.
48. Shah, P., Ramaswamy, A. V., Lazar, K., Ramaswamy, V. 2007. *Micropor. Mesopor. Mater*. 100: 210.
49. Endo, T., Doi, Y., Wakeshima, M., Hinatsu, Y. 2010. *Inorg. Chem*. 49: 10809–10814.
50. Maxim, N., Magusin, P. C. M., Kooyman, J., Wolput, J. H. M. C., Santen, R. A., Abbenhuis, H. C. L. 2001. *Chem. Mater*. 13: 2958.
51. Kirumakki, S. R., Shpeizer, B. G., Sagar, G. V., Chary, K. V. R., Clearfield, A. 2006. *J. Catalysis* 242: 319.
52. Haouas, M., Ge'rardin, C., Taulelle, F., Estournes, C., Loiseau, T., Ferey, G. 1998. *J. Chim. Phys. Phys. Chim. Biol*. 95: 302.
53. Taulelle, F., Haouas, M., Ge'rardin, C., Estournes, C., Loiseau, T., Ferey, G. 1999. *Colloid Surface A: Physicochem. Eng. Aspect*. 158: 299.
54. Arieli, D., Vaughan, D. E. W., Strohmaier, K. G., Goldfarb, D. 1999. *J. Am. Chem. Soc*. 121: 6028.
55. Burton, A. W. Z. 2004. *Z. Kristallogr*. 219: 866.
56. Shpeizer, B. G., Bakhmutov, V. I., Zhang, P. et al. 2010. *Colloid Surface A: Physicochem. Eng. Aspect*. 357: 105.
57. Nery, J. G., Giotto, M. V., Mascarenhas, Y. P., Cardoso, D., Zotin, F. M. Z., Sousa-Aguiar, E. F. 2000. *Micropor. Mesopor. Mater*. 41: 281.
58. Shames, A., Lev, O., Iozefron-Khuyavskaya, B. 1993. *J. Non-Cryst. Solids* 163: 105.
59. Bakhmutov, V. I., Shpeizer, B. G., Prosvirin, A. V., Dunbar, K. M., Clearfield, A. 2009. *Micropor. Mesopor. Mater*. 118: 78.
60. Calis, G., Frenken, P., de Boer, E., Hefni, M. A. 1987. *Zeolites* 7: 319.
61. Shah, P., Ramaswamy, A. V., Pasricha, R., Lazar, K., Ramaswamy, V. 2004. *Stud. Surf. Sci. Catal*. 154: 870.
62. Ciuparu, D., Chen, Y., Lim, S., Yang, Y., Haller, G. L., Pfefferle, L. 2004. *J. Phys. Chem. B* 108: 15565.
63. Chen, Y., Ciuparu, D., Yang, Y. H. et al. 2005. *Nanotechnology* 16: S476.
64. Poncelet, G., Centeno, M. A., Molina, R. 2005. *Appl. Catal. A: Gen*. 288: 232.
65. Kovalchuk, T. V., Sfini, H., Korchev, A. et al. 2005. *J. Phys. Chem. B*. 109: 13948.
66. Luca, V., MacLachlan, D. J., Morgan, K. 1997. *Chem. Mater*. 9: 2720.
67. Umamaheswari, V., Böhlmann, W., Poppl, A., Vinu, A., Hartmann, M. 2005. *Micropor. Mesopor. Mater*. 89: 47.
68. Peloquin, J. M., Britt, R. D. 2001. *Biochim. Biophys. Acta* 1503: 96.
69. Dikanov, S. A., Tsvetkov, Y. D. 1992. *Electron Spin Echo Envelope Modulation Spectroscopy*. Boca Raton, FL: CRC Press.

3 Features of Solid-State NMR: Diamagnetic Materials

Solid-state NMR, being widely applied in materials chemistry, is especially valuable at probing complex molecular systems, particularly composites and other heterogeneous aggregates. This method represents an ideal tool for characterizations of objects with dimensions on a nanoscale. The number of publications where solid-state NMR has been used increases constantly and very rapidly. For example, following Dybowski, just between 2005 and 2009 the solid-state NMR literature has enriched itself with more than 3300 articles [1]. In addition currently this literature includes a large number of excellent reviews that describe in detail solid-state NMR applications for supramolecular systems [2], molecular sieves catalysts [3], and zeolite structures [4]. Some of them demonstrate new solid-state NMR techniques and approaches. For example, the review by Babonneau et al. [5] describing the new techniques, aimed at precise characterizations of interfaces in sol–gel materials and hybrid interfaces, is of great interest for materials chemists because it shows how the researcher can use the NMR parameters from isotropic chemical shifts to quadrupolar and dipolar interactions.

The physical basis of NMR, briefly introduced in Chapter 1, is common for any phase state. Nevertheless, chemists, being even familiar with NMR spectroscopy for solutions, face a number of difficulties in moving to solid-state NMR, from interpreting parameters obtained by experiments, methods of signal registration, to line shapes observed in solid-state NMR spectra. This chapter, considering the features of solid-state NMR in diamagnetic solids, can help to overcome these problems. In addition, the goal of this chapter is to prevent you from being an ineffective user confessing to a "black box approach."

3.1 FROM ISOTROPIC CHEMICAL SHIFTS TO MAGNETIC SHIELDING TENSORS

Local magnetic fields induced in molecules are responsible for causing the frequency shifts in NMR spectra. Generally in solutions, researchers observe and determine isotropic chemical shifts as dispositions of sharp spectral lines and a single chemical shift value can be attributed to one sort of nucleus.

In reality, chemical shift interaction is anisotropic and chemical shift is a three-dimensional (3D) magnitude characterized by *chemical shift anisotropy*, generally invisible in solutions due to intense and high-amplitude isotropic molecular motions. Since solids show very limited motions, the aforementioned features appear in solid-state NMR spectra and chemical shift anisotropy strongly affects the shapes of observed signals. In addition, microcrystalline or amorphous samples exhibit a random distribution of crystallite orientations relative to the external magnetic field B_0. This leads to a situation when each magnetically distinct nucleus possesses a unique chemical shift anisotropy forming a unique powder NMR pattern [6]. Moreover, in contrast to solutions, in solids even determination of isotropic chemical shifts and/or an analysis of integral signal intensities can be a problem, particularly in noncrystalline materials. Such a situation occurs in amorphous silicates. Generally they are described in terms of Q(4), Q(3), and Q(2) silicon species, characterized by different isotropic ^{29}Si shifts, and ratios between these silicate tetrahedral units. Figure 3.1 shows a ^{29}Si magic angle spinning (MAS) NMR spectrum (top), recorded for a sample of glass K_2O-$2SiO_2$ that illustrates well the problem of isotropic chemical shift determination and the Q(n) analysis [7]. As seen, the resonance centered at −93 ppm is broad and not resolved. Standard deconvolution of the signal can be performed by least squares fittings to three Gaussian line shapes, corresponding to the aforementioned Q(n) units. It has been shown that the fitting procedures give an unreasonable Q(4)/Q(3)/Q(2) ratio that does not correspond to independent data. The conclusion is obvious: without additional assumptions, this deconvolution analysis is impossible and isotropic ^{29}Si chemical shifts represent distributions. Only the application of the special magic angle flipping (MAF) NMR technique (see Chapter 6) helps to avoid this problem. Figure 3.1 shows the analysis of the isotropic projection of the 2D MAF NMR dataset with the best fitting marked as dashed line and the integrated areas marked as circles. The figure shows the data obtained from the simulated 2D MAF data set with the *Gaussian* fits for each site (solid lines). First, the ^{29}Si isotropic chemical shits of the Q(n) sites correspond to Gaussian distributions that can be characterized by positions of −82.74, −91.32, and −101.67 ppm for Q(2), Q(3), and Q(4), respectively. Second, only these treatments lead to the Q(4)/Q(3)/Q(2) ratio confirmed independently.

According to quantum mechanical representations, shown in Chapter 1, the Hamiltonian, describing anisotropic chemical shift interaction, is written as

$$\hat{H}(CS) = \gamma B_0 \sigma \hat{I} \tag{3.1}$$

where

σ is the chemical shift tensor

$\hat{I}$ represents the nuclear spin operator

In turn, the chemical shift tensor σ can be expressed via a 3 × 3 matrix (1) in Scheme 3.1. Each element in this matrix characterizes the i—shielding component at field B_0 that is applied along the j-axis to be related to a measured chemical shift component expressed in Equation (2) of this scheme.

Commonly the chemical shift tensor is known to be asymmetrical. However because the antisymmetric part cannot be observed in powder patterns, the symmetrical part

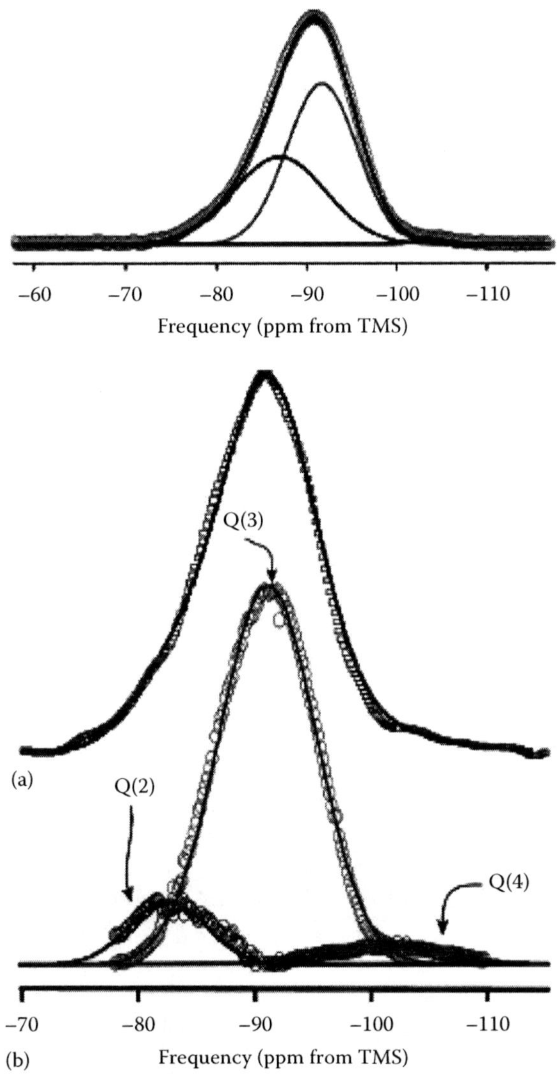

FIGURE 3.1 Top: the ^{29}Si MAS NMR spectrum, recorded for a sample of glass K_2O–$2SiO_2$ (solid line) and its best fitting (circles). The lines within, obtained by deconvolution, represent the three components: Q(4) (around−105 ppm), Q(3) (around−90 ppm), and Q(2) (around−86 ppm). Bottom: (a) the isotropic projection of the 2D MAF data set (squares) with the best fit (dashed line) and (b) the integrated areas (circles) obtained from the simulated 2D MAF data set with the Gaussian fits for each site (solid lines). (Reproduced with permission from Davis, M.C. et al., J. *Phys. Chem. A*, 114, 5503, 2010.)

can be diagonalized in the principal coordinate system where there is a set of diagonal elements referring to the principal components shown in matrix (3).

By convention, the principal components in matrix (3) are described as $\sigma_{11} \leq \sigma_{22} \leq \sigma_{33}$. It should be emphasized that component σ_{11} is taken as the smallest one. In turn, the isotropic chemical shift, noted in terms of magnetic shielding constant

$$\ddot{\sigma} = \begin{bmatrix} \sigma_{xx} & \sigma_{xy} & \sigma_{xz} \\ \sigma_{yx} & \sigma_{yy} & \sigma_{yz} \\ \sigma_{zx} & \sigma_{zy} & \sigma_{zz} \end{bmatrix} \quad (1)$$

$$\delta_{ij} = \frac{\sigma_{iso,ref} - \sigma_{ij}}{1 - \sigma_{iso,ref}} \quad (2)$$

$$\ddot{\sigma}_{PAS} = \begin{bmatrix} \sigma_{11} & 0 & 0 \\ 0 & \sigma_{22} & 0 \\ 0 & 0 & \sigma_{33} \end{bmatrix} \quad (3)$$

SCHEME 3.1 The chemical-shift tensor σ (see Equation 3.1) expressed via 3 × 3 matrix forms with the i, j elements shown in expression (2).

Three conventions are commonly used for reporting chemical shift tensors

$$\sigma_{11} \le \sigma_{22} \le \sigma_{33}$$

$$\sigma_{iso} = \frac{1}{3} \mathrm{Tr}\, \{\ddot{\sigma}\} \equiv \frac{1}{3}(\sigma_{11} + \sigma_{22} + \sigma_{33})$$

$$\Omega \equiv \sigma_{33} - \sigma_{11} \approx \delta_{11} - \delta_{33};$$

$$\kappa \equiv \frac{3(\sigma_{iso} - \sigma_{22})}{\Omega} \approx \frac{3(\delta_{22} - \delta_{iso})}{\Omega}$$

$$|\sigma_{ZZ} - \sigma_{iso}| \ge |\sigma_{XX} - \sigma_{iso}| \ge |\sigma_{YY} - \sigma_{iso}|.$$

$$|\Delta\sigma| \equiv \sigma_{ZZ} - \frac{\sigma_{XX} + \sigma_{YY}}{2}; \quad \eta = \frac{\sigma_{YY} - \sigma_{XX}}{\sigma_{ZZ} - \sigma_{iso}}$$

SCHEME 3.2 The conventional representations describing the chemical shift tensors in solid-state NMR using the frequency-ordered principal shift values, terms span Ω and skew k and the shielding anisotropy, Δσ.

σ(iso), can be defined as shown in Scheme 3.2 (top), that is, it is independent of the chosen coordination frame. Thus, the σ(iso) is identical to the isotropic chemical shift δ(iso) (δ(iso) = −σ(iso)) that is measured in solutions.

As has been emphasized in the literature, there are three conventional procedures, represented in Scheme 3.2 that describe the chemical shift tensors, measured by the solid-state NMR experiments. The first procedure uses the aforementioned formalism of the frequency-ordered principal shift values. The second one applies additional terms: span Ω and skew k. These terms describe the magnitude of the chemical shift anisotropy and degree of axial symmetry of the chemical shift tensor. Then the Ω parameter takes positive values, naturally measured in ppm, while the k parameter changes between +1 and −1, by definition.

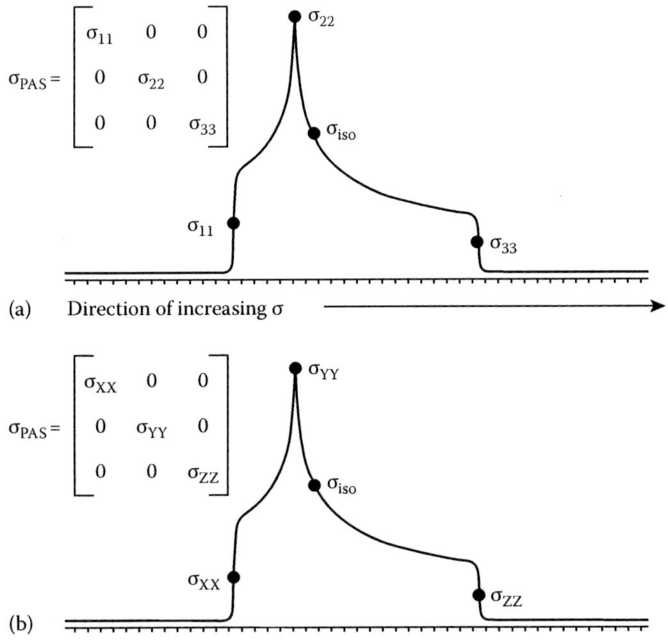

FIGURE 3.2 The shape expected for an NMR signal when the tensor is not axially symmetric. The principal components are shown in different conventional symbols corresponding to formalism (a) and (b). (Reproduced with permission from Widdifield, C.M. and Schurko, R.W., *Concepts Magn. Reson. Part A*, 34A, 91, 2009.)

Finally there is the third procedure that uses the shielding anisotropy, $\Delta\sigma$, and asymmetry parameter η, using principal components (bottom in Scheme 3.2). It is important to emphasize that the principal coordinate systems do not correspond to a molecular coordinate frame. Generally speaking, in practice, the chemical shift anisotropy and asymmetry parameters strongly affect the shape of signals that can be observed in solids.

Figure 3.2 illustrates the shape of signals that is typical of static samples when the chemical shift anisotropy is not axial symmetric.

The 3D chemical shifts as tensor magnitudes can potentially be expressed in the molecular coordinate systems. Then the components σ_{XX}, σ_{YY}, and σ_{ZZ} can be associated with chemical structures. The latter attracts a great interest because the σ_{XX}, σ_{YY}, and σ_{ZZ} measurements can be interpreted in terms of electron distributions [8]. However, generally experimental determinations of how the chemical shift tensors are oriented in the molecular frame are possible only for single crystals. Microcrystalline samples and amorphous and strongly disordered materials prevent such determinations. Nevertheless, in some cases, the orientation of the chemical shift tensor can be reasonably proposed on the basis of the available elements of the symmetry. It is possible, when, for example, rotational axes in molecules are known.

In the absence of these symmetry elements, the orientation of the chemical shift tensors in the molecular coordinate system can be obtained only by computer calculations that is considered in the following. Here, however, two points are very important. First, the calculations require independent experimental confirmations based, in turn, on experimental measurements of the tensors performed by the single crystal NMR approaches or by the recoupling NMR experiments carried out even in powder samples [6]. Second, interpretations of magnetic shielding components, themselves, in terms of the electron nature of molecules are not straight. In fact, according to computer calculations, performed for sets of similar molecules, diamagnetic shielding components change insignificantly and thus changes in the total chemical shift values (and also in the symmetry) are dictated by paramagnetic shielding components.

Taylor et al. [9] demonstrate in practice how to measure chemical shift tensors and anisotropy of chemical shifts in solids and how large they can be. Solid $HgCl_2$, studied in this work, is of great interest for three reasons. First, natural abundance of ^{199}Hg nuclei is relatively low (16.87%). Second, ^{199}Hg content in this molecule is high. Third, due to calculations, it is known that the ^{199}Hg isotropic chemical shift is very sensitive to molecular geometry: *a change in the bond length of 1 pm corresponds to a* δ(iso) variation of 50 ppm or for every 10° change in the Cl–Hg–Cl bond angle, the chemical shift changes by 100 ppm.

Figure 3.3 shows the ^{199}Hg solid-state NMR spectra, recorded for spinning and static $HgCl_2$ samples. The patterns detected at different spinning rates can be treated by standard simulation procedures (available in standard software packages of NMR spectrometers) to result in $\delta_{11} = -401$ ppm, $\delta_{22} = -442$ ppm, $\delta_{33} = -4034$ ppm, δ(iso) = −1625 ppm, and anisotropy of −2409 ppm. The data were obtained by an experiment at a spinning rate of 15 kHz. It has been noted that in all the cases the ^{199}Hg tensor has been measured as axially symmetric.

These data clearly show that the heavy nuclei, such as ^{199}Hg, can be characterized by very large chemical shift anisotropies. Their magnitudes lead to the situation where the solid-state NMR spectrum can stretch from 0 to −4000 ppm. It is obvious that this situation, in practice, is expressed as a very low NMR sensitivity. Increasing the number of scans naturally improves signal-to-noise ratios in the final spectra. However, if nuclear relaxation is long and contents of target nuclei in materials are low, the experimental time will be dramatically elongated.

In a common case, one can formulate a rule: anisotropies of chemical shifts for target nuclei are comparable to variations of isotropic chemical shifts (chemical shift scale) observed for these nuclei. Some data are listed in Table 3.1. As seen from the table, ^{1}H chemical shift anisotropy is minimal while it is maximal for ^{195}Pt nuclei. The results of direct measurements in Table 3.2 confirm this statement.

Finally, it should be noted that an axially symmetric tensor can be visualized as ellipsoid, where the largest component is situated along a chemical bond. However, a direct association of the ellipsoid with electronic density distribution is not correct.

3.1.1 Visualization of Chemical Shift Tensors and Determination of Their Orientations

As we understand now, solid-state NMR experiments in static and spinning samples lead to determination of the 3D chemical shifts. Thus, potentially, solid-state NMR can provide a larger set of information than the experiments performed in solutions.

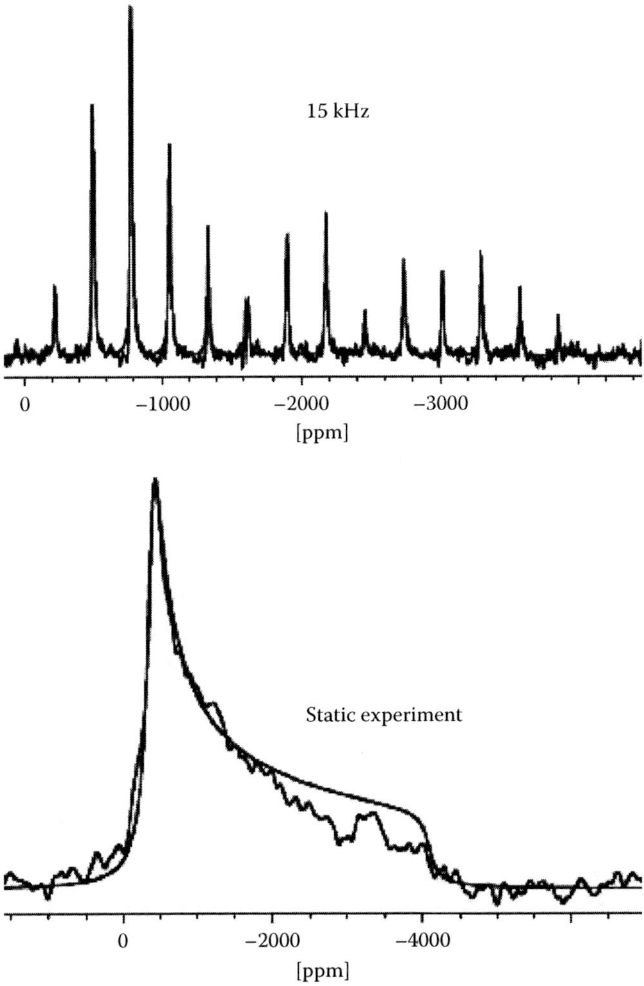

FIGURE 3.3 The ^{199}Hg solid-state NMR spectra, recorded for spinning and static $HgCl_2$ samples: solid lines illustrate the fitting procedures. (Reproduced with permission from Taylor, R.E. et al., J. *Mol. Struct.*, 930, 99, 2009.)

However this set will actually be very useful after visualization of the chemical shift tensors and determination of orientations of these components in the molecular coordinate system. Then, the shielding tensors can be associated in some sense with electrons distributed in a molecule.

According to Widdifield et al. [6], orientations of chemical shift tensors can be found experimentally in the molecular frame, when an object under investigation is a single crystal. It has been noted that generally the spherical polar coordinates are used to describe the shielding response that is parallel to the applied magnetic field B_0 and responsible for frequency shifts in the NMR spectra.

However, the contributions of different electrons to chemical shift components of the tensors can be determined only theoretically. This approach implies computing

TABLE 3.1
Chemical Shift Ranges, δ (SCS), for Different Nuclei

Nucleus	SCS (ppm)[a]
^{1}H	−50 to +20
^{11}B	−130 to +95
^{13}C	−300 to +250
^{19}F	−250 to +400
^{31}P	−125 to +500
^{15}N	−500 to +850
^{195}Pt	−1,370 to +12,000

[a] The values with "+" correspond to low-field displacements.

TABLE 3.2
Components of Screening Tensors and the Δσ Values (in ppm) for Some Nuclei, Measured in the Solid State NMR Spectra

Compound	σ_{ZZ}	σ_{YY}	σ_{XX}	Δσ
$^{14}NH_3$	237.3	278.0	278.0	−44.3
$Me^{14}NC$	370	10	10	360
$HC^{14}N$	348	−215	−215	563
$C_6{}^{19}F_6$	465.9	310.8	310.8	155.1
$Xe^{19}F_4$	528.5	58.5	58.5	470

the diamagnetic and paramagnetic shielding components represented in Figure 3.4 in the limits of Ramsey's theory.

In this figure the diamagnetic term is expressed via Equation (1), where $\sigma_{ij}(d)$ is element ij of a diamagnetic shielding tensor, ψ_0 is a ground-state wave function, $\delta_{ij}(K)$ is the delta function, and r_{kN} is the distance between electron k and nucleus N. The paramagnetic term is given by Equation (2). Here $\sigma_{ij}(p)$ is element ij of the paramagnetic shielding tensor, E_n and E_0 are the eigenvalues associated with an excited and the ground electronic state, respectively, $L^{\wedge}_{ki}$ and $L^{\wedge}_{kNj}$ are angular momentum, and the symbol "c.c." represents the complex conjugate. On the level of computing, the chemical shift components can be obtained, for example, by Amsterdam density functional software.

Generally the Amsterdam density functional methodology includes optimizations of molecular geometries and then, computing the NMR chemical shift tensors is performed on the basis of the NMR and electron paramagnetic resonance (EPR) modules using the gauge-including atomic orbitals (GIAO) method at different basis sets [6]. In turn, the density functional theory (DFT) calculations apply the Becke

$$\sigma_{ij}^{d} = \frac{e^2\mu_0}{8\pi m_e}\left\langle \psi_0 \left| \sum_k \frac{r_k \cdot r_{kN}\,\delta_{ij}^{K} - r_{kNi}\,r_{kNj}}{r_{kN}^3} \right| \psi_0 \right\rangle \qquad (1)$$

$$i, j = x, y, z,$$

$$\sigma_{ij}^{P} = \frac{e^2\mu_0}{8\pi m_e^2}\sum_n \frac{1}{E_n - E_0} \times \left[\left\langle \psi_0 \left| \sum_k \hat{L}_{ki} \right| \psi_n \right\rangle \left\langle \psi_n \left| \sum_k \frac{\hat{L}_{kNj}}{r_{kN}^3} \right| \psi_0 \right\rangle + \text{c.c.} \right] \qquad (2)$$

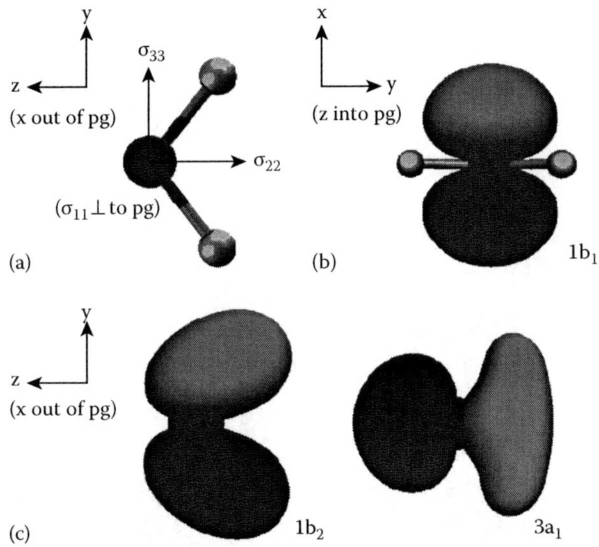

FIGURE 3.4 Ramsey's theory of the diamagnetic and paramagnetic shielding components and the chemical shift tensor calculated for oxygen in molecule H_2O in the molecular coordinate frame (a). The patterns (b) and (c) show molecular orbitals that provide main contributions to magnetic shielding. (Reproduced with permission from Widdifield, C.M. and Schurko, R.W., *Concepts Magn. Reson. Part A*, 34A, 91, 2009.)

exchange functional and the Lee correlation functional (BLYP). Detailed descriptions can be found in the review work of Widdifield and Schurko [6].

The results of the calculations carried out for ^{13}C nuclei in molecule C_2H_4 illustrate the influence of diamagnetic and paramagnetic terms on the total chemical shift tensor components. The calculations lead to the following tensor parameters: diamagnetic components σ_{11}(dia), σ_{22}(dia), and σ_{33}(dia) are computed as 247.32, 247.26, and 261.43 ppm, respectively, while the paramagnetic components σ_{11}(p), σ_{22}(p), and σ_{33}(p) are obtained as −318.51, −186.92, and −50.44 ppm.

Figure 3.4 illustrates the data obtained for the oxygen shielding in the water molecule, where the contributions of different molecular orbitals can be clearly seen.

However, on the whole, the calculations are not simple even in the case of relatively simple molecules and require a special and detailed knowledge in this field. On the other hand, the calculations can be exclusively useful particularly for porous systems, where chemical shifts of nuclei on the surface and in the matrix can be different but cannot be distinguished experimentally.

3.2 NUCLEI WITH A SPIN OF 1/2

Historically quadrupolar nuclei have played a leading role in solid-state NMR experiments performed for different materials particularly in static samples. The reason is simple: the requirement of high spectral resolution in such cases is not critical. Later due to development of the advanced high-resolution techniques, the accent has shifted toward nuclei with spins of 1/2. At the present time, solid-state NMR experiments in MAS, performed for nuclei ^{1}H, ^{19}F, ^{13}C, ^{15}N, ^{31}P, ^{29}Si, ^{117,119}Sn, ^{129}Xe, etc., play a central role in structural investigations of different materials: catalysts, polymers, glasses, zeolites, liquid crystals, natural, and pharmaceutical products. In addition, the MAS NMR techniques are very useful at probing their dynamics. Many aspects of these applications (that are not the focus of this book) can be found in original papers and reviews. For example, NMR studies of amyloid fibrils can be found in the review by Tycko [10]. ^{31}P NMR spectroscopy, applied for different object in food science, is reviewed by Spyros and Dais [11]. ^{19}F Solid-state NMR applications for biomembranes have recently been considered by Ulrich [34] and MAS NMR applied for peptides and proteins can be found in the review by Reif et al. [12].

The principal advantage of the MAS techniques is the solid-state MAS NMR spectra of nuclei with spins of 1/2 in diamagnetic solids can be well resolved as in solutions, providing, thus, detailed information on the nature and number of chemically and structurally different sites [13]. At the same time it should be emphasized that in spite of the advanced techniques, the achieved spectral resolution is strongly limited by the nature of systems. First of all, their crystallinity and molecular mobility play a major role. For example, a crystalline glycine sample, used as an NMR standard, shows the ^{13}C resonances with line widths of <5 Hz in ^{13}C CP MAS NMR spectra. Figure 3.5 shows the ^{29}Si MAS NMR experiments performed for the poorly and extremely highly crystalline samples of the zeolite ZSM-5 material (from A to C). Gies et al. [4] have performed these experiments after very careful shimming of the magnetic field, precise setting of the magic angle, and in stable sample spinning. As can be seen, the highly crystalline samples provide observation of 21 or even 22 different ^{29}Si NMR signals at a line width of 5 Hz. At the same time the resolution decreases for a poorly crystalline sample. Gies and coworkers have noted that these highly resolved silicon solid-state MAS NMR spectra can be very useful in two aspects: how to synthesize such materials and how to rationalize in detail the effects of temperature conditions and sorbed organics.

In contrast to a poorly crystalline sample in Figure 3.5, showing still different silicon units, Figure 3.6 demonstrates how dramatically the spectral resolution decreases in the ^{29}Si MAS NMR experiments performed on the strongly amorphous silica-based systems, SiO_2–Al_2O_3–ZnO, prepared by the sol–gel method. In this case, the

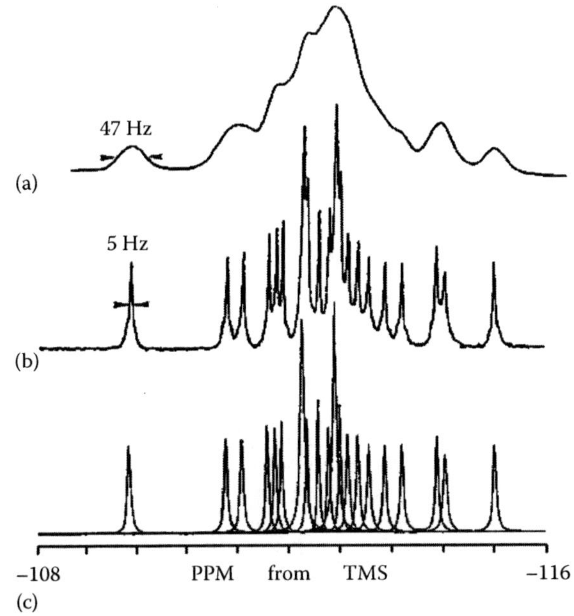

FIGURE 3.5 The ^{29}Si MAS NMR spectra recorded (a): for a poorly crystalline ZSM-5 sample, (b): for a highly crystalline ZMS sample after careful shimming coil procedures and at precise adjustments of the magic angle, (c): a simulated spectrum B. (Reproduced with permission from Fyfe, G.A. et al., *Chem. Rev.*, 91, 1525, 1991.)

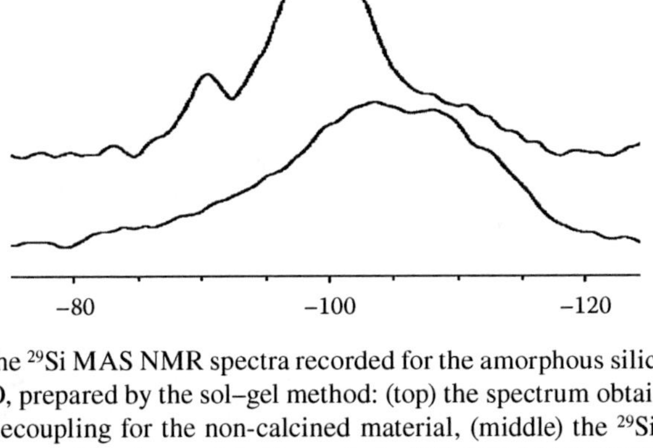

FIGURE 3.6 The ^{29}Si MAS NMR spectra recorded for the amorphous silica-based systems, SiO_2–Al_2O_3–ZnO, prepared by the sol–gel method: (top) the spectrum obtained by one-pulse sequence at ^1H decoupling for the non-calcined material, (middle) the ^{29}Si CP MAS NMR spectrum of the non-calcined material, (bottom) the ^{29}Si $\{^1H\}$ MAS NMR spectrum of the calcined material.

material shows a very broad resonance located between −90 and −120 ppm (the top spectrum, recorded for the noncalcined material by one-pulse sequence at ^{1}H decoupling), where lines coming from different silicon sites are completely overlapped and the structural information is very poor. The situation does not change critically even in the ^{29}Si CP MAS NMR spectrum or after calcination of the material. Finally, we emphasize again that such a dependence on crystallinity is the important feature of solid-state NMR in contrast to solutions, where the line widths depend on relaxation times T_1 and T_2 and shimming coils adjustments only.

Nuclei with spins of 1/2 are well known to undergo dipole–dipole homonuclear or heteronuclear interaction (e.g., proton–proton or proton–carbon interaction, respectively, which completely differs from spin–spin coupling). Therefore, their resonance frequencies will depend on the dipolar coupling constants. This coupling constant governs how the magnetic field, created by neighboring spins, will change at varying orientations of interspin vectors in the external magnetic field. The dipolar coupling constant, D, corresponding to two nuclei (I and S) can be written as

$$D = \left(\frac{\mu_0}{4\pi}\right)\frac{\hbar\gamma_I\gamma_S}{r(I-S)^3} \tag{3.2}$$

where r(I − S) is the vector between the nuclei and other magnitudes are known. In turn, as noted earlier, the dipolar interaction between the pair of nuclei depends on the orientation of the internuclear vector with respect to the direction of the static magnetic field, B_0, via factor

$$\frac{(3\cos^2(\theta)-1)}{r^3} \tag{3.3}$$

where

θ is the angle between the internuclear vector

B_0 is the direction

When molecules move rapidly and isotropically, as in solutions, this factor goes to zero and the nuclei do not feel the presence of dipolar coupling. In this connection, it would be noted that such a fast isotropic motion is also possible in solids. The isotropic motions are typical of plastic crystals. Similarly, relatively small molecular groups, for example, ions SbF_6^- and BF_4^- undergo fast isotropic rotations even in rigid solids. This example has been considered in Chapter 1, where complex $Fe(II)_5(bptz)_4(SbF_6)_{10}(CH_3CN)_{10}$ (see Figure 2.1) has been discussed. This isotropic SbF_6^- rotation leads to complete averaging of the fluorine–fluorine dipolar couplings. Only due to this motion, ^{19}F nuclei appear to be equivalent in the ^{19}F solid-state MAS NMR spectrum and therefore the corresponding ^{19}F–^{19}F spin–spin coupling constants are also invisible.

Generally internuclear dipolar coupling is not averaged in solids. Thus, in powdered samples there is a uniform distribution of internuclear orientations, and therefore the NMR spectrum will consist of a superposition of many lines, corresponding to the different dipolar couplings. The observed lines will be anisotropically broadened, inversely proportional to internuclear distances.

Following from Equation 3.2 (see the γ_I, γ_S constants), the dipolar broadening effect depends on the nature of target nuclei. For example, a ^{1}H resonance of static ice is broadened to ~10^5 Hz due to strongest H–H dipolar couplings, while liquid water gives a line broadened by only <0.1 Hz. At the same time, in contrast to ^{1}H nuclei, ^{29}Si–^{29}Si dipolar coupling is weak to result in line widths of only 2–3 kHz in diamagnetic solids [14]. This weakening effect is related to a low natural abundance of ^{29}Si nuclei.

Fast high-amplitude molecular motions (typical of solutions) can be "imitated" in rigid solids by MAS NMR experiments. Since at the angle θ of 54.4°, named "magic" by Andrew, the factor (3.3) transforms to 0, application of MAS technique eliminates, or, at least, reduces significantly the orientation-dependent effects in the solid-state NMR spectra. However it is important to understand that complete elimination of the dipolar coupling can be reached only at spinning rates that are remarkably larger than the line widths of resonances detected in the corresponding static samples. Principally this is technically possible because modern commercial NMR spectrometers can be equipped with MAS probes giving spinning rates of 30–50 kHz or even more, resulting in high-resolution MAS NMR spectra even for ^{1}H nuclei with the highest dipolar coupling constants.

In performing the MAS NMR experiments, researchers should however remember two important factors:

First, when the *homonuclear dipolar coupling* dominates and a spinning rate is smaller with respect to line widths of resonances in static samples, the MAS NMR spectra will show the isotropic resonances, accompanied by intense sideband patterns; dispositions and intensities of the sidebands depend on the spinning rate.

Second, *all the lines will still be remarkably broadened due to the remaining time-dependent terms.*

Figure 3.7 illustrates this situation and shows clearly that such spectra are very difficult to analyze. In contrast, MAS NMR spectra of molecular systems with dominant hetero- dipolar (e.g., ^{13}C–^{1}H, ^{15}N–^{1}H, ^{13}C–^{19}F, etc.) and chemical shift anisotropy interactions will show sharp lines even at intermediate spinning rates due to the absence of the aforementioned time-dependent terms. Moreover, as noted in Section 3.1, in

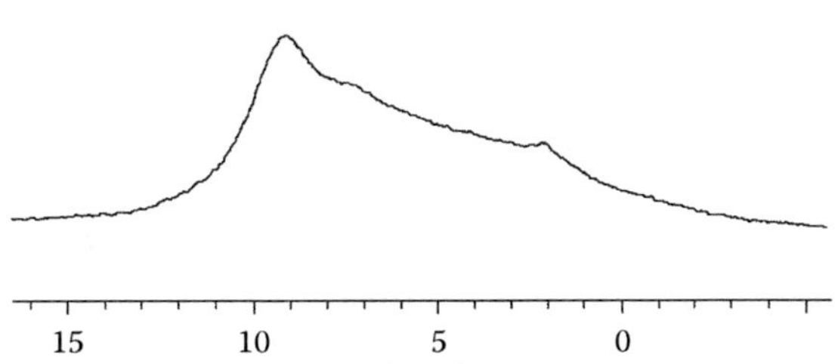

FIGURE 3.7 The ^{1}H MAS NMR spectrum of a non-calcined SiO_2–Al_2O_3 material, obtained by the sol–gel method, recorded with a single pulse sequence at spinning rate of 5 kHz. The spectrum is represented by its isotropic part showing the presence of organic template molecules.

the case of chemical shift anisotropy interactions, distributions of sidebands (see Figure 3.3) can be used to determine magnetic screening tensors (e.g., see Table 2.2). Generally for these determinations the formalism of Strub et al. [8] can be applied. It should be also added that in the case of paramagnetic solids, where the dominating dipolar electron–nucleus interactions have a "hetero" nature, isotropic resonances and their sidebands can still be remarkably broadened. However the broadening effects have another origin: short relaxation times due to small distances between nuclei and unpaired electrons.

3.3 QUADRUPOLAR NUCLEI IN SOLIDS

As noted in Chapter 1, many nuclei, for example, ^{2}H, ^{7}Li, ^{11}B, ^{17}O, ^{23}Na, ^{27}Al, ^{47}Ti, ^{51}V, and ^{93}Nb are quadrupolar; they interact with the external and local magnetic fields (the Zeeman splitting) and with electric field gradients at these nuclei. The latter causes an additional splitting of the nuclear energy levels [15]. Thus, not only isotropic and anisotropic chemical shifts but also nuclear quadrupole coupling constants and asymmetry parameters η_q can be the focus of researchers probing materials populated by such nuclei.

Theoretically, the quadrupole interactions that by their nature disturb the Zeeman energy, are generally considered as the *first*- and *second*-order energy corrections. The first-order term (being anisotropic) includes an angular dependence that is similar to that for dipole–dipole interactions. In contrast, the second-order correction is *isotropic*. At the same time, it is dependent on the Larmor frequency as

$$\frac{E_m = -(e^2qQ/(4I(2I-1))^2 m}{\omega_0} \tag{3.4}$$

where

m is the magnetic quantum number
ω_0 is the Larmor's frequency

The important conclusion is as follows: *this second-order term decreases with increasing the external magnetic field strength.*

As a function of the nuclear nature, these solid-state quadrupolar effects can become very large. In practice, the first-order broadening can spread the signal intensity over an extremely large frequency range [15]. Under this circumstance, even registration of the resonances in static powder samples or even in fast-spinning samples will be difficult particularly at low natural abundance of the nuclei, for example, ^{17}O nuclei that are important as probes for materials science. In addition, powerful and relatively long (90°) radio frequency pulses, used generally in direct excitation of spins in solids, often lead to resonances that come not only from the central (+1/2 to −1/2) transitions but also from satellite transitions typical of quadrupolar nuclei.

It is obvious from the NMR theory that the MAS NMR experiments at highest spinning rates can potentially reduce the first-order broadening due to the presence of the angular dependence. However, in spite of this MAS effect, the second-order isotropic interaction will still broaden the resonances observed in regular MAS

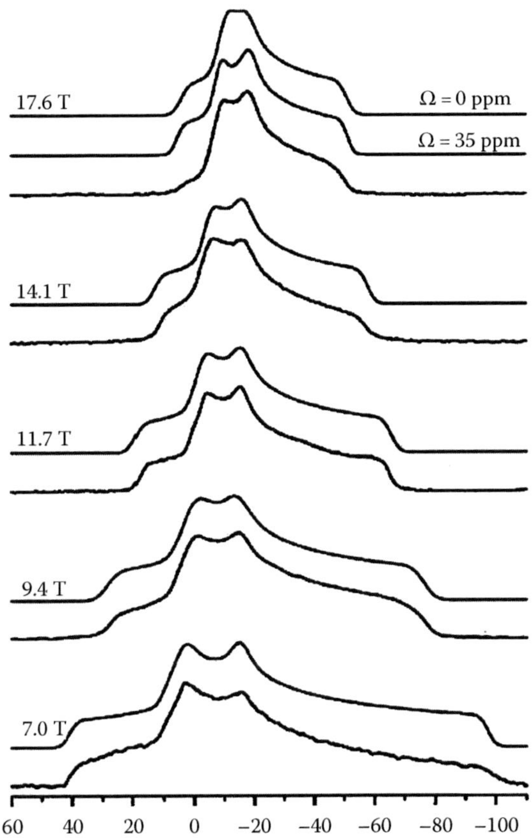

FIGURE 3.8 The ^{139}La NMR spectra recorded in a static sample of LaF_3 at different magnetic field strengths. The experimental NMR spectra are represented in the following the spectra obtained by simulations. The top spectrum corresponding to 17.6T has been computed without magnetic shielding anisotropy contribution. (Reproduced with permission from Ooms, K.J. et al., *Solid State NMR*, 28, 125, 2005.)

NMR spectra. On the other hand, from Equation 3.4, the second-order quadrupolar interactions can be reduced by the highest magnetic field strengths.

Wasylishen and coworkers demonstrate the advantages of such high-field solid-state NMR experiments on quadrupolar nuclei [16]. The measurements were carried out for ^{139}La nuclei in solid LaF_3, where static samples have been placed into the magnetic fields with strengths of 7.0, 9.4, 11.7, 14.1, and 17.6T. Figure 3.8 shows the collected data, where the "positive" influence of field strengths is evident. If at a field of 7.0T the resonance is situated between +40 and −110ppm, then at a field of 17.0T it is observed from +20 to −50ppm only.

The second important aspect demonstrated in this study is although the quadrupolar interactions strongly dominate the line shape of the ^{139}La NMR signal, nevertheless, chemical shift anisotropy also provides a contribution that is not negligible. Thus, Wasylishen and coworkers have shown that highest magnetic fields can be successfully used for the measurement of the La magnetic shielding tensors.

As can be seen, the line shape in Figure 3.8 is typical of the asymmetry parameter, η(Q) that is close to 1. The breadth of the central transition spectrum recorded at the magnetic field B_0 of 7.0T is approximately equal to 140kHz. However this magnitude reduces to 56kHz observed at a field of 17.6T. Then by standard simulation procedures, available in the software package, the quadrupole coupling constant can be determined as 15.9MHz at asymmetry parameter η(Q) of 0.82. In addition, an analysis of the highest magnetic field measurements leads to the ^{139}La magnetic shielding tensor (see the following) and determination of the isotropic chemical shift equal to −135ppm, in reference to $LaCl_3$ water solution.

It should be emphasized that the ^{139}La magnetic shift anisotropy plays an insignificant role in the breadth of all the spectra in Figure 3.8. Nevertheless, the signal intensities and separation of the two horns, observed in the magnetic field B_0 of 17.6T, are still affected by small anisotropies. Using this observation, the authors have performed simulation procedures for the intensities of the horns detected at the different magnetic fields to obtain finally a nonaxially symmetric shielding tensor characterized with a skew of 0.6ppm and a span of 35ppm. Even visually the spectrum at the highest magnetic field B_0 of 17.6T, calculated in the absence of shielding anisotropy (see top in Figure 3.8), and the experimental spectrum at this field differ remarkably. At the same time, the contribution of the chemical shift anisotropy to the line shape becomes practically negligible when the NMR spectra are recorded in the fields below 14.1T. Expedience of applying the highest magnetic fields for such quadrupolar nuclei is obvious.

3.3.1 Two Categories of Quadrupolar Nuclei

Quadrupolar nuclei can be divided into two important categories that show different spectral behavior. The first includes nuclei with integer spins:

$$^{2}\mathrm{H} \quad ^{6}\mathrm{Li} \quad ^{10}\mathrm{B} \quad ^{14}\mathrm{N} \quad ^{50}\mathrm{V} \quad ^{138}\mathrm{La} \tag{3.5}$$

The second includes only nuclei with noninteger spins:

$$^{7}\mathrm{Li} \quad ^{11}\mathrm{B} \quad ^{27}\mathrm{Al} \quad ^{23}\mathrm{Na} \quad ^{35}\mathrm{Cl} \quad ^{37}\mathrm{Cl} \quad ^{47}\mathrm{Ti} \quad ^{49}\mathrm{Ti} \quad ^{51}\mathrm{V} \quad ^{81}\mathrm{Br} \quad ^{93}\mathrm{Nb} \quad ^{121}\mathrm{Sb} \quad ^{133}\mathrm{Cs} \quad ^{139}\mathrm{La} \tag{3.6}$$

The noninteger spin nuclei do not undergo first-order broadenings for their central transitions. Therefore these nuclei are convenient in a spectroscopic sense and can be investigated even in static NMR experiments. It should be added that the central transitions of quadrupolar nuclei are generally generated in the experiments as most intense resonances in static powder samples while the satellite transitions are spread over a frequency range of many MHz. Therefore their special detection is difficult by regular pulse techniques [15]. At the same time, the central transitions of these nuclei remain broadened, for example, by tens of kHz due to the strong second-order quadrupolar effects. As a result, generally the static powder samples display NMR spectra, where resonances belonging to different sites can overlap. The MAS NMR technique cannot help dramatically in this case because the second-order quadrupolar broadenings are reduced only partially under these conditions. A higher resolution in such MAS NMR spectra can be reached by alternative approaches that are based on two-dimensional (2D) experiments, complex sample reorientations during the experiments, and excitation of multiple quantum transitions considered in the following. In contrast, the MAS techniques are very successful for nuclei of the first category.

Although the majority of solid-state NMR experiments, performed on noninteger spin nuclei, concentrates generally on the central transitions (thus providing a structural analysis of materials), in some cases, observation of the satellite transitions by the MAS technique is also important. There are two aspects in this connection. First, these MAS NMR experiments on the noninteger spin nuclei can be used to adjust accurately the magic angle, because the width of the spinning sideband patterns is very sensitive to the angle. Second, the satellite transitions in MAS NMR spectra are preferable to observe, particularly for nuclei with relatively weak quadrupolar interactions. For example, the symmetrical AlO_6 sites in some materials show the too-small quadrupolar coupling constants that are difficult to characterize quantitatively by registration of the central transitions. However, when the magic angle is accurately adjusted, the MAS NMR spectra exhibit sets of relatively sharp sidebands that reproduce well the static powder patterns. Then, these sidebands from the satellite transitions can be simulated by conventional procedures to result in the quadrupole coupling constants and asymmetry parameters characterizing the observed sites. It has been shown the quadrupole coupling constants up to 2 and 3.3 MHz for spins of 3/2 and 5/2, respectively, can be successfully measured by this method [15]. It is obvious that for amorphous molecular systems the change in the extent of the sideband manifolds can be applied for estimations of the average coupling constants and their distributions only.

Finally, it should be emphasized that interactions of nuclear spins with electric field gradients located at these nuclei are strongest and therefore magnitudes of the quadrupole coupling constants and the asymmetry parameters, η_q, contribute to the line shapes of the central and satellite transitions in the static and MAS NMR spectra even for paramagnetic systems particularly at relatively small concentrations of paramagnetic centers.

3.4 DETECTION OF NMR SIGNALS IN SOLIDS: COMMON ASPECTS

Due to the rapid high-amplitude isotropic molecular motions, resonance lines in NMR spectra of liquids are generally very sharp while lines in solids are strongly broadened. Because a primary source of this broadening in diamagnetic samples is the magnetic dipolar nucleus–nucleus interaction, governed by the dipolar factor (3.3) and/or quadrupolar effects, the measurements can be carried out in static or spinning samples.

3.4.1 WIDE-LINE NMR

The broadening effect by dipolar or quadrupolar interaction is particularly large for static powder samples. Detection and investigation of such resonances is an object of solid-state wide-line NMR and the experiments are generally performed on wide-bore superconducting magnets with a bore diameter of 89 mm. These experiments need high-power solid-state NMR probe heads to excite uniformly the very large frequency diapason and the radio frequency coils should have good quality factors Q.

Interpretation of the wide lines in the structural context is generally based on an approach suggested by Van Vleck in 1948 [17]. In the frameworks of the Van Vleck approach, an observed line is characterized by the *dipolar second moment* (measured as *square width*). The second moment of a line is calculated in terms of sum (3.7) over all nuclei pairs in crystals or sum Σr_{jk}^6 in a polycrystalline sample.

$$\frac{\sum (3\cos^2\theta_{jk} - 1)^2}{r_{jk}^6} \tag{3.7}$$

Since the r_{jk}^6 factor dominates strongly for nearest dipole–dipole contacts, structural information about molecules, groups, or ions can be obtained by computing and comparison of the experimental and calculated wide-line NMR spectra.

A *single crystal* sample can be oriented in the external magnetic field, B_0. In this case, a wide-line NMR spectrum of the crystal can be analyzed with better accuracy. For example, a single crystal of $CaSO_4 \cdot 2H_2O$, where protons are relatively isolated, exhibits the Pake doublet, providing accurate determination of the proton–proton distance [18].

Strongly amorphous and nonordered systems give rise to the wide-line solid-state NMR spectra, where directional information "nucleus–nucleus" is lost and line widths mask the presence of chemically and structurally different nuclei. Therefore, at present, the wide-line solid-state NMR experiments are rather applied to extract dynamic information in solids from molecular (or group) motions to phase transitions by relaxation time measurements and/or a line-width analysis. It should be emphasized that in this context the wide-line experiments are even preferable because they can be performed in a very large temperature range versus MAS experiments where fast spinning leads to additional problems in temperature control (see the following). This factor is very important in characterizations of materials of all kinds: amino acids, polypeptides, proteins, polymers, rubber, metabolites, steroids, etc. In solid proteins, for example, it is possible to identify and characterize energetically a number of motions contributing to relaxation rates such as methyl-group reorientations, segmental motions, side-chain motions, and bound-water reorientations. Here, the static solid-state ^2H NMR spectroscopy plays a central role.

Finally, it should be noted that at the present time, methods of detection and excitation of nuclear spins are based on different pulse sequences. The pulse sequences are too numerous to be covered in this text. In addition, they often depend on the nature of target nuclei, properties of materials, and formulated tasks. Nevertheless they can be represented by two groups: direct and indirect excitation.

3.4.2 Direct Excitation

A simple experiment based on direct excitation of nuclei has been shown in Chapter 1: the RF pulse in the single-pulse sequence acts on a nuclear spin, exciting the necessary range of frequencies. Since commercial NMR spectrometers routinely produce the powerful radio frequency pulses capable of an excitation area of 50–100 kHz, this diapason is sufficient to detect resonances in most of the diamagnetic materials containing nuclei with spins of 1/2, even in the presence of strong dipolar coupling or chemical shift anisotropy interaction.

For quadrupolar nuclei with spins ≥1, where very strong quadrupolar interactions are dominating to produce inhomogeneous broadenings on the order of *several MHz*, such an excitation is not enough to collect the undistorted FIDs. In fact, when the pulse is applied at the Larmor frequency, only central transitions of these nuclei are effectively irradiated while satellite transitions are off resonance. This situation leads to significant distortions of the resonance shapes observed after Fourier transformation. The problem can be avoided by decreasing pulse lengths and increasing pulse powers or by selective irradiation by relatively long pulses, details of which can be found in Smith's review [15]. For example, aluminum-containing materials are often studied by ^{27}Al solid-state MAS NMR with relatively short radio frequency pulses. Under this condition, excitation of the satellite transitions for ^{27}Al nuclei is practically absent.

A similar problem appears in detection of ^{2}H NMR signals in solids. To obtain undistorted solid-state NMR spectra extremely short radio frequency pulses should be applied. However in practice this is often impossible to realize. Reichert and Schneider [19] have suggested the ^{2}H NMR experiments with composite and shaped pulses. These pulses allow reducing the phase and amplitude distortions typical of this case. It has been emphasized that this approach does not require additional spectrometer hardware as well as additional data manipulations.

Another approach is based on the frequency-stepped adiabatic half passage (FSAHP), described in detail in the work of Kentgens et al. [20]. According to Kentgens, when a spin system is irradiated far *off resonance*, the effective field is oriented practically parallel to the static magnetic field B_0. Then, if this frequency is *slowly* varied to the resonance condition, the magnetization will follow the effective field that becomes perpendicular to the field B_0 under the resonance condition. Then, switching the radio frequency irradiation off at this point, named as a half passage, allows the collection of the FID. Kentgens emphasizes that this FSAHP technique shows a large advantage versus the regular short-pulse experiments because it results in NMR spectra where peak intensities are increased and less dependent on the RF field strength.

As we have noted in Chapter 1, a relatively long dead time of NMR spectrometers can complicate registration of the quadrupolar nuclei, excited directly by a single pulse sequence. In fact, when a macroscopic magnetization, created by a radio frequency pulse, decays rapidly with respect to the spectrometer dead time, Fourier transformation converts the FIDs into NMR spectra, where signal intensities are strongly reduced and a baseline undergoes very strong distortions. Generally this technical problem can be minimized in the NMR experiments based on Hahn-echo pulse sequences:

$$(90^{\circ}_{X} - \tau - 180^{\circ}_{Y} - \tau - \text{FID}) \tag{3.8}$$

or solid echo pulse sequences:

$$(90^\circ_X - \tau - 90^\circ_Y - \tau - \text{FID}) \tag{3.9}$$

where τ is echo-delay time. In order to see the full Hahn-echo FIDs, the data collection can be started immediately after 180° pulses. It should be noted that these NMR experiments, performed in spinning samples, require echo-delays that should be synchronized with rotor periods to reach maximal peak intensities [21]. In the absence of the synchronization or due to a poor stability of spinning rates the intensities strongly decrease. As we show in the following, the echo-based experiments can be successfully applied for nuclei with spins of 1/2 even in paramagnetic materials, where the Hahn-echo NMR experiments at different carrier frequencies can be performed to represent the resulting NMR spectra by their summing after Fourier transformation. Unfortunately these experiments can be accompanied by dephasing effects that can lead to the incomplete focusing of the echo signals.

Routine solid-state NMR experiments, based on the direct excitation of nuclei, generally require large numbers of scans to reach good signal-to-noise ratios in the NMR spectra. In addition, they generally imply long relaxation delays. The best result can be obtained when the acting angle of RF pulses is optimized, in accordance with the Ernst rule mentioned earlier [22]. This rule is particularly important for materials with low-abounded nuclei, for example, ^{29}Si nuclei in glasses or silica-based materials, which relax particularly slowly (T_1 of 100 s or longer) and the recorded ^{29}Si NMR spectra are needed for quantitative treatments in terms of integral intensities.

3.4.3 Cross-Polarization

Solid-state MAS NMR experiments with single-pulse direct excitation are a large research field in materials chemistry particularly at high natural abundance of target nuclei. However this technique is practically useless at probing the naturally abundant ^{13}C samples, such as materials containing organic components or bioorganic solids. It is known that, for example, the solid-state ^{1}H MAS NMR has not been so important as solid-state ^{13}C NMR because of the serious problems appearing in the ^{1}H NMR spectra caused by the strong interproton dipolar coupling, on the one hand, and the efficient spin diffusion, on the other hand. Due to the spin diffusion process, considered in Chapter 7, the interproton dipolar interactions become homogeneous through the sample. In addition, such homogenous dipolar interactions cannot be often averaged away by simple magic angle spinning.

Cross-polarization (CP) from abundant spins (^{1}H, ^{19}F) to dilute spins (^{13}C, ^{15}N, or ^{29}Si) solves two important problems in solid-state NMR. *First*, the CP improves strongly signal-to-noise ratios in NMR spectra of nuclei with a low natural abundance and low γ values. *The second* is connected with spin lattice relaxation times of dilute spins. They are very long in diamagnetic solids varying, for example, from tens of seconds for ^{13}C or ^{15}N nuclei to several minutes for ^{29}Si nuclei in silicate glasses in the absence of paramagnetic impurities. Under these conditions long relaxation delays lengthen sharply the time needed for collection of NMR data.

The task of a cross-polarization pulse sequence in (3.10) is a transfer magnetization from the ^1H spins with a high natural abundance to the ^{13}C (or ^{15}N or ^{29}Si) spins via the dipolar coupling between them, which is based on their close space proximity.

$$\begin{array}{ll} ^{1}\mathrm{H} & 90^{\circ}_{\mathrm{X}} - \text{Contact pulse}_{-\mathrm{Y}} \quad {}^{1}\mathrm{H}\ \text{Decoupling} \\ ^{13}\mathrm{C} & \quad - \text{Contact pulse}_{-\mathrm{Y}} - {}^{13}\mathrm{C}\ \text{FID} \end{array} \tag{3.10}$$

The amplitudes of two contact pulses, operating at proton and carbon frequencies, should be adjusted to achieve the Hartmann–Hahn conditions [23]:

$$\gamma_H B_1(^1H) = \gamma_C B_1(^{13}C) \tag{3.11}$$

When this condition is reached, the energy gaps between the ^1H and ^{13}C rotating-frame spin states become equal and a transition requiring energy on a ^1H spin can be exactly compensated by a transition releasing energy on a ^{13}C spin. This pulse sequence in a combination with the MAS technique leads to rapid accumulations of the ^{13}C NMR spectra because the recycling delay is dictated now by the ^1H (but not ^{13}C) relaxation.

The efficiency of the CP NMR technique depends remarkably on three factors: accuracy in adjustment of the Hartmann–Hahn conditions (here a critical role belongs to accurate calibration of 90° pulses and application of a correct carrier frequency for a ^1H channel), distances between coupled nuclei, and molecular mobility in the solid state [24]. Figure 3.9 shows the kinetics of the cross-polarization ^1H–^{31}P measured in a standard sample $CaHPO_4 \cdot 2H_2O$ at spinning rate of 3 kHz. As we can see, the CP ^{31}P intensity increases at the beginning on increasing the contact time and then it decreases, illustrating that any CP experiment should be optimized in

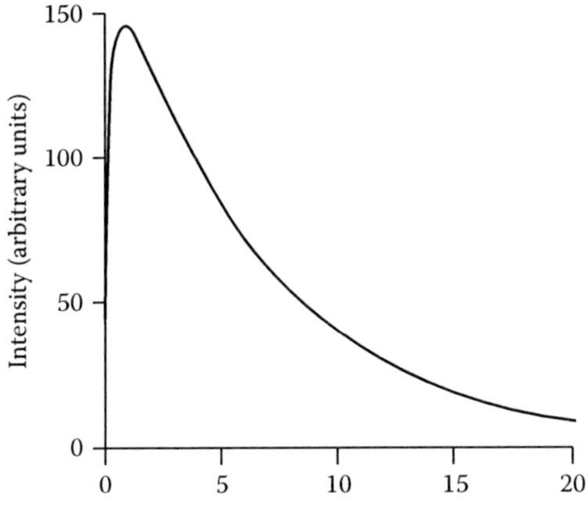

FIGURE 3.9 Kinetics of the cross polarization ^1H–^{31}P collected for a standard sample of $CaHPO_4 \cdot 2H_2O$ at spinning rate of 3 kHz. (Reproduced with permission from Kolodzieski, W. and Klinowski, J., *Chem. Rev.*, 102, 613, 2002.)

terms of contact times to make the measurements quantitative. This optimization is particularly important in studying complex materials, for example, the large-pore alumino-phosphate molecular sieves.

The curve shown in Figure 3.9 and other similar CP kinetic curves can be treated quantitatively via Equation 3.12:

$$I(t) = I_0\left(1 - \frac{T(IS)}{T(I)_{1\rho}}\right)^{-1}\left[\exp\left(\frac{-t}{T(I)_{1\rho}}\right) - \exp\left(\frac{-t}{T(IS)}\right)\right] \tag{3.12}$$

where

T(IS) is the cross-polarization constant

$T(I)_{1\rho}$ is spin-lattice relaxation time in the rotating frame, for ^1H nuclei

It should be emphasized that spin-lattice relaxation times in the rotating frame are remarkably shorter than those in the laboratory coordinate systems due to the relative weakness of the applied radio frequency field. The CP time constants T(IS) in Equation 3.12 are specific. They characterize functional groups involved into the study. The T(IS) constants are governed by the internuclear distances (via factor of $1/r^3$), the number of neighboring protons, and the mobility of these groups. In fact, for a *rigid* lattice of a solid, the CP rate constant can be expressed as

$$\frac{1}{T(IS)} = \frac{3}{2}\left(\frac{2\pi M_2(IS)}{5(M_2(II))^{1/2}}\right)^{1/2} \tag{3.13}$$

where $M_2(IS)$ and $M_2(II)$ are the second moments that are dependent on the strength of the I–S and I–I dipolar interactions. The cross-polarization constant becomes smaller when the second moments decrease due to raising the more intense molecular (or group) motions.

According to Peersen et al. [25], accurately setting the amplitudes for the Hartmann–Hahn matched spin-lock fields can be achieved by *variable amplitude* cross-polarization experiments where the cross-polarization buildup data are collected when the ^1H pulse in standard sequence (3.10) is replaced for a pulse train with incrementing ^1H pulse time changing from 5 μs to 11 ms. As we show in the following, this variable amplitude pulse sequence can be very successfully applied to distinguish between two phases in different materials and to characterize, for example, their crystallinity [26].

The common aspects of cross-polarization experiments performed on different nuclei and their features can be found in the review by Wasylichen and coauthors [28]. Table 3.3 lists a set of the standard samples, generally used for adjustments of the Hartmann–Hahn conditions. Here ratios $\gamma(^1H)/\gamma(s)$ represent the factors of signal enhancement expected in the ^1H–S cross-polarization experiments for nuclei S with spins of 1/2.

The theory of cross polarization shows, that contact time, providing polarization transfer, elongates on lowering the γ constants. For example, the ^{109}Ag CP MAS

TABLE 3.3
Enhancement Factors Expected for Cross-Polarization from Proton to S Where S Are Nuclei with Spins of 1/2

Nucleus	$\gamma(^1H)/\gamma(S)$	Standard
^{13}C	3.98	Glycine
^{15}N	−9.86	NH_4NO_3
^{29}Si	−5.03	$(CH_3)_3Si(CH_2)_2COONa$
^{31}P	2.47	$NH_4H_2PO_4$
^{77}Se	5.22	$(NH_4)_2SeO_4$
^{109}Ag	−21.4	$[CH_3CHC(OH)COO]Ag$
^{113}Cd	−4.48	$Cd(NO_3)_2$
^{119}Sn	−2.67	$Sn((C_6H_{11})_4$
^{125}Te	−3.14	$Te(OH)_6$
^{195}Pt	−4.58	$(NH_4)_2PtCl_6$
^{199}Hg	5.52	$(NBu_4)_2[Hg(SCN)_4]$
^{207}Pb	4.79	$PbPH_4$

NMR experiments can be realized in a contact time up to 50 ms versus 1.5 or 6 ms generally needed for ^{13}C and ^{29}Si nuclei, respectively.

The cross-polarization, for example, from ^1H nuclei to ^{13}C nuclei, can be modified by combining the standard cross-polarization section with polarization inversion [27]. This cross-polarization–polarization-inversion experiment helps to distinguish in the MAS NMR spectra between the groups ^{13}CH and $^{13}CH_2$, helping in the signal assignments. The cross-polarization also plays an important role in recording the solid-state NMR spectra (MAS or static) for systems containing nuclei with very large chemical shift anisotropies. For example, the CP can help in registration of ^{199}Hg nuclei or $^{119,117}Sn$ nuclei in solids.

3.4.4 Cross-Polarization with Quadrupolar Nuclei

Generally the cross-polarization experiments performed with participation of quadrupolar nuclei require specific Hartmann–Hahn matching conditions. These conditions can be realized when *the energy of splitting in the rotating coordinate systems of both nuclei is the same.* In other words, only a *small fraction* of the quadrupolar nuclei can have a Hartmann–Hahn match with the other nuclei. Therefore under this condition the signal-to-noise ratios in the NMR spectra will be poor. In such cases, direct excitation is preferable, particularly for nuclei with a high natural abundance.

The CP(^1H–^{27}Al) MAS NMR experiment is shown in Figure 3.10, where the signal-to-noise ratio decreases remarkably in going from direct excitation to the ^1H–^{27}Al cross-polarization. Nevertheless in the context of signal assignments the importance of the CP experiment is obvious.

In fact, the figure shows a remarkable increase in the relative intensity of the signal belonging to the extra framework aluminum.

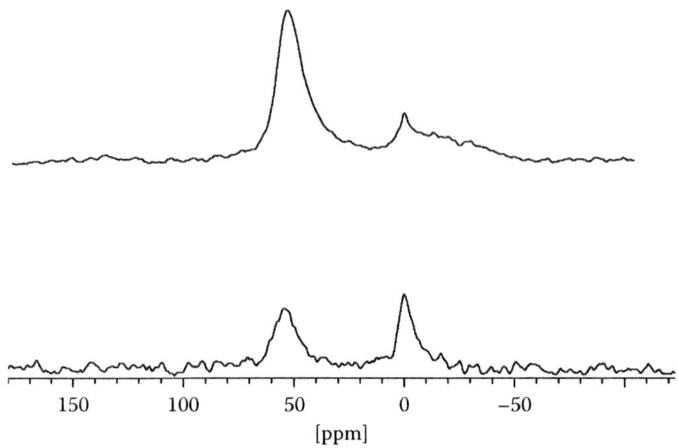

FIGURE 3.10 The ^{27}Al MAS NMR spectra recorded in a sample of microporous system SiO_2 Al_2O_7 at a spinning rate of 8 kHz: (top) direct excitation showing an intense resonance at 55 ppm belonging to tetrahedral AlO_4 species in the silica matrix; ^1H–^{27}Al cross-polarization showing the second intense resonance of extra framework six-coordinated Al–O units.

The solid-state ^2H NMR spectra are difficult to observe even under MAS conditions particularly in low enriched samples, when a system investigated has low molecular mobility and hence a very large quadrupolar splitting is expected. In spite of the expected relatively small efficiency of the cross polarization ^1H–^2H, the CP MAS NMR experiments can help to collect the NMR pattern with a less distorted baseline (Figure 3.11).

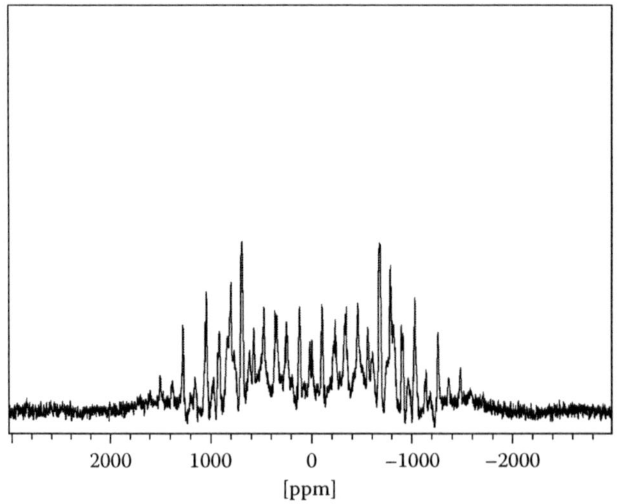

FIGURE 3.11 The CP ^1H–^2H NMR spectrum recorded with a standard 7 mm prove head at spinning rate of 6 kHz for a sample of Ph_3P labeled at para-position. The spectrum was obtained with a contact time of 10 ms.

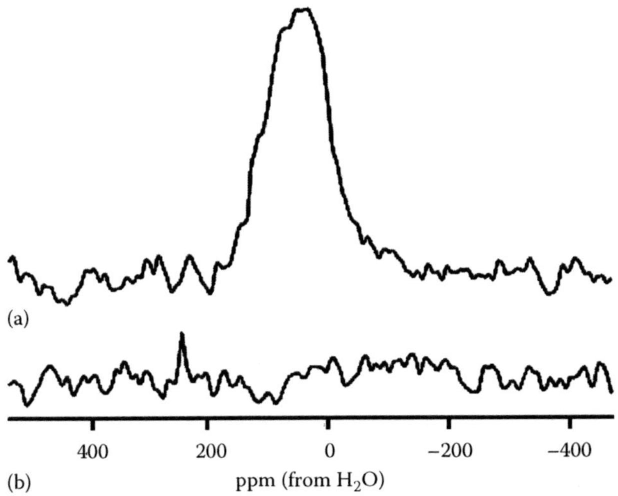

FIGURE 3.12 The ^{27}Al–^{17}O cross-polarization MAS NMR spectrum recorded in a static sample of α-alumina (a) and the NMR spectrum obtained by ^{17}O direct excitation (b). (Reproduced with permission from Bakhmutov, V.I., *Chem. Rev.*, 111, 530, 2011.)

In spite of this principal limitation, the cross-polarization NMR technique in the case of some combinations of quadrupolar nuclei can be very successfully applied for probing different materials [15]. Figure 3.12 compares the ^{27}Al–^{17}O cross-polarization NMR spectrum recorded in a static sample of α-alumina and the NMR spectrum obtained by the ^{17}O direct excitation. As we can see, the cross-polarization from abundant spins ^{27}Al to dilute spins ^{17}O illustrates a very large increase in a signal-to-noise ratio. As noted in [15], only very weak radio frequency fields can be used and the RF field strengths should be the same (≈10 kHz) for both nuclei.

Multiple resonance NMR techniques, including cross-polarization sections, indirect detection (TRAPDOR), and/or nutation are also available. Their applications for quadrupolar nuclei can be found in Smith's review [15].

Finally, it should be emphasized that the efficiency of the cross-polarization transfer depends on T_1 relaxation times of abundant spins. One can expect that strong shortening of the ^1H T_1 times can violate the conditions of polarization transfers. Therefore *due to the fast relaxation in paramagnetic materials, the efficiency of the CP experiments is dramatically reduced.* Nevertheless, as is shown in the following, sensitivity enhancement is still possible in the ^{13}C solid-state NMR spectroscopy of some paramagnetic systems.

3.5 HIGH-POWER DECOUPLING

The MAS conditions lead to narrowing the lines, if spinning rates are larger than line widths in the static NMR spectra. However, actual high resolution (commonly less than 5 Hz in crystalline samples) cannot be reached by simple MAS, when target nuclei in solids undergo the very strong dipolar interactions with neighbors at small distances between them. This situation, typical of rare nuclei such as ^{13}C or ^{15}N etc.

at short *bond* distances ^{1}H–^{13}C or ^{1}H–^{15}N or ^{13}C–^{19}F, etc., requires effective suppressing of the very strong ^{1}H–X dipolar coupling. The suppressing effect can be reached by the high-power decoupling technique that produces ^{1}H or ^{19}F irradiation, when the ^{13}C or ^{15}N solid-state NMR FIDs are recorded. The decoupling section can be included in a cross-polarization or a direct excitation experiment.

Technically the decoupling (requiring the presence of the second channel in an NMR probe) is realized by continuous irradiation (CW) at a power of 100–1000 W or by special multipulse sequences operating at proton or fluorine frequency [21]. These pulse trains are called to reach two goals: first, they should reduce the time-averaged dipolar interactions to zero and second, they should avoid the problems related to a finite pulse length and field inhomogeneity. Cycles of 3, 4, 6, or even 52 pulses can be applied to exceed the width of the ^{1}H (or ^{19}F) NMR spectra. Among them, sequences two pulse phase modulation (*TPPM*) (providing heteronuclear X-{^{1}H} decoupling), *WAHUHA*, and *MREV-8* (effective in homonuclear, e.g., proton–proton decoupling) are relatively short and most applicable and generally included in the software of commercial spectrometers. All these experiments are routine and very effective when NMR probes are well tuned.

The theoretical and practical aspects of heteronuclear decoupling in solid-state NMR include different factors determining decoupling performance. They are carefully considered by Hodgkinson [29]. One of the practically important results of these considerations is the fact that shapes of lines, as well as their widths can be strongly dependent on the decoupling method. These effects are illustrated in Figure 3.13. The figure shows the ^{13}C {^{1}H} MAS NMR spectra, that is, recorded by direct ^{13}C excitation in the presence of a decoupling field. It is clearly seen that the resonance of $^{13}C(\alpha)$-enriched alanine changes remarkably in going from the CW decoupling method to the pulse-based decoupling approaches. In addition, the TPPM and XiX

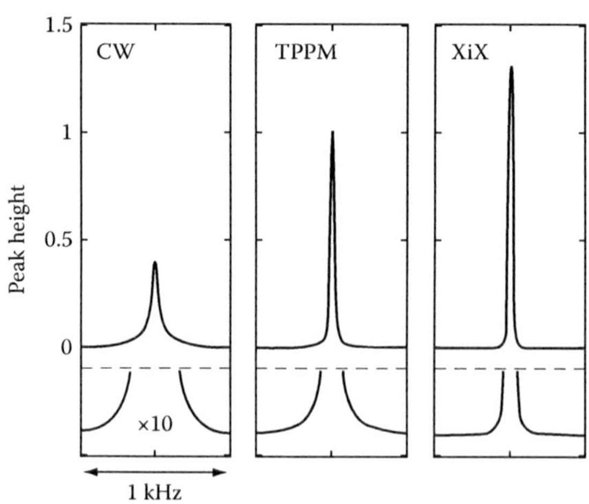

FIGURE 3.13 The C(α) resonance line, detected in a sample of polycrystalline ^{13}C-labeled alanine, recorded at a spinning rate of 30 kHz with a spectrometer working at 150.8 MHz (^{13}C) and 600 MHz (^{1}H) under different proton-decoupling methods. (Reproduced with permission from Hodgkinson, P., *Prog. Nucl. Magn. Reson. Spectros.*, 46, 197, 2005.)

(X-Inverse-X-Decoupling) *full-widths-at-half*-maxima are very similar while the height of the XiX-decoupled signal is significantly greater. At the same time, the baseline expansions illustrate that the TPPM-decoupled signal has a significantly wider foot. In practice, these effects should be taken into account in an analysis of integral intensities particularly for overlapped resonances when deconvolution is necessary.

Since dipolar and spin–spin couplings between other nuclei, other than ^{1}H and ^{19}F, are generally less significant, the corresponding heteronuclear decoupling experiments are not popular. First, the heteronuclear spin–spin coupling magnitudes can be observed directly in the NMR experiments (see Chapter 1) and they can be informative even in solids. Second, the much larger chemical shift anisotropies for such nuclei (e.g., ^{31}P) complicate technically their decoupling. Third, the couplings can be refocused by spin-echo pulses [15].

Finally it should be emphasized that the presence of the ^{1}H decoupling field is not always useful. Wasylishen et al. have carefully analyzed results of proton decoupling applied for some paramagnetic compounds [30]. They have found a number of negative effects. Nevertheless it has been demonstrated that proton decoupling in observation of nuclei ^{13}C in paramagnetic complexes formed by amino acids is still possible. The authors have shown what decoupling approaches are most effective in this case.

3.6 MULTI-QUANTUM NMR EXPERIMENTS

Following Andrew and Szczesniak [31], ensembles of nuclear spins are nonlinear systems and therefore their important features cannot be completely described by the linear response that appears in the single-quantum NMR spectra. Theory of NMR shows that one can realize excitation of the multiple-quantum transitions that are normally forbidden. Already the first NMR experiments performed with the continuous-wave spectrometers have demonstrated that multiple-quantum transitions appear on increasing the radio frequency power to drive the nuclear spins beyond the range of their linear response, that, nevertheless, were difficult to rationalize [30]. Later due to the rapid development of 2D Fourier NMR methodology, the information about multiple-quantum transitions became available. To be precise, this information is encoded in the *one-quantum* NMR spectrum, when it is registered during the detection period t(2), but results from excitation of the forbidden multiple-quantum coherences earlier during the spin evolution period t(2). Then, varying the t(1) duration allows us to probe the behavior of the forbidden coherences in time. For example, if number of nonequivalent 1/2 spins is N, then, the number of possible P quantum transitions takes values from N to N–P, where $1 \leq P \leq N$. For this reason, a smaller amount of lines appear in the same spectral width and the multiple-quantum spectrum is considerably less complicated and much better resolved. Moreover, such spectrum gives enough information about dipolar couplings that are important in terms of molecular structure determinations. This explains why the multiple-quantum technique a powerful tool in investigating dipolar couplings particularly in oriented phases.

As we now know, the single-quantum spectra NMR of quadrupolar nuclei in solids are dominated by very strong quadrupolar interactions. As a result, the quadrupolar interactions often mask more weak interactions, such as chemical shift anisotropies, important, in turn, for description of many materials. Because the multiple-quantum spectra are not affected by quadrupolar broadenings, they can be successfully used to extract the CSA information. For example, a combination of deuterium double-quantum NMR with magic angle spinning leads to accurate determinations of the chemical shift anisotropy and measurements of the isotropic shifts for ^2H nuclei in powdered samples [31,32].

Smith et al. [15] show that there are different schemes to realize the 2D multiquantum MAS NMR experiments effective for studies of quadrupolar nuclei in solids. Figure 3.14a illustrates a simple experiment, where a multiquantum transition is excited by the action of a single and high-power radio frequency pulse that evolves the multiquantum coherence during a time t(1). Then, a second pulse is applied to convert this coherence into coherence p = −1, observed, in turn, during time t(2). One can rationalize that the signal is recorded after the second pulse action and then, the echo signal will be formed at a time t(2) equal to the quadrupole anisotropy multiplied by time t(1). Both of the pulses in this experiment are nonselective to excite, thus, all coherences. Finally, after 2D Fourier transformation, the resonances show up as ridges located along the quadrupolar anisotropy axis and the isotropic

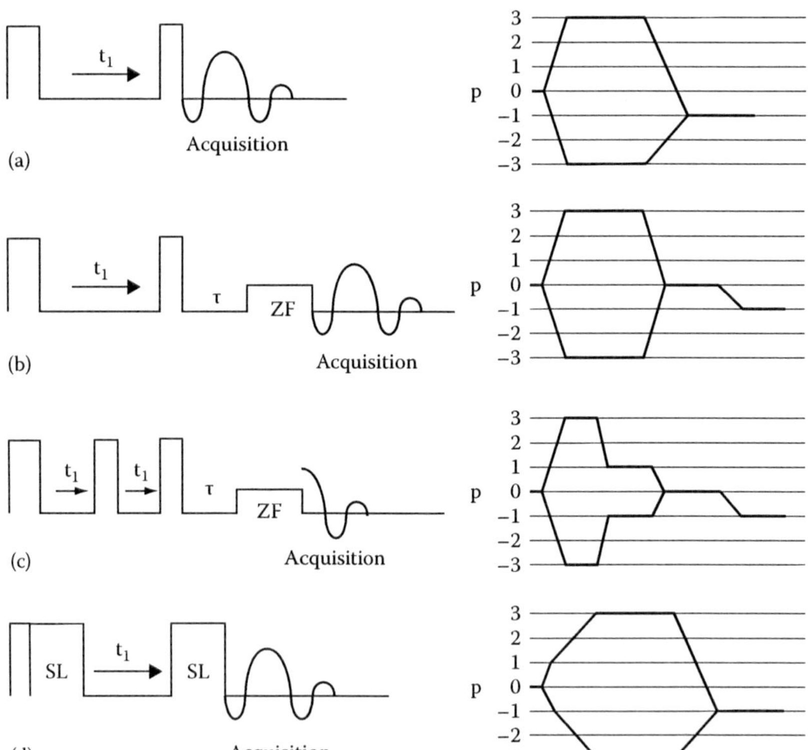

FIGURE 3.14 The schemes of different multiquantum MAS NMR experiments (see symbols and conditions of experiments a–d in the text). (Reproduced with permission from Smith, M.E. and van Eck, E.R.H., *Prog. Nucl. Magn. Reson. Spectrosc.*, 34, 159, 1999.)

spectrum appears as projection of the entire 2D spectrum. Figure 3.14b illustrates an experiment with the so-called Z-filter pulse. The pulse is selective and capable of symmetrical coherence transfer pathway in contrast to asymmetrical one in the earlier discussed simple experiments. A more complex t(1) split experiment is illustrated in Figure 3.14c. Here the isotropic spectrum is obtained directly in the absence of shearing the data. However, it should be emphasized that due to wide distributions in chemical shifts typical of amorphous materials, a part of signal intensity will be lost. The latter is obviously the disadvantage of the t(1) split experiment. Finally, Figure 3.14d shows the pulse sequence, where the excitation and the conversion of the coherences are performed by a spin-locking at the duration Tr/4. This experiment has a great advantage in collection of quantitative information. Some practical applications of these experiments is considered in Chapter 4.

3.7 DIPOLAR RECOUPLING NMR TECHNIQUES

Due to the coherence, which averages the chemical shift anisotropy interactions, the MAS NMR experiments result in high-resolution NMR spectra in solids to finally show isotropic δ values as in solutions. In addition, since the dipole–dipole interactions are characterized by the traceless second-rank tensors, the MAS experiments reduce also the dipolar interactions to zero narrowing thus the observed resonances. This is a great advantage of solid-state MAS NMR. On the other hand, due to the MAS conditions, important structural information is being lost. In fact, it is impossible to determine the dipolar coupling constant D (see Equation 3.2) directly leading to the internuclear separations r. Therefore many efforts have been expended to find the conditions for the MAS NMR experiments that recover the dipole–dipole interaction and provide, thus, calculations of distances between target nuclei.

At the present time there are numerous and various dipolar recoupling techniques for measuring the dipolar coupling under MAS conditions, based on different echo-pulse sequences. In spite of the variety, they can be split into two fields, based on the homonuclear and heteronuclear dipolar recoupling in MAS NMR.

Dipolar homonuclear recoupling implies generally the double-quantum (DQ) MAS NMR spectroscopy that applies pulse sequences back-to-back (BABA), dipolar recovery at the magnetic angle (DRAMA), dipolar recoupling with a windowless multi-pulse irradiation (DRAWS), and homonuclear rotary resonance (HORROR), recently considered in Gullion and Vega's review [33]. The ^{1}H–^{1}H BABA MAS NMR spectroscopy can be commonly represented in Figure 3.15, showing the scheme of the pulse sequence and the coherence transfer pathway diagram. The experiment consists of the following steps: the double-quantum coherence is excited and evolves during an incremented time, t_1. Then, the double-quantum coherence is converted into the detectable single-quantum coherence, collected during an acquisition time, t_2. It has been noted that in order to select the concrete coherence transfer pathways, a phase cycling scheme should be employed. When the increment, t_1, becomes equal to a rotor period, then the rotor-synchronized 2D NMR spectrum can be recorded. In turn, reducing t_1 and therefore increasing the DQ spectral width, leads to the observation of a DQ MAS spinning sideband pattern that can be analyzed quantitatively in terms of coupling and internuclear distances.

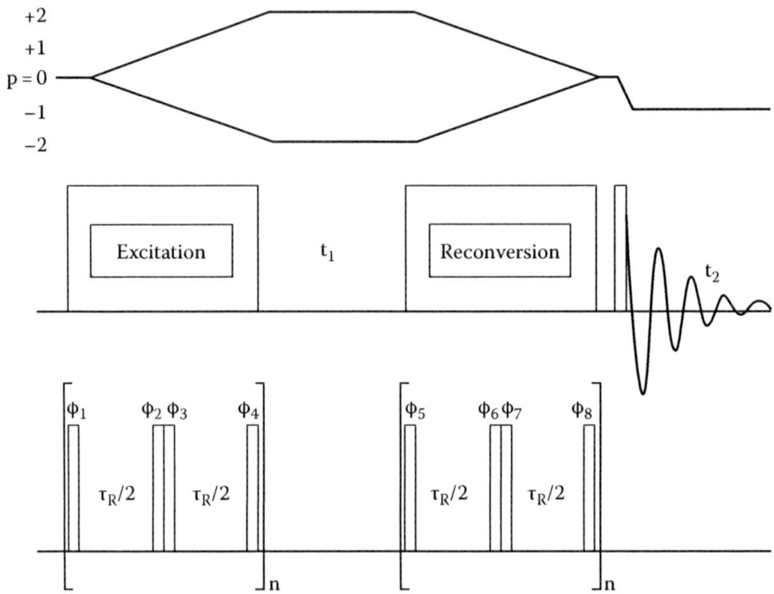

FIGURE 3.15 The diagram of the pulse sequence and coherence transfer pathways operating in ^{1}H DQ BABA MAS NMR experiments. (Reproduced with permission from Gullion, T. and Vega, A.J., *Prog. Nucl. Magn. Reson. Spectros.*, 47, 123, 2005.)

By analogy, recovering the dipolar heteronuclear coupling, for example, between pairs of spins of 1/2, such as ^{13}C–^{15}N, etc., can potentially lead to measurements of distances between them, when the rotational echo double resonance (REDOR) MAS NMR experiments are performed. Such experiments are not simple, particularly with participation of quadrupolar nuclei, and require detailed considerations on the theoretical and experimental levels. Both of them are given by Gullion and Vega for spin pairs I = 1/2 and S > 1/2, one of which is quadrupolar [33].

The Hamiltonian for spin pairs I = 1/2 and S > 1/2 in a MAS sample, placed into the external magnetic field, is commonly expressed via Equation (1), shown in Scheme 3.3.

In this equation, the last two terms correspond to the interaction appearing in the presence of the radio frequency field, while the two first terms represent the *orientation-independent components* of the chemical shift interactions for the I, S spins, which correspond to the isotropic chemical shifts in the external magnetic field (see Chapter 1). The *orientation-dependent* component (see Equation 2) of the chemical shift interaction is characterized by an anisotropy, noted here as ω_Z, and an asymmetry parameter η_C. The angle-dependent term, noted as C(t), and anisotropy ω_Z are given by Equations (3 and 4). Here the Z component of the chemical shift tensor in the principal coordinate system is named as σ_{ZZ}^{PAS} and the angles θ_C and ϕ_C take the instantaneous values of the polar angles of the field B_0 in the principal coordinate system of the chemical shift tensor.

When the sample is spinning, these angles change and then, the time-dependent term, reflecting the behavior of the chemical shift interaction, can be given by $C(t) = \omega_Z g((t)$ with function g(t). expressed via Equation (5). In this expression, ω_R is the

$$H = H_{iso,I} + H_{iso,S} + H_{csa,I} + H_{csa,S} + H_Q + H_D + H_{rf,I} + H_{rf,S} \quad (1)$$

$$H_{csa} = C(t)\hat{I}_z \quad (2)$$

$$C(t) = \frac{1}{2}\omega_z (3\cos^2\theta_c - 1 - \eta_c \sin^2\theta_c \cos 2\phi_c) \quad (3)$$

$$\omega_z = -\omega_0 \left(\sigma_{zz}^{PAS} - \sigma_{iso}\right) \quad (4)$$

$$g(t) = g_1 \cos(\omega_R t + \psi_1) + g_2 \cos(2\omega_R t + \psi_2) \quad (5)$$

$$H_Q = \frac{1}{2} Q(t) \left[\hat{S}_z^2 - \frac{1}{3} S(S+1)\hat{I}\right] \quad (6)$$

$$g'(t) = g_1' \cos(\omega_R t + \psi_1') + g_2' \cos(2\omega_R t + \psi_2') \quad (7)$$

SCHEME 3.3 The Hamiltonian and elements of the theory for the REDOR MAS NMR experiments performed for spin pairs I = ½ and S > ½ placed into the external magnetic field.

angular rotation rate and functions g_1, g_2, ψ_1, and ψ_2 describe the initial orientation of the crystallites of the sample in the rotor frame and the initial orientation of the asymmetry parameter. Finally, the similar expressions can be written for nuclear spins S.

Quadrupole interactions are expressed via the Hamiltonian (6). Again, they have the time-dependent term that can be given as $Q(t) = \omega_Q(t)\, g'(t)$, where function $g'(t)$ (see Equation (7)) is similar to those in Equation (5).

The heteronuclear dipolar interaction for isolated I–S pairs is written via Equation (1) in Scheme 3.4. The equation contains the terms d(t) and ω_D that are given by Equations (2) and (3), respectively. In turn, the angle θ_D is the angle formed by the I–S vector and the external magnetic field, which changes due to sample spinning. As we know, the dipolar coupling for the spins distanced by the r is expressed as $D = \gamma_I\gamma_S\hbar/2\pi r^3$. The time-dependent term $d(t) = \omega_Q\, g''(t)$ corresponds to function $g''(t)$ in expression (4). Finally the Hamiltonian, for example, describing spin I at application of radio frequency pulses is $H_{rf,I} = \omega_{1,I}\hat{I}_{X,Y}$ with $\omega_{1,I} = \gamma_I B_{1,I}$ on applying the radio frequency field $B_{1,I}$.

Gullion and Vega have considered the practical consequences of the application of the basic REDOR pulse sequence illustrated in Figure 3.16a. Here there are two rotor periods with 180° pulses operating in channel I at times t = 0 and T_R and in channel S at times $T_R/2$ and $3T_R/2$. One can show that in the presence of these four radio frequency pulses, two Hamiltonians $\bar{H}_{CSA,I}$ and $\bar{H}_{CSA,S}$, corresponding to chemical shift anisotropy interactions, are averaged to 0 in cycles (5). The same conclusion can be deduced for the quadrupolar interactions (see cycles (6)). However, in contrast to the aforementioned interactions, the hetero-dipolar Hamiltonian will be *not zero* and defined by $-\bar{H}_D$. Thus, the REDOR experiment will correspond to the average Hamiltonian in expression (8). It would be added that the REDOR MAS NMR experiments can include the cross-polarization section to provide the corresponding signal enhancement as shown in Figure 3.16b.

$$H_D = d(t)\hat{I}_z\hat{S}_z \quad (1)$$

$$d(t) = -\omega_D(1 - 3\cos^2\theta_D) \quad (2)$$

$$\omega_D = 2\pi D \quad (3)$$

$$g''(t) = g_1''\cos(\omega_R t + \psi_1'') + g_2''\cos(2\omega_R t + \psi_2'') \quad (4)$$

$$-\bar{H}_{csa,I},\ \bar{H}_{csa,I},\ \bar{H}_{csa,I},\ -\bar{H}_{csa,I} \quad (5)$$

$$\bar{H}_Q,\ -\bar{H}_Q,\ \bar{H}_Q,\ -\bar{H}_Q \quad (6)$$

$$-\bar{H}_D,\ -\bar{H}_D,\ -\bar{H}_D,\ -\bar{H}_D \quad (7)$$

$$\bar{H}_D = \left(\frac{2\sqrt{2}}{\pi}\omega_D \sin 2\beta \sin\alpha\right)\hat{I}_z\hat{S}_z \quad (8)$$

SCHEME 3.4 The heteronuclear dipolar interactions for isolated spin pairs I – S (I = ½ and S > ½) in REDOR MAS NMR experiments.

Finally the REDOR experiment can be performed at different numbers of rotor cycles between the excitation and detection sections. Generally two REDOR experiments are carried out: the first one without pulses in channel S (as a control step), while the second experiment produces the dipolar dephased signal of intensity S, which is analyzed in terms of ΔS/S magnitudes.

Applications of the multiple-quantum REDOR MAS NMR experiments resulting in determination of dipolar coupling ^{27}Al–^{31}P, important for some materials and many other recoupling experiments implying the presence of intense quadrupolar interactions, such as transfer of polarization in double resonance (TRAPDOR), rotational echo adiabatic passage double resonance (REAPDOR), and REDER are also available and can be found in the review by Gullion and Vega [33].

Since ^2H nuclei are very attractive for solid-state NMR studies of different materials particularly in probing and characterizing their molecular mobility, one of the approaches to ^2H NMR can be illustrated here. It is well known that uniform irradiation of the deuterium spectral range is technically difficult because of the very large deuterium quadrupolar frequency area. Generally speaking, this circumstance complicates the REDOR applications. However special composite radio frequency pulses can help to avoid this problem. For example, Figure 3.17 shows a pulse sequence that has been applied in the REDOR.

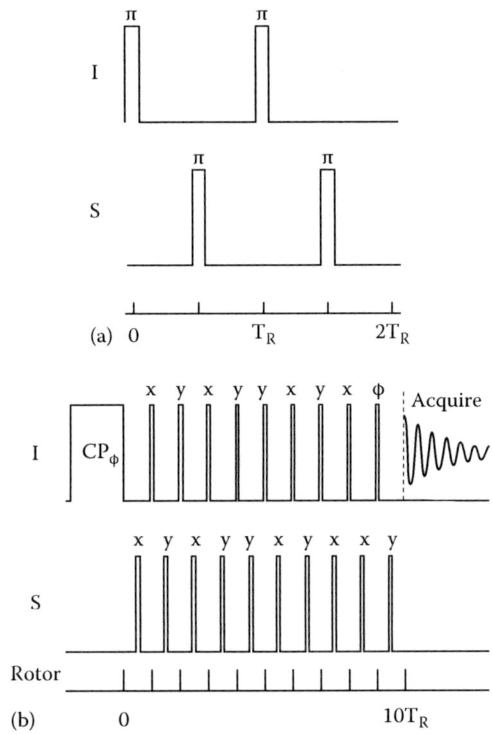

FIGURE 3.16 (a) The representative building block for the REDOR experiment, containing two rotor periods with 180° pulses operating in channel I and channel S. (b) The REDOR experiment, where the cross-polarization section is added. (Reproduced with permission from Gullion, T. and Vega, A.J., *Prog. Nucl. Magn. Reson. Spectrosc.*, 47, 123, 2005.)

NMR experiments performed for samples containing the enriched ^{13}C–^{2}H pairs. In the absence of intense molecular motions, the quadrupolar frequency of ^{2}H nuclei located at chemical C–H bonds in solids covers a region of 240 kHz. When the application of only a single dipolar recoupling pulse in the deuterium channel becomes unsatisfactory, then, instead of the 180° pulse, a single ^{2}H composite 90° pulse can be applied as a more effective one. It is important that the pulses in the observed channel are applied at the midpoint and at the end of each rotor cycle. As we can see, the distance between these two nuclear labels is determined by the dihedral angle ψ and the solid curve has been obtained after the calculations carried out for dipolar coupling of 72 kHz. The latter, in turn, corresponds to internuclear distance r of 3.99 Å.

To conclude this section we emphasize that solid-state recoupling NMR cannot be considered as a replacement for the regular diffraction methods in spite of a good NMR accuracy. A combination of the different methods is obviously more preferable. For example, the data on single-crystal x-ray structures already known for some systems can be used for quantitative interpretations of double-quantum MAS NMR spectra, recorded for related systems, where the diffraction data are absent. Nevertheless sometimes, as noted by Ulrich [34], the REDOR MAS NMR spectroscopy becomes a single physical method capable of determining the internuclear

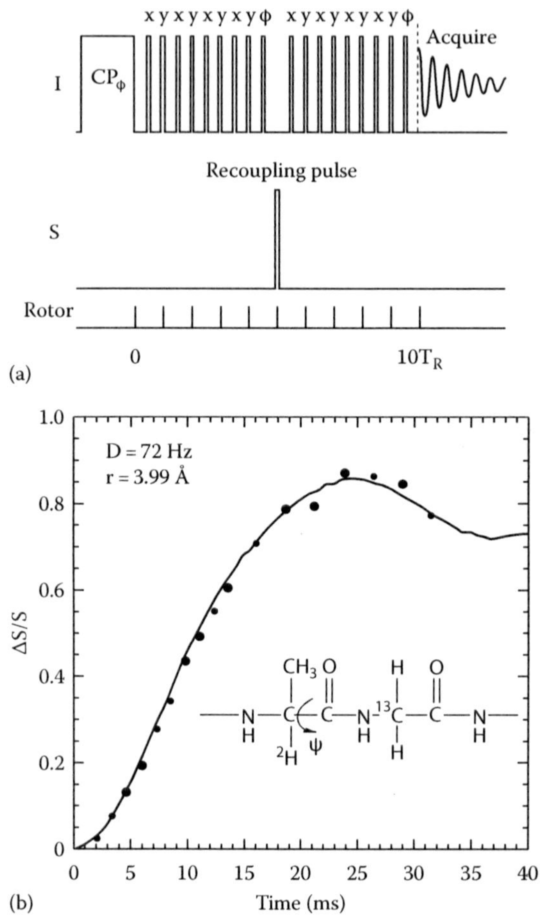

FIGURE 3.17 (a) The REDOR MAS NMR experiments containing a single recoupling pulse in the deuterium channel. (b) The evolution of the ^{13}C–^{2}H REDOR $\Delta S/S$ fraction, observed in time for a ^{13}C labeled molecule $(Ala\text{-}Gly)_{15}$. (Reproduced with permission from Gullion, T. and Vega, A.J., *Prog. Nucl. Magn. Reson. Spectrosc.*, 47, 123, 2005.)

distances in solids. The latter has been demonstrated by ^{19}F REDOR MAS NMR experiments performed in solid ^{19}F-labeled peptides. It has been noted that the accuracy of the resulting 3D peptide structures, based on the internuclear distances, is not a consequence of the errors coming from NMR, itself, but it is rather related to the conformational assumptions taken for the structural analysis.

3.8 SIDEBAND MANIPULATIONS

Static solid-state NMR of polycrystalline compounds or amorphous solids shows a broad powder pattern even for isolated nuclei with spins of 1/2 or a Pake's powder pattern that is typical of quadrupolar nuclei. In both cases the observation is accompanied by remarkable loss in peak sensitivity. When systems contain many structurally and/or chemically nonequivalent sites, the static NMR spectra are difficult to analyze because they mask the presence of different signals. The MAS NMR in the fast spinning regime reduces or completely removes all anisotropic interactions of the second rank: chemical shift, dipolar, or quadrupolar interactions. However, at intermediate spinning rates, which are comparable with the spread in Larmor frequencies, the MAS NMR spectra show sets of sidebands separated by the distances (measured in Hz) equal to the spinning rates, and "centered" at the isotropic shift frequencies. The theoretical details of this phenomenon can be found in Antzutkin [35].

If a material, investigated by solid-state NMR, is supposed to have different chemical and/or structural sites, the corresponding MAS NMR spectrum could show a number of sideband sets and isotropic resonances. According to the theory [35], these sidebands and isotropic resonances can be distinguished in the MAS NMR spectra recorded at various spinning rates. This is a simple approach often used in practice of MAS NMR. However the approach to running NMR at various spinning rates risks making the experiments very long particularly at relatively long relaxation times of target nuclei. To avoid this problem, the different pulse sequences can be applied as shown in Figure 3.18.

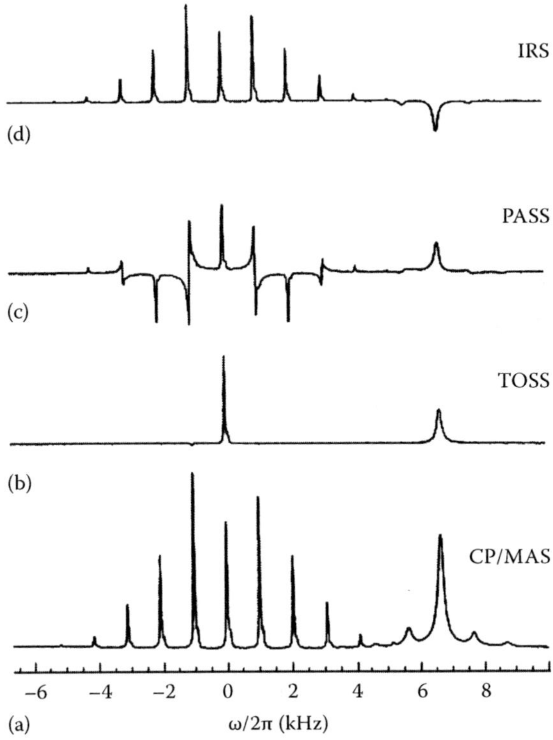

FIGURE 3.18 (a) The regular CP ^{13}C {^{1}H} NMR spectrum recorded for a sample of glycine at spinning rate of 1030 Hz, (b) The TOSS experiments, (c) the PASS experiment, (d) the IRS experiment performed under the same conditions. (Reproduced with permission from Antzutkin, O.N., *Prog. Nucl. Magn. Reson. Spectrosc.*, 35, 203, 1999.)

The spectra illustrate three widely applied approaches aimed at manipulations of spinning sidebands that are based on actions of strong 180° pulses with precise timing.

The sidebands in a MAS NMR spectrum can be completely eliminated by the total suppression of spinning sideband pulse sequence (TOSS), suggested by Dixon. The TOSS pulse sequence results in the MAS NMR spectra exhibiting only the signals corresponding to the isotropic chemical shifts (see Figure 3.18b). Thus this approach provides observation of solid-state NMR spectra similar to those in solutions.

The second approach implies an additional phase-shift that is specific for each sideband:

$$\phi k = -k\theta \tag{3.14}$$

Here θ represents the pitch of the so-called PASS sequence (see Figure 3.18c). It has been noted that the PASS is the best approach in a 2D sideband separation scheme. In fact, for systems with many chemical sites, sideband patterns overlap in the regular MAS spectra while the PASS experiment separates all sidebands by order.

The third approach creates conditions where a sideband pattern can be in phase with the corresponding center resonance, while the whole sideband manifold, corresponding to another site, will be inverted by the isotropic rotation sequence (IRS) (see Figure 3.18d). The IRS can be used to prepare a state of pure difference polarization regardless of the orientations of the crystallites in a powder sample. It has been emphasized that under the IRS conditions the spin system effectively evolves only under the isotropic chemical shifts. This effect can be applied for 2D versions of solid-state NMR to separate isotropic and anisotropic chemical shifts. Different modifications of these main pulse sequences, for example, the SELTIC sequence, are also available and can be found in the original papers.

3.9 ZERO-FIELD SOLID-STATE NMR EXPERIMENTS

To observe an NMR signal, an initial nuclear polarization is generally created by the strong external magnetic field, B_0. On the other hand, the nuclei can be polarized even in the absence of the B_0 by the *internal* magnetic fields that appear in a sample due to its nature. Andrew and Szczesniak [31] have noted that such zero-field NMR experiments are possible and the first zero-field NMR spectrum has been recorded by Reif and Purcell (1953) (see reference [37] in the review [31]). for a solid hydrogen sample due to very strong proton–proton dipolar interactions. Similarly, even in the absence of an external magnetic field, the internal magnetic fields polarize ^{57}Fe and ^{59}Co nuclei in ferromagnetic iron and cobalt again resulting in the zero-field NMR spectra.

Other physical principles make a basis for the zero-field NMR experiments, where the main efforts are directed to enhancement of spectral resolution in NMR spectra of polycrystalline samples. An initial nuclear polarization by these experiments is induced in a sample, usually by its placing into a strong external magnetic field. Then, this highly polarized sample is transferred mechanically into a magnetic field of an intermediate strength. At the same time, this intermediate field should still be notably larger than local fields induced in the sample due to strong dipole–dipole interactions. By the next step, this intermediate field is rapidly switched off. Under this condition, the nuclear spins undergo an evolution, during time τ_1, which is initiated obviously by the dipole–dipole forces. After the evolution time, the intermediate field is again turned on. Then, the sample is again placed into the high magnetic field. Finally, a high-field NMR signal is recorded as the FID as function of τ_1. Fourier transformation of the FID(τ_1) data gives, thus, a zero-field NMR spectrum.

The summary of this zero-field experiment is as follows the manipulations lead to the high-magnetic-field NMR spectrum, recorded naturally at a high sensitivity and due to an evolution of the nuclear spin system in the absence of the external high magnetic field. It is important that during the evolution period all directions are equivalent and numerous crystallites in a polycrystalline sample will behave identically. Under this condition, the resulting NMR spectrum will show only sharp lines. Andrew notes [31] that even powder samples will exhibit dipolar couplings with high resolution, which can be usually reached only in field-oriented single crystals. Figure 3.19 illustrates three proton NMR spectra obtained in a sample of the barium chloride monohydride molecule. The spectra have been obtained in

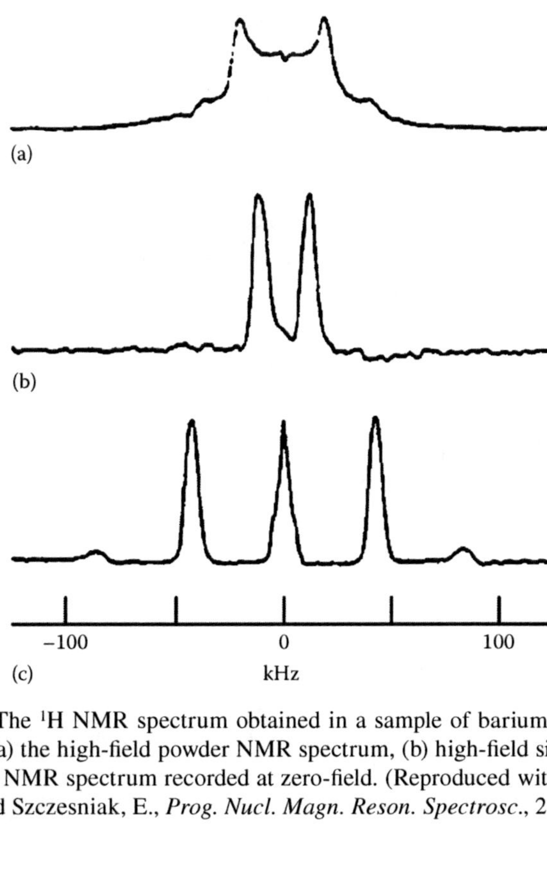

FIGURE 3.19 The ^{1}H NMR spectrum obtained in a sample of barium chlorate monohydrate molecule: (a) the high-field powder NMR spectrum, (b) high-field single crystal NMR spectrum, (c) the NMR spectrum recorded at zero-field. (Reproduced with permission from Andrew, E.R. and Szczesniak, E., *Prog. Nucl. Magn. Reson. Spectrosc.*, 28, 11, 1995.)

a high external magnetic field, where the molecule shows a typical powder line shape (Pake's doublet) (a); in a high external magnetic field using a crystal orientation (b); and in a zero-field using a powder (c). As can be seen, the spectra result in observation of dipolar couplings in powder samples at high resolution. Due to the absence of orientation broadenings, these zero field experiments open the way to perform determination of structures even in the absence of single crystals or oriented samples.

The limitations in applications of this method are also obvious: the zero-field technique can be successfully used when the spin-lattice relaxation time, T_1, in a low magnetic field is long enough to complete the experimental cycle. Finally the technical limitation is related to special equipment needed for such experiments. It should be added that the zero-field NMR experiments can be performed on spinning samples. However acting pulses and rotation rates should be synchronized in such cases.

Finally the detailed theory of the zero-field NMR experiments [36] and also acceleration of experiments by replacing the mechanical transfer with electrical switching [37] can be found in the original papers.

3.10 SOLID-STATE NMR IMAGING

The strategy of improving the spectral resolution in solid-state NMR experiments is obviously very important in a structural analysis of various materials. It is well known that high-resolution NMR spectra can be obtained, when the strong external magnetic field is maximally homogenous and capable of keeping the resonance condition in each point through the volume of a sample investigated. This is a key statement important for studies of liquids and solids. However in the case of large sizes of heterogeneous samples, the resonance condition cannot be satisfied often throughout the whole object. However, the external magnetic field could be homogenized in a small volume of the sample and thus a high-resolved NMR signal can still be recorded from this volume. Then, this small volume can be "spatially shifted" to collect the next signal. Finally, repeating these experiments can result in observation of target nuclei distributed through the sample volume. This is the principal idea in NMR imaging that was realized by Lauterbur in 1973 [38].

Currently, solid-state NMR imaging (often called the spatially resolved solid-state NMR) is widely used as an important instrument in probing spatial chemistry and molecular dynamics of different materials. For example, measurements of catalyst porosity, characterizations of polymer materials, and studies of their morphology or the ceramic processing can be in focus of such applications.

Spatial resolution can be achieved by different techniques based on application of strong oscillatory field gradients, and/or multiple-quantum coherences evolution or line-narrowing approaches such as magic echoes, magic angle sample spinning and multiple-pulse excitation. The basic principles of these experiments can be found in the review presented by Demco and Blumichm [39]. The authors have classified the domains for applications of spatially resolved NMR as *imaging*, *microscopy*, and *volume localized spectroscopy* of materials.

Spatial resolution, named as ΔX^{-1}, by frequency encoding is related by the relationship (3.15)

$$\Delta X = \frac{2\pi\Delta\nu}{\gamma(G_X)} \tag{3.15}$$

where

$\Delta\nu$ is the full width at half height of NMR spectrum

G_X is the field gradient strength applied along the x direction

This relation implies that the spatial resolution can be improved by increasing the gradient strength or reducing the line width. The latter can be realized by manipulations of spins.

Since the NMR imaging techniques are various and numerous, it is reasonable to be limited, for example, by multiple-quantum coherences that are widely applied for static solids. Generally these experiments include excitation, evolution, reconversion, and detection (by 2D NMR) sections that are represented in Figure 3.20. As one can see, the three-pulse sequence is applied in combination with 180° refocusing pulses to excite the multiple-quantum coherences in the presence of strong dipolar coupling. The same pulse scheme can be applied for quadrupolar nuclei. The key

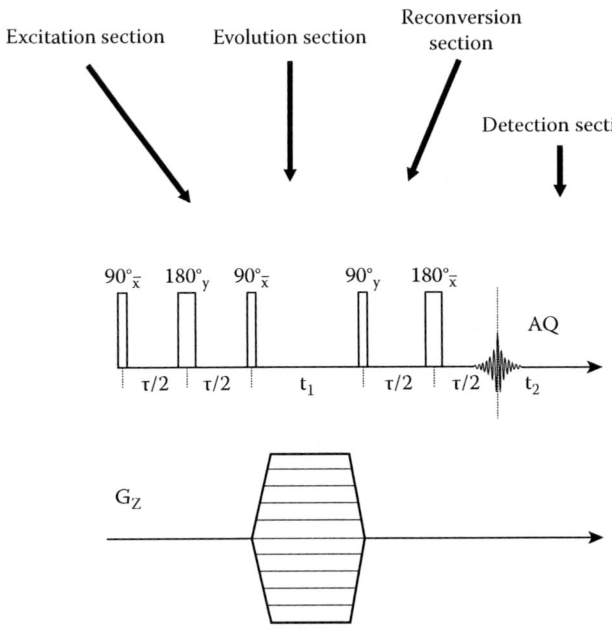

FIGURE 3.20 Excitation, evolution, reconversion, and detection time sections represented in the three-pulse sequence applied in a combination with 180° refocusing pulses to excite and detect the multiple-quantum coherences. (Reproduced with permission from Demco, D.E. and Blumichm, B., *Concepts Magn. Reson. A*, 12, 188, 2000.)

part of the experiments is: the strong field gradient is used during the time t_1 for the phase encoding of space information.

Demco and Blumichm have emphasized that quadrupolar nuclei, such as ^{2}H in different polymer materials, or ^{27}Al nuclei, situated, for example, in ceramics, are the best candidates for studies by the spectroscopic multiple-quantum imaging. In fact, these experiments can give information on the local electron symmetry and molecular mobility and orientation. When the multiple-quantum coherences of quadrupolar nuclei are replaced for the multiple-quantum transitions of the dipolar-coupled nuclei, the same experiments can be performed for ^{1}H nuclei [40].

To conclude, it should be added that the solid-state NMR imaging experiments can imply application of the MAS techniques to reduce the dipolar, chemical shift anisotropy and quadrupolar interactions. However, spinning rates and the field-gradient pulses should again be synchronized to obtain best results.

REFERENCES

1. Dybovski, C., Bai, S. 2008. *Analyt. Chem.* 80: 4295.
2. Brown, S. P., Spiesss, H. W. 2001. *Chem. Rev.* 101: 4125–4155.
3. Klinowski, J. 1991. *Chem. Rev.* 91: 1459.
4. Fyfe, G. A., Feng, Y., Grondey, H., Kokotailo, G. T., Gies, H. 1991. *Chem. Rev.* 91: 1525–1543.
5. Bonhomme, C., Coelho, C., Baccile, N., Gervais, C., Azais, T., Babonneau, F. 2007. *Acc. Chem. Res.* 40: 738.
6. Widdifield, C. M., Schurko, R. W. 2009. *Concepts Magn. Reson. Part A.* 34A: 91–126.
7. Davis, M. C., Kaseman, D. C., Parvani, S. M. et al. 2010. *J. Phys. Chem. A* 114: 5503–5508.
8. Strub, H., Beeler, A. J., Grant, D. M., Michel, J., Cutts, P. W., Zilm, K. W. 1983. *J. Am. Chem. Soc.* 105: 3333–3334.
9. Taylor, R. E., Carver, C. T., Larsen, R. E., Dmitrenko, O., Bai, S., Dybowski, C. 2009. *J. Mol. Struct.* 930: 99–109.
10. Tycko R. 2003. *Prog. Nucl. Magn. Reson. Spectr.* 42: 53–68.
11. Spyros, A., Dais, P. 2009. *Prog. Nucl. Magn. Reson. Spectrosc.* 54: 195–207.
12. Hologne, M., Chevelkov, V., Reif, B. 2006. *Prog. Nucl. Magn. Reson. Spectrosc.* 48: 211–232.
13. Lesage, A. 2009. *Phys. Chem. Chem. Phys.* 11: 6876.
14. Bakhmutov, V. I. 2009. Nuclear magnetic resonance spectrometry. In *Encyclopedia of Applied Spectroscopy*, D. L. Andrews (ed.). Weinheim, Germany: Wiley.
15. Smith, M. E., van Eck, E. R. H. 1999. *Prog. Nucl. Magn. Reson. Spectrosc.* 34: 159.
16. Ooms, K. J., Feindel, K. W., Willans, M. J., Wasylishen, R. E., Hanna, J. V., Pike, K. J., Smith, M. E. 2005. *Solid State NMR* 28: 125–134.
17. Van Vleck, J. H. 1948. *Phys. Rev.* 74: 1168–1183.
18. Pake, G. E. 1948. *J. Chem. Phys.* 16: 327–336.
19. Reichert, D., Schneider, H. 1993. *Meas. Sci. Technol.* 4: 1376–1379.
20. Kentgens, A. P. M., Bos, A., Dirken, P. J. 1994. *Solid State NMR* 3: 315.
21. Duer, M. J. 2002. *Solid-State NMR Spectroscopy: Principles and Applications*. Oxford, U.K.: Blackwell Science.
22. Ernst, R. R., Anderson, W. A. 1966. *Rev. Sci. Instrum.* 37: 93–102.
23. Hartmann, S. R., Hahn. E. L. 1962. *Phys. Rev.* 128: 2042–2053.
24. Kolodzieski, W., Klinowski, J. 2002. *Chem. Rev.* 102: 613.

25. Peersen O. B., Wu, X., Kustanovich, I., Smith, S. O. 1993. *J. Magn. Reson. A* 104: 334–339.
26. Park, S., Johnson, D. K., Ishizawa, C. I., Pariila, P. A., Davis, M. F. 2009. *Cellulose* 16: 641.
27. Wu, X., Zilm, K. W. 1993. *J. Magn. Reson. A* 102: 205.
28. Bryce, D. L., Bernard, G. M., Gee, M., Lumsden, M. D., Eichele, K., Wasylishen, R. E. 2001. *Can. J. Anal. Sci. Spectrosc.* 46: 46–82.
29. Hodgkinson, P. 2005. *Prog. Nucl. Magn. Reson. Spectrosc.* 46: 197–222.
30. Willans, M. J., Sears, D. N., Wasylishen, R. E. 2008. *J. Magn. Reson.* 191: 31.
31. Andrew, E. R., Szczesniak, E. 1995. *Prog. Nucl. Magn. Reson. Spectrosc.* 28: 11–36.
32. Eckman, R., Mtiller, L., Pines, A. 1980. *Chem. Phys. Lett.* 74: 376.
33. Gullion, T., Vega, A. J. 2005. *Prog. Nucl. Magn. Reson. Spectrosc.* 47: 123–136.
34. Ulrich, A. S. 2005. *Prog. Nucl. Magn. Reson. Spectrosc.* 46: 1–21.
35. Antzutkin, O. N. 1999. *Prog. Nucl. Magn. Reson. Spectrosc.* 35: 203–266.
36. Hennel, J. W., Stachurovam, M. 1985. *Z. Phys. B* 60: 49.
37. Sun, B. Q., Pines, A. 1994. *J. Magn. Reson.* 109: 157.
38. Lauterbur, P. C. 1973. *Nature* 242: 190–193.
39. Demco, D. E., Blumichm, B. 2000. *Concepts Magn. Reson. A* 12: 188–206.
40. Bakhmutov, V. I. 2011. *Chem. Rev.* 111: 530–562.

4 Practice of Multinuclear NMR in Diamagnetic Solids: General Information and Examples of Applications

NMR is applicable to any nuclei that possess nonzero magnetic moments. Nevertheless, results of NMR applications will depend strongly on the nature of materials and target nuclei, and adequacy of formulated tasks. Although NMR can be applied, in principle, for most of the nuclei in the periodic table, in reality the range of nuclei is remarkably limited by NMR sensitivity and sensitivity of measured NMR parameters to environments of nuclei. Convenience of experiments and simplicity of their interpretations also play an important role. Generally such nuclei as ^{1}H, ^{2}H, ^{19}F, ^{13}C, ^{29}Si, ^{17}O, ^{31}P, ^{51}V, ^{93}Nb, ^{181}Ta, and the halogen nuclei are mostly attractive and more often used as probes in studying different materials. Sometimes even the simple well-recorded NMR spectra are sufficient to describe complex molecular systems, for example, the immobilized phosphine linkers situated on the surface of silica-based matrix [1]. Figure 4.1 illustrates two types of such linkers that exhibit the very sharp ^{31}P resonances observed at 22.4 and −3.9 ppm for groups $[R'PR_2Et]^+$ and $[R'PR_2]$, respectively.

However even simple NMR experiments can be interpreted only on the basis of common information on target nuclei and their applications. This is the aim of this chapter, which can be useful or even used as a guide, particularly by beginners.

4.1 STRUCTURAL INFORMATION BY INCREASING THE SPECTRAL RESOLUTION IN MULTINUCLEAR SOLID-STATE NMR

Probing and describing structures of materials by solid-state multinuclear NMR generally focuses on determination of the number of different sites, their chemical identification, and relative contents as structurally forming units. Various target nuclei can be involved into MAS NMR experiments to describe better matrix of materials, for example. The latter is particularly important for amorphous solids, when the single-crystal x-ray method cannot help and the x-ray powder diffraction patterns can be simulated only in the presence of these independent NMR data. It is obvious that increasing the structural information can be achieved by the higher spectral

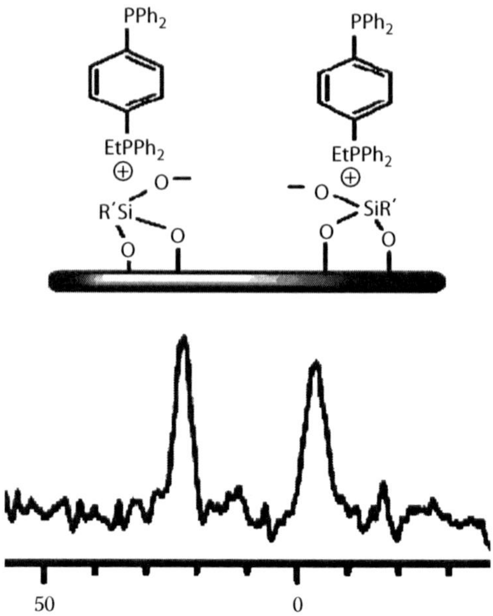

FIGURE 4.1 Top: The silica surface, containing the immobilized phosphine linkers; Bottom: the ^{31}P MAS NMR spectrum characterizing the groups $[R'PR_2Et]^+$ and $[R'PR_2]$. (Reproduced with permission from Yang, Y. et al., *J. Am. Chem. Soc.*, 130, 3771, 2008.)

resolution and/or by development of new NMR approaches. Both of the aspects are important at performing solid-state NMR experiments for nuclei with spins of 1/2 [2] and quadrupolar nuclei [3].

4.1.1 Spectral Resolution for Nonquadrupolar Nuclei

The first group of the nuclei, particularly ^{1}H, can be detected at ultra fast magic angle spinning (MAS) rates because line widths of nuclei ^{1}H, governed by very strong dipolar couplings, decrease approximately linearly with increasing the MAS frequency. Remarkable improvements in MAS technology have been achieved by applications of small NMR rotors (2.5 mm diameter or less) in commercially available NMR probes. At the present time one can routinely reach spinning frequencies up to 65 kHz. This very fast MAS condition, itself, can show the high-resolution ^{1}H NMR spectra of organic solids, mesoporous silica nanoparticles, microcrystalline proteins, etc. [2].

The progress in increasing the spectral resolution can be seen in the 1H MAS NMR experiments performed at spinning rates between 10 and 65 kHz for the silica matrix, where an organic compound is grafted inside the pores. As seen in Figure 4.2, at 10 kHz the resonances are still broadened and overlapped while at 60 kHz they are already well resolved providing the reliable spectral assignments. It is obvious also that the fast spinning is accompanied by increasing the NMR sensitivity due to redistribution in the signal intensities by reducing the sidebands.

As noted in Chapter 3, the homonuclear dipolar interaction ^{1}H–^{1}H, particularly strong in rigid solids, still broadens resonance lines in the ^{1}H MAS NMR spectra,

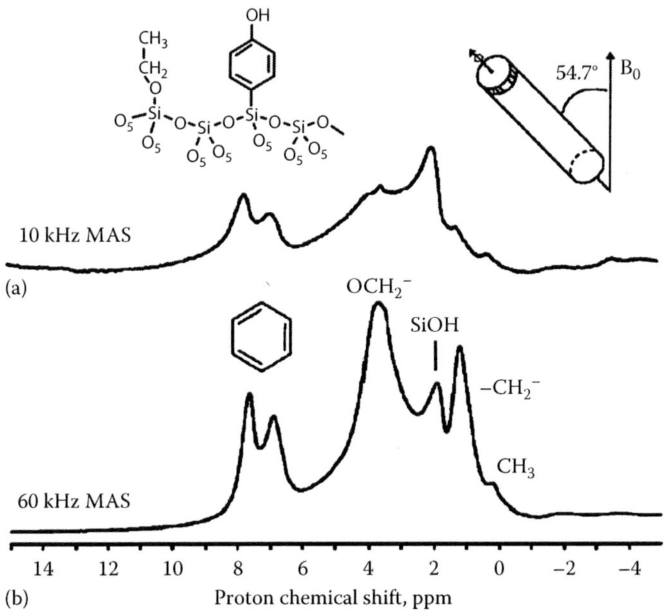

FIGURE 4.2 The ^1H MAS NMR spectra recorded at spinning rates of 10 (a) and 60 (b) kHz in a sample of the mesoporous silica matrix with the shown organic compound, which is grafted in pores. The spectra are obtained in a 1.3 mm NMR rotor. (Reproduced with permission from Lesage, A., *Phys. Chem. Chem. Phys.*, 11, 6876, 2009.)

recorded even at high spinning rates. Such broadenings can be significantly weakened or eliminated at fast spinning in combination with homonuclear dipolar decoupling techniques combined rotation and multiple pulse spectroscopy (CRAMPS), pioneered by Gerstein [4].

The principal ideas of CRAMPS imply (a) the MAS rotation in *the coordinate space* leading to $(1 - 3\cos^2\theta) = 0$ and (b) the coherent averaging of the dipolar coupling in spin space via resonant cyclic, and periodic multiple-pulse excitations over cycle times that are short relative to the inverse of the homogeneous dipolar coupling. Thus, for example, in a one-dimensional (1D) CRAMPS experiment, the resulting signal is registered in the windows between pulses in the cyclic and periodic excitations, which attenuate homogeneous dipolar interactions. Generally this technique provides the best effect for resonances of nuclei with spins of 1/2, which are broadened due to homogeneous dipolar homo- coupling and shielding anisotropies.

Currently the technique of improving the resolution, involves new pulse sequences, such as decoupling using mind-boggling optimization (DUMBO) or phase-modulated-Lee-Goldburg (PMLG) [2] The DUMBO sequence can narrow the ^1H resonances up to a few Hz, as shown by the ^1H MAS NMR spectra in Figure 4.3. 2D ^1H–^1H double-quantum (DQ) MAS experiments increasing the resolution are also available. However, it should be emphasized here that all of these ^1H NMR approaches are particularly successful for crystalline samples because the nature of line broadenings in strongly amorphous systems is also related to chemical shift distributions.

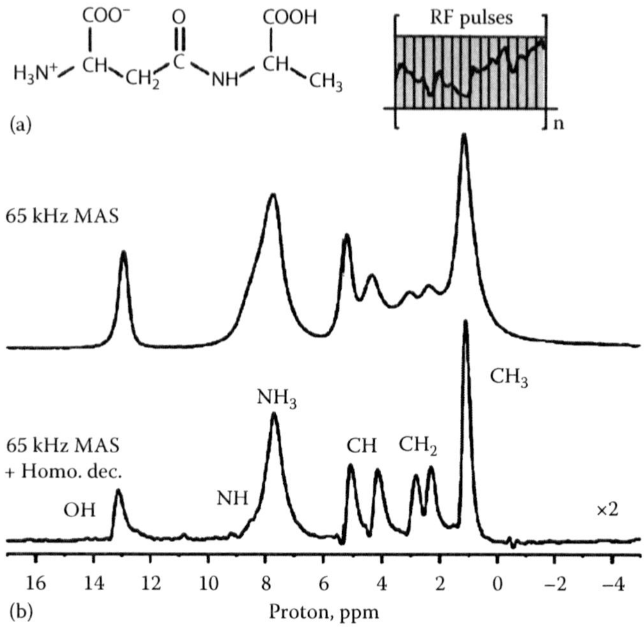

FIGURE 4.3 Comparison of the ^{1}H MAS NMR spectra recorded in the solid dipeptide b-L-Asp-L-Ala. (a) without homonuclear decoupling; (b) with decoupling performed by a DUMBO pulse sequence. (Reproduced with permission from Lesage, A., *Phys. Chem. Chem. Phys.*, 11, 6876, 2009.)

^{1}H NMR signals can be sharpened by isotopic dilution with deuterium. This approach obviously requires previous chemical treatments or even quite expensive synthetic procedures. Nevertheless it is very promising especially for biomolecular systems. For example, deuteration of groups NH or OH leads to the ^{1}H NMR spectra, where resonances are broadened only by a 100 Hz even at relatively small spinning rates and in the absence of homo-decoupling techniques [2]. Finally, a combination of cross-polarization, MAS, and ^{1}H-decoupling can be routinely applied to obtain high sensitivity and high resolution NMR spectra for nuclei of 1/2 other than ^{1}H for both crystalline and amorphous materials.

4.1.2 Signal Assignments

Structural information can be reliable only with reliable assignments of the NMR signals. Strategy in assignments of resonances can be based on *the homo- and/or heteronuclear correlation spectroscopy*. For nuclei with spins of 1/2, other than ^{1}H, for example ^{13}C nuclei, the signal assignments can be carried out by the homonuclear correlation spectroscopy, where the NMR data are collected either by dipolar-based transfers or via J-based techniques. Double-quantum (or zero-quantum) homonuclear dipolar recoupling MAS techniques (see Chapter 3) can be applied for probing the ^{29}Si–^{29}Si or ^{31}P–^{31}P dipolar interaction, as an important part of structural information in solids [2].

The structure of zeolites can be accurately solved by the ^{29}Si DQ dipolar recoupling NMR experiments, which lead to internuclear silicon distances. The 2D ^{29}Si DQ technique is very suitable to solve structures of zeolites and mesoporous silica-based materials. For example, the 2D ^{29}Si incredible natural abundance double quantum transfer (INADEQUATE) experiments are very powerful tools for unambiguous determination of *interconnectivities* between the silicon atoms. These multidimensional and multinuclear MAS NMR correlations are useful for elucidation of structural details about inorganic and hybrid systems [2,5]. Practice of solving the local structure can be illustrated here by the NMR data described by Lesage in the review [2]. The well-resolved ^{29}Si double-quantum dipolar recoupling NMR spectra of a zeolite sample have been obtained at natural ^{29}Si abundance. First, the spectra show the silicon–silicon correlations needed for signal assignments (Figure 4.4) and

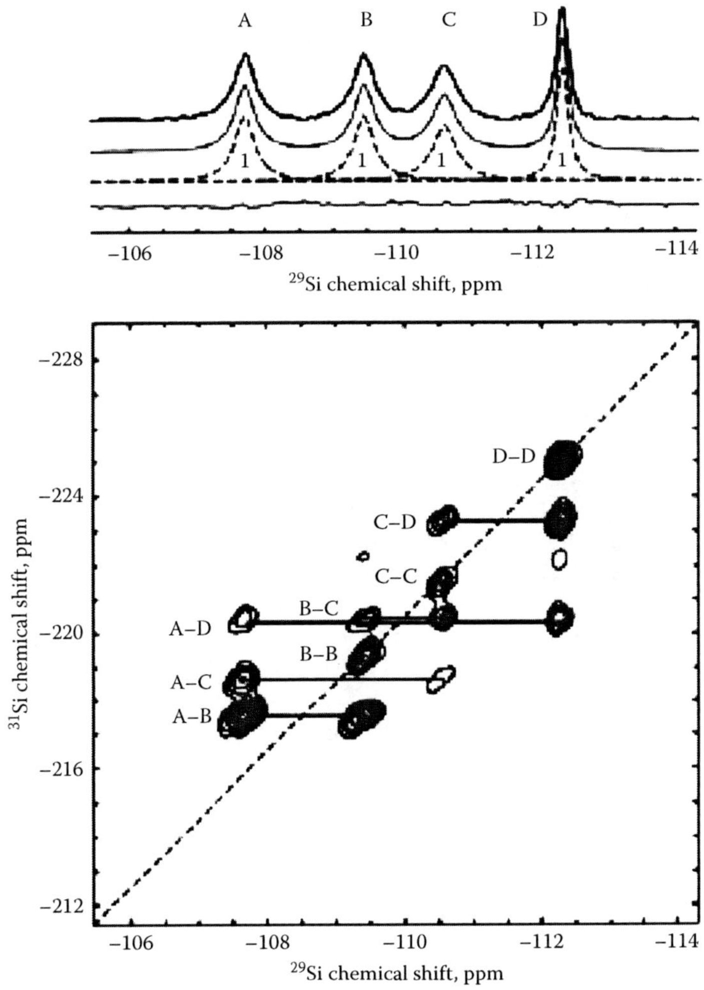

FIGURE 4.4 The ^{29}Si DQ MAS NMR spectra obtained in a sample of a purely siliceous zeolite: (top) 1D ^{29}Si DQ correlation spectrum; (bottom) 2D ^{29}Si DQ correlation spectrum. The corresponding silicon atoms are assigned in Scheme 4.1. (Reproduced with permission from Lesage, A., *Phys. Chem. Chem. Phys.*, 11, 6876, 2009.)

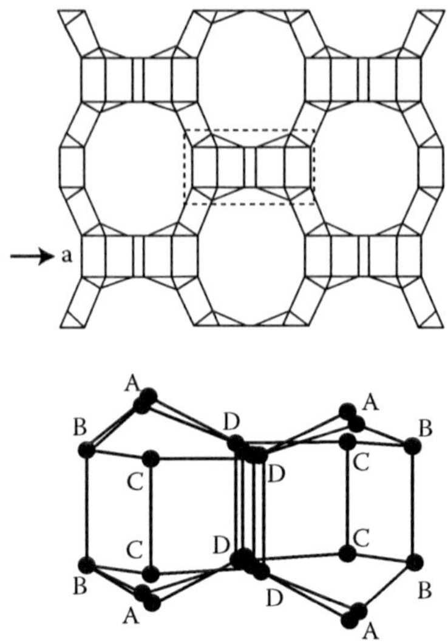

SCHEME 4.1 Structural motives in a zeolite system corresponding to the NMR spectra in Figure 4.4.

second, the experiments reveal the internuclear silicon distances. When the unit cell parameters and space group are known from the powder XRD data, then the structure of the zeolite can be completely solved (Scheme 4.1). In the following we will show that the accurate structure of frameworks in many materials can be obtained by a combination of the solid-state NMR data with x-ray diffraction data and first principle calculations.

Isotropic spin–spin coupling (for coupling mechanisms see Chapter 1) is also a significant important element in signal assignments. These couplings, particularly important for spins of 1/2, can provide local structural information, which is complementary to that extracted via dipolar and chemical shift interactions. The coupling constants can be measured in the solid state by different methods. For example, the J-resolved NMR spectroscopy is used for accurate measurements of even small scalar couplings $^{13}C–^{13}C$, $^{15}N–^{15}N$, and $^{29}Si–^{29}Si$. Lesage [2] has noted that homonuclear scalar couplings, such as J(P–P) and J(C–C) can be measured even in disordered samples by a combination of spin-echo experiments with the 2D INADEQUATE spectroscopy. These NMR experiments naturally result in distributions of the coupling constant in noncrystalline systems. For example, the solid-state ^{31}P NMR study of *N,N*-bis(diphenylphosphino)-*N*-((*S*)-amethylbenzyl)-amine has led to the average $^{2}J(P–P)$ constant in the P–N–P fragment that takes values between 22.6 and 27.9 Hz. According to Emsley et al. [6], the $^{2}J(P–N–P)$-coupling constants show a clear dependence on the local structural parameters. Hence they can be used as a probe for the structural distortions.

4.1.3 Resolving Quadrupolar Nuclei

Chapter 3 has demonstrated the main problem in solid-state NMR experiments on quadrupolar nuclei, related to the second-order quadrupolar broadening. The effect cannot be completely removed by simple MAS techniques even at high spinning rates. These second-order quadrupolar broadenings mask the presence of structurally different sites, limiting, thus, actual spectral resolution.

However, the broadening effect can be strongly reduced by sample rotations around two different angles or by double rotation, DOR. The DOR technique, considered carefully by Ashbrook [3], has great advantages leading to the high-resolution NMR spectra of the 1D NMR experiment that are recorded in real time. However, a special design of DOR NMR probes (see following) obviously limits the application of this technique.

In contrast, the 2D multiquantum MAS experiments can be performed with regular commercial NMR probes. Therefore they are much more popular. These experiments combine the regular MAS rotation of a sample and manipulations with the nuclear spins, thus providing a high resolution and helping in signal assignments and a structural analysis as well. The results of one of the experiments are represented in Figure 4.5. As can be seen, the one-pulse ^{27}Al MAS NMR spectrum obtained for the $AlPO_4$-14 material exhibits a very poor spectral resolution obviously due to the

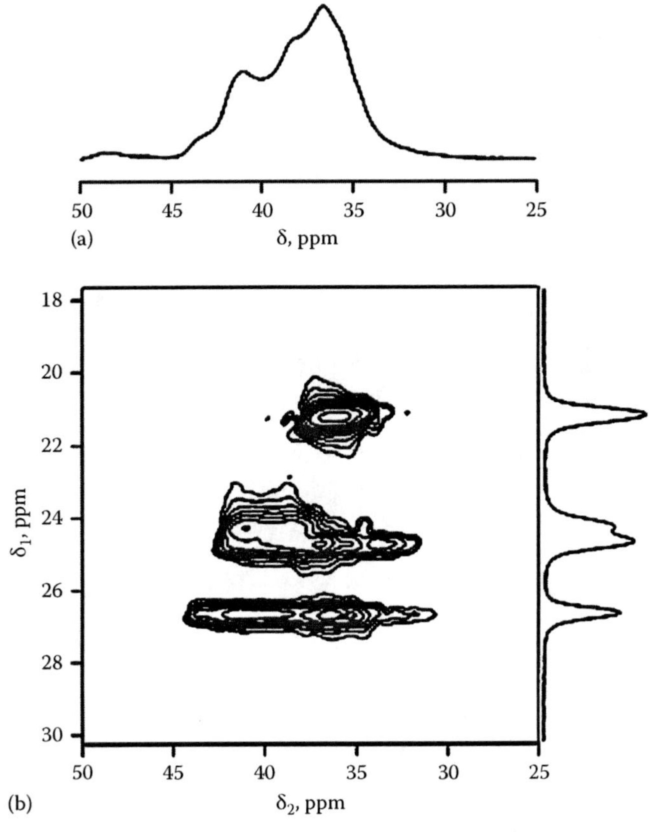

FIGURE 4.5 The ^{27}Al (a) MAS and (b) 2D triple-quantum MAS NMR spectrum and corresponding isotropic projection recorded for a material of the $AlPO_4$-14 family. (Reproduced with permission from Ashbrook, S.E., *Phys. Chem. Chem. Phys.*, 11, 6892, 2009.)

second-order quadrupolar broadening. The observed ^{27}Al resonances are strongly overlapped and their quantitative analysis is difficult. At the same time, this broadening effect is completely suppressed in the two-dimensional (2D) triple-quantum MAS NMR spectrum, where all the different aluminum species are well resolved providing thus more detailed structural information.

Alternative ways of enhancement of spectral resolution in solid-state NMR of quadrupolar nuclei have been created by the satellite-transition MAS NMR experiments at a very precise magic angle setting [3], or by the experiments at highest magnetic field strengths. These experiments can be found in works of Ooms et al. [7].

Frydman et al. [8] have suggested the other strategy in enhancing the spectral resolution of chemically or structurally nonequivalent sites in multinuclear NMR spectra of powdered and/or disordered materials. This strategy is based on relaxation behavior of nuclei in a combination with signal-enhancement methodologies, such as the quadrupolar Carr–Purcell–Meiboom–Gill pulse trains. It has been shown that the relaxation-assisted separations can be very successful for characterizations of materials by the high-resolution solid-state NMR spectra recorded for low-γ nuclei. In addition, the authors [8] have formulated the limitations of these 2D NMR methodologies and their potential. Generally speaking, even the simple partly relaxed NMR spectra collected by inversion-recovery pulse sequences can potentially resolve the overlapped lines, when they have different relaxation T_1 times.

4.2 PROBING PROXIMITIES OF NUCLEI AND MEASUREMENTS OF INTERNUCLEAR DISTANCES IN DIAMAGNETIC SOLIDS

Terms "proximity" and "distance" imply mutual locations of nuclei interpreted on qualitative and quantitative levels, respectively. Internuclear proximity plays an important role in line assignments of NMR spectra as the first step in any structural study of solids.

Probing proton–proton proximities by solid-state ^1H DQ MAS NMR can be very successful even in disordered and heterogeneous materials. Brown [9] considers this NMR technique applied for different materials: silica ZSM-12 zeolites, described via the poly(ethylene oxide)/cross-linked silicone networks, template mesoporous silicate films, polymer functionalized carbon nanotubes, potassium aluminum phosphate glasses, and hydrated cements.

The NMR study of a silica-based material, modified by zirconium species on the matrix surface, as schematically shown in Scheme 4.2, allows us to trace the spectroscopic logic in signal assignments.

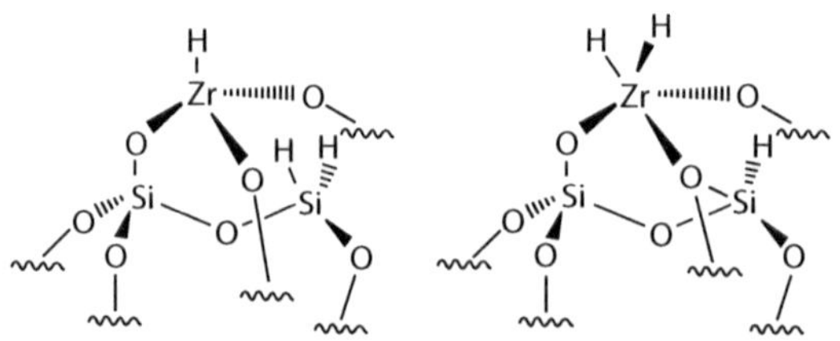

SCHEME 4.2 Structural elements of a silica-based material modified by the Zr species.

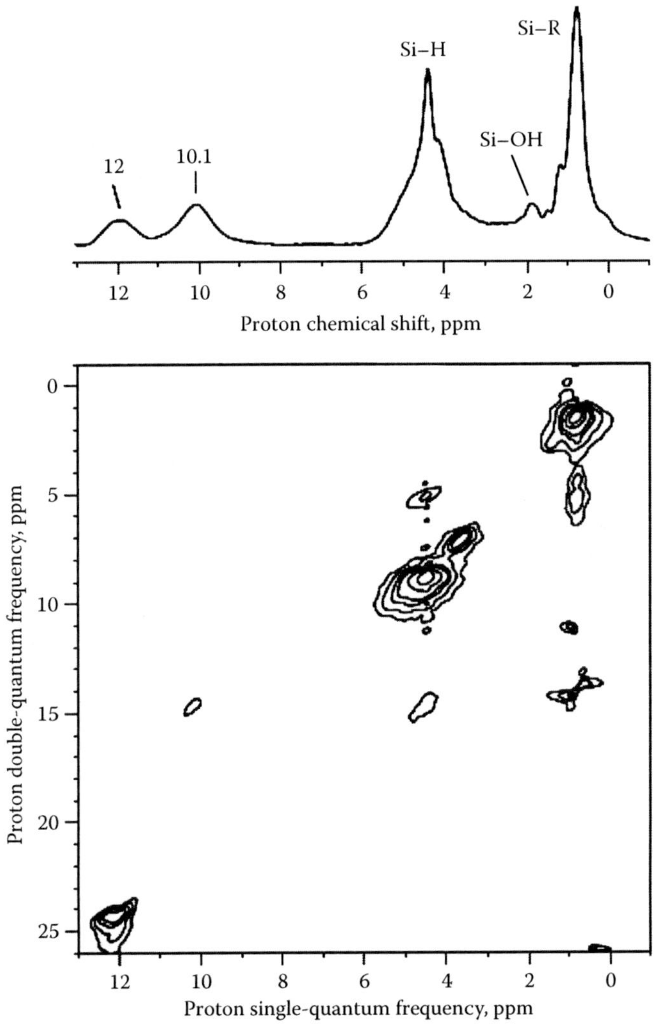

FIGURE 4.6 The proton (500 MHz) MAS and DQ MAS (30 kHz) spectra recorded for a silica material containing the zinc organic species on the silica surface. One rotor period of BABA recoupling has been applied for the excitation and reconversion of double quantum coherence. (Reproduced with permission from Brown, S.P., *Prog. Nucl. Magn. Reson. Spectrosc.*, 50, 199, 2007.)

Figure 4.6 shows the ^{1}H NMR data collected for this material. sample, allowing us to trace the logic in the signal assignments. As can be seen, the single quantum ^{1}H MAS NMR spectrum exhibits all of the protons situated on the silica surface. The high-field resonances observed between 4.5 and 1.5 ppm can be well assigned to proton–silica units, marked as Si–H and Si–OH on the basis of the well-known spectral–structural relationships. Then, two resonances in a low field with chemical shifts of 10.1 and 12 ppm obviously belong to different zirconium hydride surface species (Scheme 2.1). Their further discrimination can be carried out on only the basis of the double quantum proton–proton proximity experiments. Following from the 2D ^{1}H NMR pattern, only one resonance, observed at δ of 12 ppm, exhibits a diagonal peak. Thus this signal corresponds to the ZrH_2 species. Then the resonance with δ of 10.1 ppm can be well assigned to the isolated ZrH groups. In addition, it is remarkable that the ZrH groups show a weak correlation at a DQ frequency of 14.5 ppm with the SiH resonance, observed at 4.4 ppm.

The theoretical grounds, considered in Chapter 3, allow us to accept solid-state NMR as a powerful tool for high-precision measurements of interatomic distances in ordered solids. In fact, large magnitudes of long-range dipole–dipole interactions that, on the one hand, prevent the observation of high-resolution NMR spectra, and on the other hand contain sufficient information on distances between nuclei. This is particularly important when regular diffraction methods are not applicable.

The classical Pake's powder NMR spectrum [10], typical of two diluted nuclei (e.g., protons) in a static sample, is shown in Figure 4.7. The shape of the resonance is formed by superposition of two subspectra (shown as dashed lines) and the distance between the two singularities corresponds to the dipolar coupling constant ω_D. The ω_D splitting characterizes the static constant, if the solid is rigid. Then, recalculation of coupling constant ω_D gives a distance between the nuclei. A weak point in this simple NMR approach is obviously related to a relatively low NMR sensitivity in the case of rare target nuclei and the fact that the registered nuclei are supposed to be strongly coupled only in the pair. Other couplings should be negligible.

Currently the high potential of NMR techniques capable of *direct* determination of proton–proton dipolar coupling and, thus, measurement of ^{1}H–^{1}H distances is well recognized and widely used for increasingly complex solids [2]. Among various approaches, the 2D proton–proton double-quantum MAS NMR experiments play a major role. This technique has been applied by Spiess and coworkers, cited by Lesage [2], for organic molecular systems. Leskes and Vega [11] have applied the

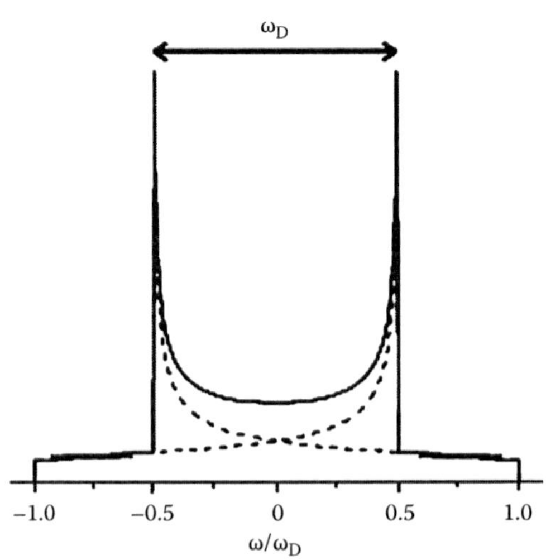

FIGURE 4.7 The classical Pake dipolar powder spectrum for two diluted nuclei in a static sample. (Reproduced with permission from Ooms, K.J. et al., *Solid State NMR*, 28, 125, 2005.)

more advanced triple-quantum ^{1}H NMR experiments to observe the correlations between three protons.

Ulrich [12] has considered the application of the Carr–Purcell–Meiboom–Gill (CPMG) pulse sequence for determinations of ^{19}F–^{19}F distances in fluorine-labeled biomolecules. These pulse trains refocus the CSA and heteronuclear couplings, while the homonuclear ^{19}F–^{19}F interactions still remain unchanged. They are evolving during the echo period. Then, Fourier transformation produces a pure dipolar spectrum that provides, in turn, measurements of F–F dipolar coupling and the corresponding ^{19}F–^{19}F distances.

Less direct NMR methods, requiring, however, more advanced techniques, such as separated local field (*SLF*), polarization inversion spin exchange at magic angle (*PISEMA*) polarization inversion time averaged nutation spin exchange at magic angle (*PITANSEMA*) and transferred echo double resonance (*TEDOR*), simplify determinations of dipolar couplings and hence internuclear distances. Their physical basis and applications for high-precision measurements of internuclear distances have recently been represented by Lee and Khitrin [13].

This review formulates the main features of the modern approaches as: (a) the NMR technique uses 2D single-echo experiments enhanced by adiabatic cross-polarization; (b) the 2D NMR experiments result in *unperturbed dipolar powder* spectra and can be performed even for diluted (1% and lower) nuclear spins; (c) the technique helps to resolve structures of *amorphous* or *disordered* systems. However, it has been emphasized that the measurements are more effective and accurate for crystalline molecular systems.

Lee and Khitrin have suggested the NMR technique especially developed for experiments in static samples. When molecules are isotopically labeled, for example, labeled glycine, this NMR technique helps to achieve accuracy in measuring internuclear distances C–C and C–N, which is well compared with that generally realized in the x-ray and neutron-diffraction methods. For example, distance Cα–N in the glycine molecule is determined as 1.496 ± 0.002 Å by solid-state NMR and 1.476 ± 0.001 Å by neutron diffraction [13].

Since new solid-state MAS NMR methods create an opportunity for three-dimensional (3D) structural determinations, these methods can be used in probing absorbed organic molecules in microporous materials that can form molecular complexes. Organic species can be absorbed within frameworks of zeolites and their surface localizations play a major role in understanding catalytic properties of zeolite systems.

Following Fyfe et al. [14], usefulness of the ^{19}F–^{29}Si CP, REDOR, and TEDOR NMR experiments can be illustrated in practice by studies of the silicate octadecasil. These experiments have been performed to localize precisely the F anions situated within the 3D framework of this compound. As has been emphasized, such problems are often difficult to solve by traditional x-ray diffraction because of many objective reasons: the single crystals can be too small or twinned, for example.

The framework of octadecasil is depicted in Figure 4.8. Thus the task consists of accurately determining the Si–F distances. Because the ^{29}Si and ^{19}F resonances are generally well detected in the MAS NMR spectra, the determination of internuclear distances can be based on the ^{19}F–^{29}Si dipolar couplings observed experimentally.

According to the principles and the protocol of REDOR MAS NMR considered in Chapter 3, the experiments include the radio frequency pulses applied to reintroduce

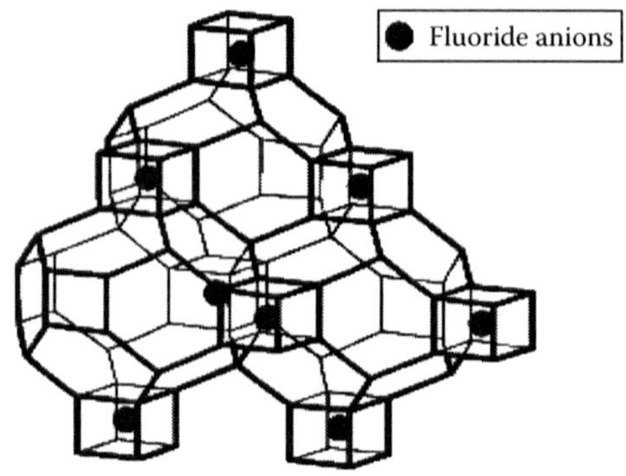

FIGURE 4.8 Schematic representation of the framework structure of octadecasil. There is the packing arrangement of the two types: the large and small cages. (Reproduced with permission from Fyfe, C.A. et al., J. *Am. Chem. Soc.*, 119, 12210, 1997.)

the heteronuclear dipolar couplings that have been eliminated by MAS. Two different experiments should be performed. The first requires dephasing while the second should be done without the dephasing. Thus, the echo intensity, S_0, observed in the absence of dephasing, and echo S_F registered in the presence of dephasing to reintroduce heteronuclear dipolar couplings, can be recorded. Then, the difference in intensities, $\Delta S = S_0 - S_F$, that appear due to the effect of heteronuclear dipolar couplings, should be normalized as ratios $\Delta S/S_0$. Figure 4.9 illustrates the data collected. The curves show the typical nonperiodic oscillations, when the $\Delta S/S_0$ values are plotted as a function of the number of rotor cycles n or the evolution time or the dimension parameter, representing $nD\tau_r$ where the τ_r is the time of a complete rotor evolution.

Then, the experimental curves can be fitted to unknown ^{29}Si–^{19}F dipolar couplings. As we can see, the curves are well fitted to the ^{29}Si–^{19}F dipolar couplings of 1150 Hz to account for a full occupancy by ions F^- (the dashed line) and a 85%

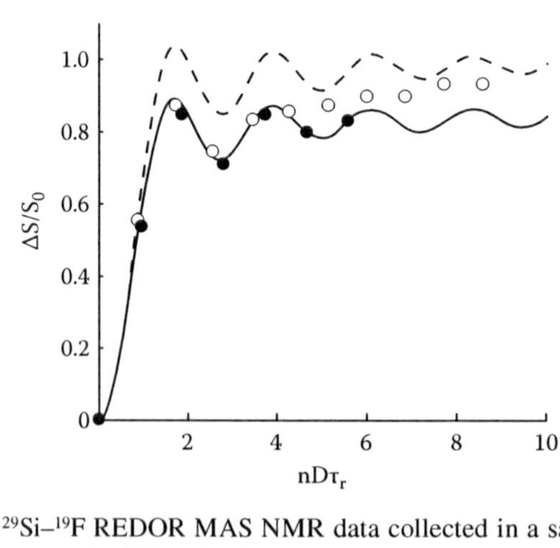

FIGURE 4.9 The ^{29}Si–^{19}F REDOR MAS NMR data collected in a sample of octadecasil. The filled circles correspond to ^{19}F dephasing pulses used at a spinning rate of 2.480 kHz and the open circles show the results obtained a spinning rate of 2.675 kHz. (Reproduced with permission from Fyfe, C.A. et al., *J. Am. Chem. Soc.*, 119, 12210, 1997.)

occupancy (the solid line). Finally, accurate Si–F distances of 2.69 ± 0.04 Å can be calculated in excellent agreement with the value of 2.63 Å determined from the single crystal x-ray structure.

Fyfe emphasizes that these MAS NMR experiments can be successfully applied for other materials. For example, ^{19}F–^{31}P interatomic distances can be well determined in aluminophosphate molecular sieves ($AlPO_4$'s) with fluorine-containing guests or other framework/sorbate host–guest complexes.

4.3 ^{31}P SOLID-STATE NMR

Generally NMR users, even beginners or researchers, working in solutions, are familiar with ^{13}C and ^{1}H NMR spectra, chemical shift ranges of ^{1}H and ^{13}C nuclei, and the dependence of their chemical shifts on the nature of molecules investigated. Therefore the following sections will consider nuclei, other than protons and carbons, which are important for NMR studies of materials.

^{13}P nuclei have 100% naturally abundant spins of 1/2, providing excellent NMR sensitivity 390 times larger than ^{13}C nuclei, for example. Chemical shifts of these nuclei cover a range between 500 and −200 ppm, referring to H_3PO_4 as an external standard in Figure 4.10. This range demonstrates the large sensitivity of isotropic shift $\delta(^{31}P)$ to the nature of phosphorous atoms and their environments. For example, phosphorous with valence of 5 in phosphates can be seen between 20 and −40 ppm while the trivalent phosphorous compounds show lines between 100 and 200 ppm.

A substituent on the oxygen atoms, for example, in the phosphates, affects the electron density in the phosphorous atoms and thus the presence of metal cations in the phosphate groups can be recognized by ^{31}P NMR. Thus there is a basis for discrimination between chemically different phosphorous groups or groups that are similar chemically but differ structurally. All these features are important for applications of solid-state ^{31}P NMR in materials chemistry as a powerful analytical tool [15].

The analysis of materials by ^{31}P NMR obviously implies recording the highly resolved NMR spectra under MAS conditions. Then, a pivotal moment is determination of number and isotropic chemical shifts for the ^{31}P resonances. For example, metal phosphates and metal phosphonates are of great interest as catalysts [16]. Therefore appearance of a single or two phosphorous resonances in the spectra of metal phosphates is an important criterion for their structural characterization.

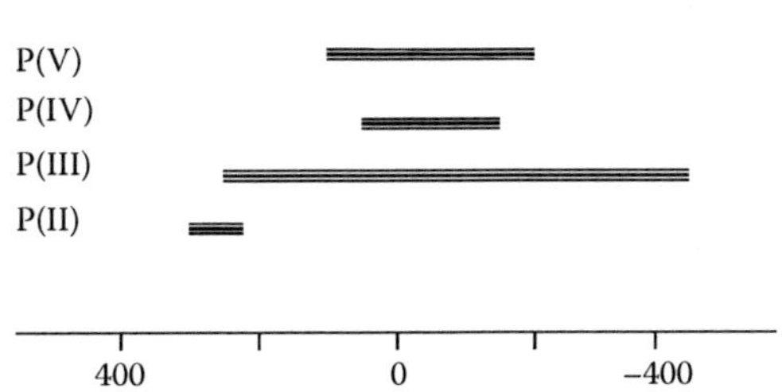

FIGURE 4.10 ^{31}P Chemical shifts referring to phosphoric acid.

Moreover the number of different phosphorous sites can be used for calculations of x-ray powder diffraction patterns.

Besides the simple ^{31}P MAS NMR experiments, aimed at measurements of isotropic chemical shifts, determination of ^{31}P magnetic shielding tensors also attracts a significant interest in probing the hybrid metal phosphonates. These tensors characterize the connectivity, which in turn describes different types of the coordination in PO_3 groups. This is particularly important, for example, for layered metal phosphonates, where a PO_3/metal inorganic core is separated by the organic moiety bound to phosphorous [17]. According to the classification suggested by Massiot et al. the PO_3 groups without protons on the oxygen atoms show three types of coordination:

1. Each of the three oxygen atoms of the phosphonate unit is coordinated to only one metal atom, corresponding to the (111) connectivity.
2. One of the oxygen atoms is coordinated to two metal atoms, while the two other atoms are bonded only to one metal atom, the (112) connectivity.
3. There are two bridging oxygen atoms with the (122) connectivity.

It has been established that the type of the connectivity strongly affects the parameters of the ^{31}P chemical shift tensors that can be used as a structural criterion. Figure 4.11 illustrates the validity of this statement: the chemical shift tensor parameters dictate the shapes of ^{31}P resonances in static samples of zinc phosphonates. As can be seen, the shapes change dramatically as a function of the connectivity.

It should be added that such (or MAS) ^{31}P NMR spectra can be treated to reveal the asymmetry parameter and magnitudes δ_{11}, δ_{22}, and δ_{33} as important characteristics for structurally describing the phosphorous centers. Table 4.1 lists these data showing a relatively small δ(iso) change versus significant variations in values δ_{11}, δ_{22}, and δ_{33}.

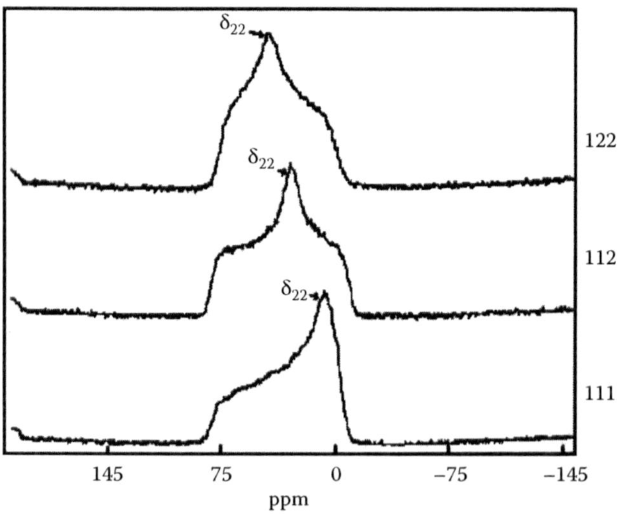

FIGURE 4.11 Static ^{31}P NMR spectra recorded in static samples of zinc phosphonates showing different connectivity, determined independently. (Reproduced with permission from Massiot, D. et al., *Chem. Mater.*, 9, 6, 1997.)

TABLE 4.1
^{31}P Magnetic Shielding Tensors (Measured in ppm) and Asymmetry Parameter η Determined for Some Zinc Phosphonates

Comp/Connectivity	δ(iso)	η	δ_{11}	δ_{22}	δ_{33}
$Zn(O_3PC_2H_4NH_2)$/111	27.0	0.4	−7.1	10.6	77.5
$Zn(O_3PC_2H_4CO_2H)3/2H_2O$/112	32.4	0.8	−9.8	27.7	79.3
$Zn(O_3PC_2H_4CO_2)2$/122	35.9	0.9	68	37.6	2.1
$Zn(O_3PC_2H_5)2H_2O$/122	37.7	0.9	71.7	39.5	1.9

Source: Massiot, D. et al., *Chem. Mater.*, 9, 6, 1997.

The divalent cadmium phosphonate hybrid material $Cd_2Cl_2(H_2O)_4(H_2L)$, synthesized from the ethylenediamine-*N,N*-bis(methylenephosphonic acid), marked as H_4L, has shown unexpected spectroscopic results in solid state ^{31}P NMR studies that can be interesting in the methodical context. The ^{31}P MAS NMR spectra of a $Cd_2Cl_2(H_2O)_4(H_2L)$ sample have revealed the ^{31}P isotropic chemical shift of 8.7 ppm, which is very similar to that measured in initial H_4L (10.5 ppm). Obviously such $\delta(^{31}P)$ changes are not sufficient to prove even formation of the material with the binding ligands [18]. However in contrast to the ^{31}P isotropic shifts, the static ^{31}P NMR spectra, shown in Figure 4.12, demonstrate how strongly the ^{31}P chemical shift tensor parameters depend on the chemical coordination.

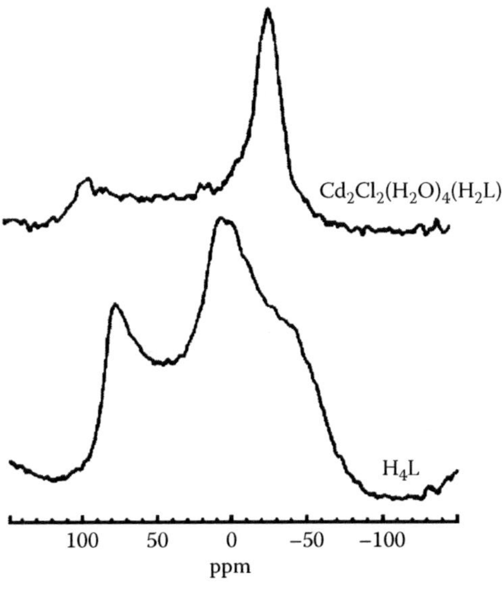

FIGURE 4.12 Static ^{31}P CP NMR spectra of aminophosphonic acid H_4L (bottom): δ(iso) = 10.5 ppm, δ_{11} = −50 ppm, δ_{22} = 3 ppm, δ_{33} = 78 ppm, η = 0.8; and $Cd_2Cl_2(H_2O)_4(H_2L)$ (top): δ(iso) = 8.7 ppm, δ_{11} = −32 ppm, δ_{22} = −32 ppm, δ_{33} = 90 ppm, η = 0.0. (Reproduced with permission from Bakhmutova-Albert, E.V. et al., *Inorg. Chem.*, 43, 1264, 2004.)

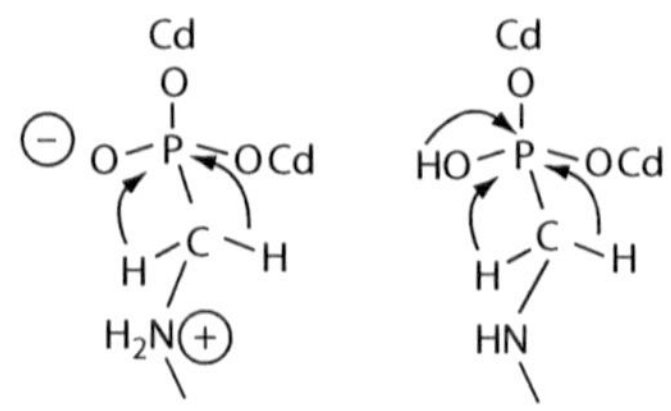

1a **1b**

FIGURE 4.13 Possible proton–phosphorous dipole–dipole interactions bearing on cross-polarized ^{31}P NMR spectra.

A molecular or zwitterionic nature, depicted in Figure 4.13, is the second important aspect in structural characterization of this material. The problem could be solved by the accurate localization of the protons in fragments **1a** or **1b** that is difficult or impossible by the regular structural methods due to a poor crystalline character of such systems. However magnitudes of the cross-polarization rates (see Chapter 3), calculated from the cross-polarization ^{1}H–^{31}P kinetics in the MAS NMR experiments, will depend on the number of protons involved in the cross-polarization process. Thus, potentially, one of these structures can be distinguished: three protons, closest to ^{31}P (CH_2 and POH), are involved in magnetization transfer in structure **1b** versus two protons (CH_2) in **1a** (see the corresponding arrows).

Due to the natural abundance and the high magnetogyric ratio, ^{31}P nuclei are very attractive for MAS NMR studies based on dipolar ^{31}P–^{31}P correlation experiments, considered by Schnell [19]. In fact, to account for these strong homonuclear dipole–dipole interactions between ^{31}P nuclei, correlation experiments are possible at spatial ^{31}P–^{31}P proximities within a distance range of 0.4–0.5 nm, dependent on the chosen recoupling times and conditions. A simple experiment is demonstrated in Figure 4.14. The ^{31}P–^{31}P double-quantum and single-quantum NMR spectra were recorded in a sample of potassium-intercalated $CdPS_3$ at a spinning rate of 15 kHz. As seen from the NMR pattern, there are two pairs of DQ peaks located on either side of the diagonal. Thus the resonances, observed at 102.0/102.3 and 97.4/96.3 ppm, obviously belong to chemically nonequivalent phosphorous atoms in the same $P_2S_6^{4-}$ group. It has been noted that in potassium-intercalated $CdPS_3$ the dipolar couplings are realized via an intramolecular way over a P–S–P bond with a distance of 0.35 nm.

Wide applications of ^{31}P solid-state NMR from the simple single-pulse 1D MAS experiments to complex pulse sequences, including 2D MAS-J-INADEQUATE ^{31}P–^{31}P experiments or heteronuclear correlation experiments, can be found in the recent review by Smith and Hanna [20]. The authors have emphasized an important role of ^{31}P NMR in biomaterials science and first principles calculations of ^{31}P NMR parameters that are often used to identify, for example, structurally different units, and carry out signal assignments. At the same time, the authors have shown that such calculations are not always conclusive. For example, the connectivity of the PO_4 units in the phosphate molecule β-$Ca(PO_3)_2$ (that is principally important for probing bio-glasses), cannot be clarified by the first principles calculations. However, this connectivity can be well determined by the MAS-J-INADEQUATE spectrum, represented in Figure 4.15. The 2D ^{31}P NMR spectra show directly how the PO_4 units

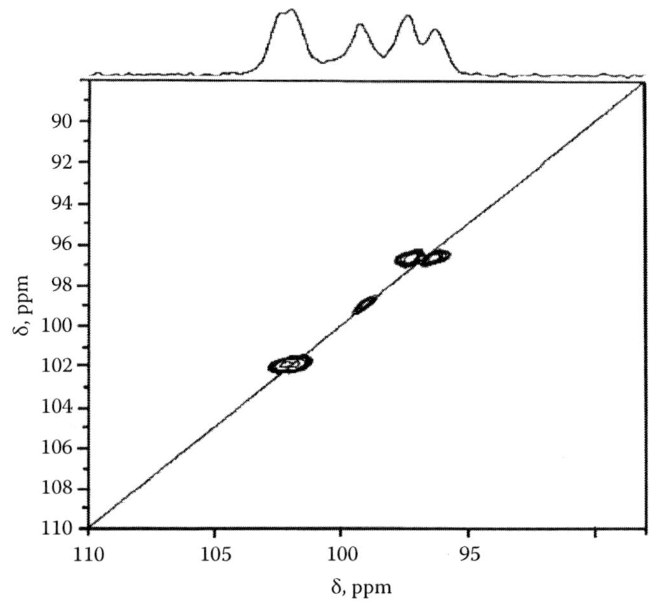

FIGURE 4.14 The ^{31}P–^{31}P double-quantum NMR spectrum recorded in a sample of potassium-intercalated $CdPS_3$. The spinning rate was 15 kHz and duration for excitation and reconversion in a dipolar recoupling was 533 ms. (Reproduced with permission from Schnell, I., *Prog. Nucl. Magn. Reson. Spectrosc.*, 45, 145, 2004.)

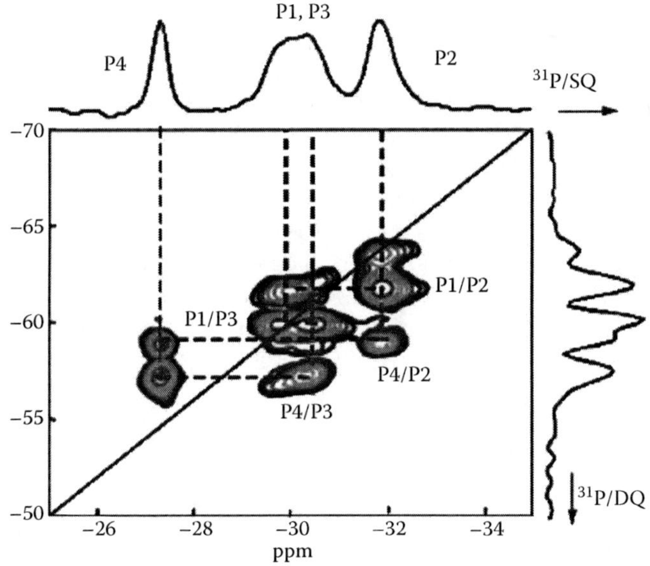

FIGURE 4.15 The ^{31}P MAS-J-INADEQUATE spectrum recorded in a sample of β-$Ca(PO_3)_2$ showing the connectivity of the PO_4 units. (Reproduced with permission from Hanna, J.V. and Smith, M.E., *Solid State NMR*, 38, 1, 2010.)

are connected in this solid. Other NMR approaches, such as the ^{31}P 2D refocused INADEQUATE or INADEQUATE spin-echo NMR experiments, are also available providing determination of the connectivity and ordering in bio-glasses.

In spite of difficulties in the theoretical approaches, there are situations when the *ab initio* calculations of ^{31}P NMR parameters become a unique instrument in studying amorphous or poorly crystalline materials. Recent studies by Chappell et al. [21], focused on structure and properties of the surface in hydroxyl-apatite, have used first principles calculations of ^{31}P NMR parameters. The calculations have been carried out in the limits of the density functional theory (DFT) by using the plane-wave pseudopotential method. The magnetic shielding tensors for ^{31}P nuclei were calculated and compared with the experimental spectral frequencies, which were measured relative to a reference material. Finally, additional calculations were performed on single molecules. It has been established that the DFT-calculated ^{31}P isotropic chemical shifts and shielding tensors for ^{31}P nuclei, situated on the surface, differ from the nuclei in the bulk crystals. The chemical shift difference is not significant but it is consistent with the ^{31}P NMR data obtained from bone and synthetically prepared crystalline samples of hydroxyl-apatite. Thus, the high potential of ^{31}P NMR in combinations with an *ab initio* modeling in order to describe the surface structure of the materials is obvious.

4.4 SOLID-STATE NMR OF HALOGEN NUCLEI AND OXYGEN

The spectral parameters of halogen nuclei $^{35,37}Cl^-$, $^{78,81}Br$, and ^{127}I are very sensitive to their local environments and they can be successfully used as excellent probes in studies of materials. The general problem in detections of chlorine, bromine, and iodine resonances is connected with their pronounced quadrupolar nature resulting in large quadrupole coupling constants, which can reach 22, 49, or 30 MHz, respectively.

It is obvious that the collection of NMR data is always aimed at a good signal-to-noise ratio in the resulting NMR spectra. On the one hand, the natural abundance of the halogen nuclei is high. On the other hand, at nonsymmetric environments of $^{35,37}Cl^-$, $^{78,81}Br$, and ^{127}I nuclei, typical of many materials, interactions between the quadrupole moments and the electric field gradients at these nuclei will strongly perturb the Zeeman's splitting and lead to quadrupolar broadenings as the dominant contributions to the observed NMR line widths. Even the central transitions, which are generally better observed for most of the quadrupolar nuclei, can be strongly broadened up to several MHz. All these effects dramatically reduce NMR sensitivity, particularly for powder samples. In addition, the large quadrupolar interactions in excitation of nuclei with a single pulse sequence lead to very short FIDs and the very broad resulting NMR spectra. In turn, these effects can be coupled with the presence of acoustic ringing in the receiver coils. Technically it greatly weakens the NMR signal.

Nevertheless, in spite of these objective difficulties, the $^{35,37}Cl^-$, $^{78,81}Br$, and ^{127}I solid-state NMR experiments can be successfully applied for characterizations of materials especially at highest magnetic fields. This positive influence of the B_0 has been considered in the previous chapter.

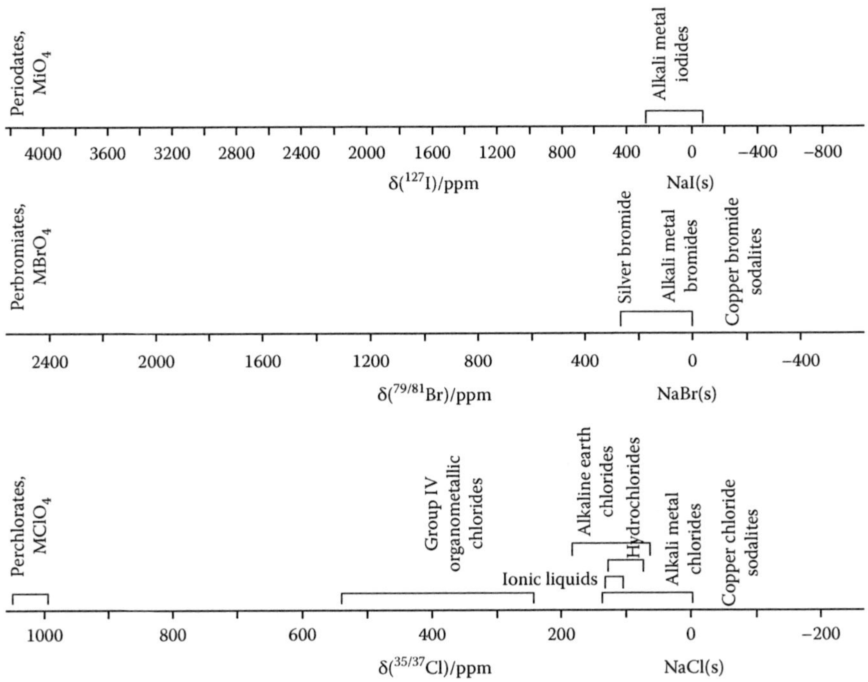

FIGURE 4.16 The isotropic chemical shift ranges for chlorine, bromine, and iodine nuclei. The chemical shift values refer to the external standard NaCl, NaBr, and NaI taken as δ = 0 ppm. (Reproduced with permission from Chapman, R.P. et al., *Prog. Nucl. Magn. Reson. Spectrosc.*, 55, 215, 2009.)

Chapman et al. [22] show how the solid-state chemical shifts of $^{35,37}Cl^-$, $^{78,81}Br$, and ^{127}I nuclei change in some types of solids (see δ(iso) scales in Figure 4.16).

Wasylishen et al. [23] recommend applications of the multiple pulse spin-echo experiments for increasing the sensitivity in NMR spectra, which are recorded in static samples containing halogen nuclei. Scheme 4.3 illustrates the common quadrupolar Carr–Purcell–Meiboom–Gill (QCPMG) pulse train and the modified QCPMG sequence, application of which transforms the powder patterns into a series of sharp spikelets depicted in Figure 4.17. The spectra were recorded at the high external magnetic field of 11.75 T.

The ^{127}I NMR spectra of a static KIO_4 sample are characterized by quadrupolar coupling constant C_Q of 20.66 MHz at a zero asymmetry parameter. As can be seen, the modified QCPMG pulse train leads to 80% increasing sensitivity. It was also noted that a decrease in the intensity of the spikelets in the 320–350 kHz area is explained by the imperfect excitation.

Other pulse sequences and 2D NMR experiments aimed at enhancement of signal-to-noise ratios are also available and can be found in Chapman et al. [22]. Among them, the MQ MAS NMR experiments, performed in chlorine-containing systems, are capable of discrimination between two different chlorine sites even with the very small differences in chlorine chemical shifts. When the line widths of the central transitions

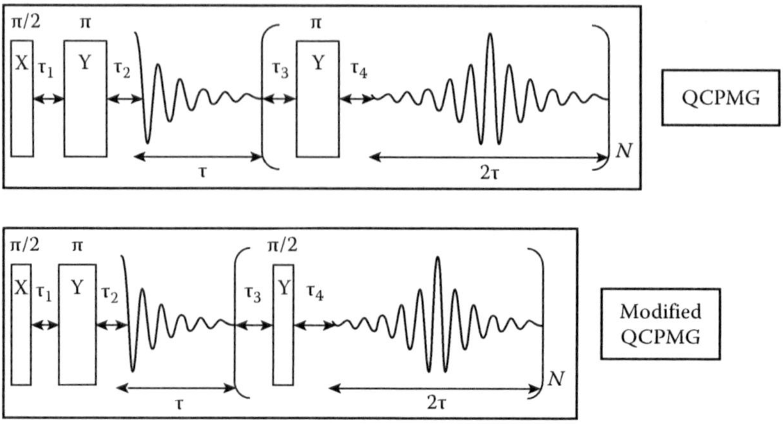

SCHEME 4.3 The pulse sequences operating in the spin-echo experiments.

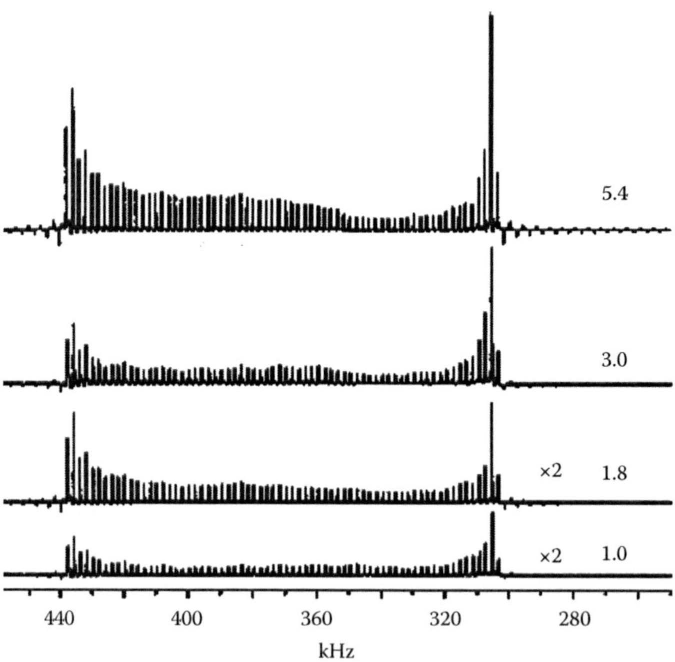

FIGURE 4.17 The ^{127}I NMR spectra recorded in a static powdered sample of KIO_4 from top to bottom: converging DFS with modified QCPMG pulse sequence, converging DFS with QCPMG pulse sequence, modified QCPMG pulse sequence and classical QCPMG pulse sequence. (Reproduced with permission from Siegel, R.S. et al., *Concepts Magn. Reson. Part A*, 26A, 62, 2005.)

are too large and the transitions cannot be uniformly excited by radio frequency pulses, then one can use the variable-offset experiments to record full NMR spectra.

The chlorine, bromine, and iodine solid-state NMR experiments in materials can be generally performed under MAS conditions. Due to MAS, the isotropic resonance is not affected by any CSA interactions and the δ(iso) values can be found. The second-order quadrupolar interactions are not averaged in MAS spectra, providing, thus, determination of quadrupolar coupling constants and the asymmetry parameters. They can be used in a structural analysis. Finally, the broadening quadrupolar effects can be decreased by the experiments at highest magnetic fields in combination with highest spinning rates. Currently, the highest magnetic field available for NMR is 21.84 T, while the highest commercially available MAS frequency is 70 kHz at using 1.3 mm rotors.

The practice of NMR studies, where halogen nuclei are targets, can be well illustrated by the NMR data collected by Stebbins and coworkers for samples of Cl-containing glasses [24]. The MAS NMR experiments have been performed on samples of silicate and alumino-silicate glasses in the magnetic fields of 14.1 and 18.8 T. It has been demonstrated that quadrupolar coupling constants of ^{35}Cl nuclei are measured between 3.0 and 4.4 MHz in silicate glasses and between 2.9 and 3.6 MHz typical of aluminosilicate glasses.

The ^{35}Cl MAS NMR spectra of the Ca-containing glasses are shown in Figure 4.18. As can be seen, both of the spectra, recorded in the magnetic field of 18.8 T, show a single broad peak that is centered at ≅2 ppm. It is remarkable that this line is observed at −57 ppm in the field of 14.1 T. Comparing the NMR spectra shows clearly that the line shapes (and also chemical shits) remain very similar in going from Al-free glasses to glasses containing aluminum. Hence chlorine ions are not binding to the aluminum centers. Thus, even simple MAS NMR experiments can be very useful and valuable for structural conclusions. The authors [24] have noted that the fully relaxed ^{35}Cl MAS NMR spectra can be collected at relaxation delays of 5 s.

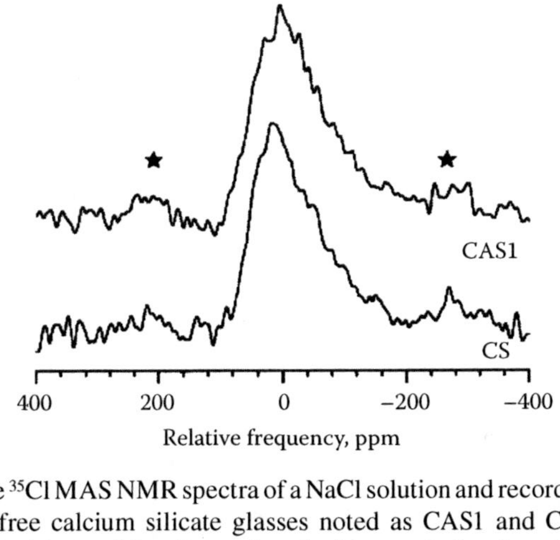

FIGURE 4.18 The ^{35}Cl MAS NMR spectra of a NaCl solution and recorded for Cl-containing aluminous and Al-free calcium silicate glasses noted as CAS1 and CS, respectively. The sidebands are noted with asterisks. (Reproduced with permission from Sandland, T.O. et al., *Geochim. Cosmochim. Acta*, 68, 5059, 2004.)

However the NMR spectra with a higher signal-to-noise ratio, which are still suitable for quantitative analysis, can be recorded with shorter delays of 0.05 s.

Finally it should be emphasized that at the present time, there are numerous spectral structural relationships that can be used in a structural analysis of systems containing halogen nuclei. Among them, the correlations between the chlorine chemical shift and the metal–Cl bond distance seem to be most important.

Following Gerothanassis [25], solid-state ^{17}O NMR has a number of advantages that can be formulated briefly as

1. ^{17}O chemical shifts cover a very large region between 1600 and 0 ppm, referring to a liquid water sample, demonstrating, thus, a high analytical potential of ^{17}O NMR.
2. Similarly to other quadrupolar nuclei with noninteger spins, ^{17}O nuclei show the central transitions that are not affected by first-order broadenings and thus they can be observed as relatively sharp resonances.
3. The satellite transitions are often unobservable under MAS conditions and the ^{17}O MAS NMR spectra show only the central transitions.
4. The line shapes of the central transitions are dominated by second-order quadrupolar effects.

Thus all these features underline that static and/or MAS ^{17}O NMR can be a powerful tool for characterizations of solids, describing the systems in terms of the nuclear quadrupolar coupling constants and shielding tensors. Moreover, as noted in Gerothanassis [25], if single crystals are available, then orientations of ^{17}O shielding tensor components can be determined relative to the local molecular symmetry.

The main problem of observing the ^{17}O NMR signals in solids is a very low natural abundance of these nuclei "multiplied" by relatively small amounts of oxygen atoms in different materials. This problem can be avoided completely by NMR experiments performed on ^{17}O-enriched compounds. Partly this problem can be solved by ^{1}H–^{17}O cross-polarization that enhances sensitivity even in static spectra. However the CP enhancement effect is not too high because only a small fraction of the quadrupolar nuclei will have a Hartmann–Hahn match (see Chapter 3). Nevertheless cross-polarization still has the advantage and it is demonstrated here in Figure 4.19. This figure shows the static ^{1}H–^{17}O CP NMR experiments carried out on samples of a hydrous albite glass compared with single pulse excitation NMR experiments. As the spectra show, the Al–O–Si signal increases remarkably in intensity by cross-polarization. Besides this effect, the CP data correspond to a structural situation where the Al–O–Si sites are located closer to protons than Si–O–Si units.

Other approaches to increasing the ^{17}O sensitivity in the MAS spectra are also available. For example, in silica gels the sensitivity can be increased by fast amplitude modulated pulses using a rotor assisted population transfer (RAPT). Smith and coworkers have communicated that S/N ratios can be enhanced by a factor of 5 [26].

A relatively poor spectral resolution in the ^{17}O solid-state NMR spectra is a second problem, complicating their application for a structural analysis. As noted earlier, the second-order quadrupolar interactions strongly broaden ^{17}O lines coming from the central transitions and this effect cannot be removed under MAS conditions.

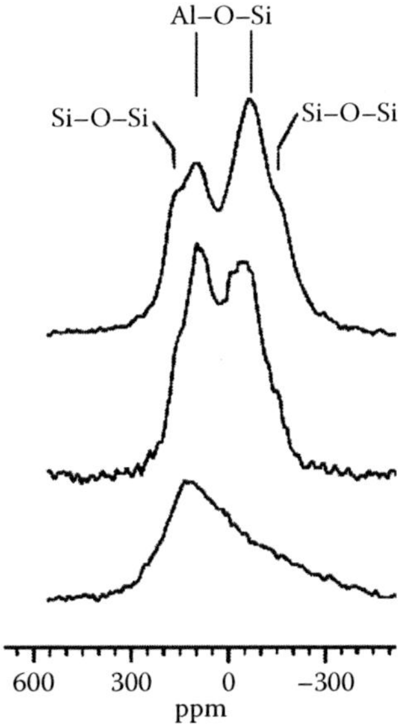

FIGURE 4.19 Comparison of the static 54.2 MHz ^{17}O NMR experiment performed with a single pulse sequence (bottom) and NMR spectra obtained by the CP at a contact time of =20 ms (middle) and 0.05 ms (top). ^{1}H and ^{17}O radio frequency field strengths were 31.3 and 10.4 kHz, respectively. (Reproduced with permission from Gerothanassis, I.P., *Prog. Nucl. Magn. Reson. Spectrosc.*, 57, 1, 2010.)

However, again, the second-order quadrupolar interactions can be minimized by the ^{17}O MAS experiments at strong magnetic fields.

Much better improvements in spectral resolution can be achieved for ^{17}O nuclei by complex mechanical motions of NMR rotors, for example, by DOR and dynamic angle spinning (DAS) [25]. It is obvious however that this technique requires a special design of NMR probes and NMR rotors. The design of the DOR NMR experiment is schematically shown in Figure 4.20: a sample is spinning

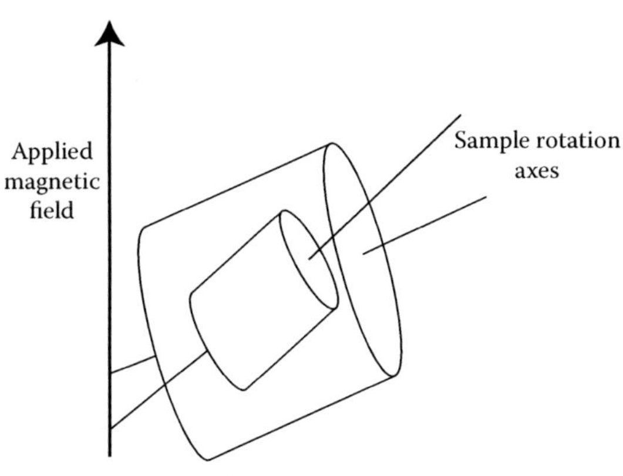

FIGURE 4.20 Schematic representation of the double rotation NMR experiment. (Reproduced with permission from Gerothanassis, I.P., *Prog. Nucl. Magn. Reson. Spectrosc.*, 57, 1, 2010.)

rapidly inside a sample-containing small inner rotor, which is embedded inside a large outer rotor. Under these conditions, the inner rotor is spinning along an axis that is inclined at an angle of 30.56° or 70.1° with respect to the spinning axis of the outer rotor. Simultaneously, the outer rotor spins along an axis that is set at the magic angle.

An alternative way can be based on ^{17}O multiple quantum NMR, for example, used in studies of silicate glasses, characterized by different Si/Al ratios. One of the important aspects in these studies is determination of different bridging oxygen sites, obviously requiring the well-resolved ^{17}O NMR spectra. As it is seen in Figure 4.21, the 3Q ^{17}O NMR experiments, performed for samples of $NaAlO_2$–SiO_2, lead to direct observation of Al–O–Al and Si–O–Al sites.

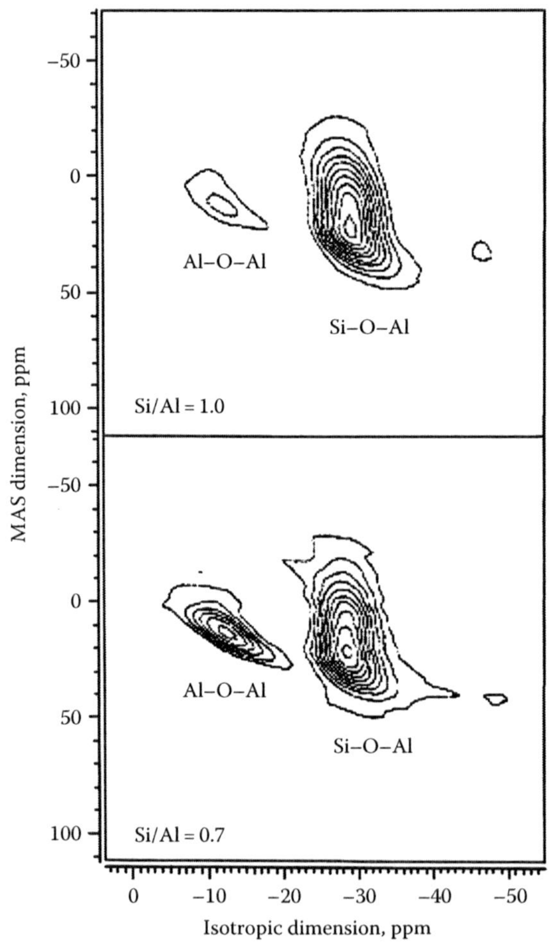

FIGURE 4.21 The ^{17}O 54.2MHz 3Q MAS NMR spectra recorded in two aluminous glasses of the $NaAlO_2$–SiO_2 formula obtained at different Si/Al ratios. (Reproduced with permission from Gerothanassis, I.P., *Prog. Nucl. Magn. Reson. Spectrosc.*, 57, 1, 2010.)

4.5 SOLID-STATE NMR OF OXIDE MATERIALS: NUCLEI ^{51}V, ^{93}NB, AND ^{181}TA

Vanadium, niobium, and tantalum oxide materials are widely used in chemistry and also in the chemical industry, electronics, and metallurgy. Vanadium-based systems are used as catalysts in selective oxidation of hydrocarbons, or reduction of nitrogen oxides with ammonia. Recent years have revealed increasing interest in the chemistry of niobium-based oxide systems. They also show improved catalytic properties in many industrial processes, where they can be applied as niobium oxide itself, and mixed oxides such as Nb_2O_5–SiO_2, Nb_2O_5–Al_2O_3, Nb_2O_5–TiO_2, Nb_2O_5–V_2O_5, etc. Therefore characterization of the oxide catalysts by direct observation of these nuclei was also of great interest during the last decade.

Lapina et al. [27] have noted that in the context of NMR properties, $^{51}V(7/2)$, $^{93}Nb(9/2)$, and $^{181}Ta(7/2)$ nuclei are very similar and show the energy level diagrams typical of quadrupolar nuclei: the spectral patterns are perturbed by the first-order and the second-order quadrupolar interactions.

Generally, ^{51}V nuclei exhibit moderate quadrupolar coupling constants that change between 2 and 6MHz. Only in some cases can they reach a magnitude of 10MHz. The magnetic shielding anisotropy for ^{51}V nuclei is as large as $\leq$1000ppm. However typical values are within a range of 100–500ppm as a function of the coordination environments. The isotropic ^{51}V NMR chemical shifts in solid vanadium-based materials change in a range of 1200ppm, where the diapason of 500ppm is more typical. Finally, compound $VOCl_3$ is generally used as an external reference in ^{51}V NMR spectra.

The ^{93}Nb solid-state NMR spectra, dominated by the quadrupolar interactions, show variation in isotropic ^{93}Nb chemical shifts in a range of 4000ppm, generally referring to $NbCl_3$ in a CH_3CN solution. The second-order quadrupolar interactions lead to typical characteristic powder patterns corresponding to the central transitions from +1/2 to −1/2, while satellite transitions are spread far from the central Larmor's frequency. Generally speaking, for nuclei with spins of 9/2 the satellite transitions are typically situated closer to the central transition. Therefore the resulting spectra show not only central transitions, but also several satellite transitions that are routinely observed.

In contrast to ^{51}V and ^{93}Nb nuclei, ^{181}Ta nuclei are less popular among materials chemists due to "inconvenient" spectroscopic properties. They have a very large quadrupolar moment (317×10^{-30} m^2 versus -5.2×10^{-30} m^2 for ^{51}V), fast quadrupolar relaxation, causing large line broadenings, and a low-gamma constant reducing NMR sensitivity. Nevertheless, Lapina et al. [27] have noted that when the ^{181}Ta NMR spectra are still observed, they are similar to those obtained for ^{51}V and ^{93}Nb nuclei and show sets of central and satellite transitions. The ^{181}Ta isotropic chemical shift range covers 3450ppm referred to $K[TaCl_6]$.

In spite of the complicity in registration of nuclei with pronounced quadrupolar properties, modern commercial FT solid-state NMR spectrometers are generally well equipped in order to observe ^{51}V, ^{93}Nb, and ^{181}Ta NMR spectra in solid materials. Minimal requirements for the spectrometers are the reasonably high magnetic fields

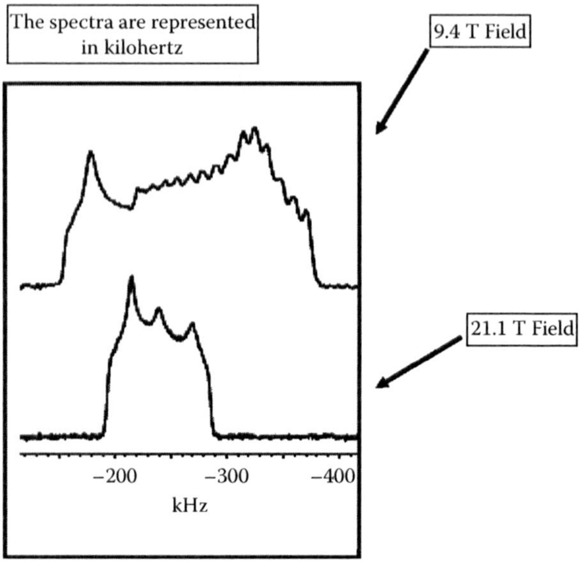

FIGURE 4.22 Effects of the magnetic field strengths on the ^{93}Nb NMR spectra recorded in a static sample of compound La_3NbO_7: (top) the field of 9.4 T, (bottom) the field of 21.1 T. (Reproduced with permission from Lapina, O.B. et al., *Prog. Nucl. Magn. Reson. Spectrosc.*, 53, 128, 2008.)

of ≥9 T, spinning rates of ≥15 kHz, extremely short radio frequency pulses, generally less than 0.5 μs, and very fast digitizing rates.

The advantage of high magnetic fields is obvious for two reasons: increasing the NMR sensitivity and dramatic effects on the line shapes for quadrupolar nuclei with half-integer spins. The latter is illustrated in Figure 4.22, where the static powder pattern of La_3NbO_7 undergoes significant narrowing in the highest magnetic field.

Currently there is a large set of NMR techniques that can be applied for structural characterization of materials where ^{51}V, ^{93}Nb, and ^{181}Ta nuclei are targets. Hahn- and solid-echo pulse sequences or a QCPMG pulse train, remarkably increasing NMR sensitivity, can be used to perform NMR experiments in a static sample.

Technical progress in the MAS NMR technique and a special design of NMR rotors help now to reach spinning rates up to 70 kHz. However, the choice of the rates potentially applied in MAS experiments will strongly depend on the nature of investigated systems and on the applied magnetic field. This statement can be clarified by the following example. In order to record a MAS ^{51}V NMR spectrum with typical vanadium isotropic chemical shifts and chemical shift anisotropies (see preceding text), which is free from sidebands, the spinning rate should be more than 40 kHz, when the applied magnetic field strength is of 9.4 T. In fact, at lower spinning rates, interpretation of very complex vanadium spectral patterns will be difficult. At the same time, one can show that at a higher magnetic field strength, for example, of 21 T, the chemical shifts and anisotropies, particularly, expressed in frequency units, will increase, requiring for a sideband-free spectrum a spinning rate of 90 kHz. However, this rate is impossible technically.

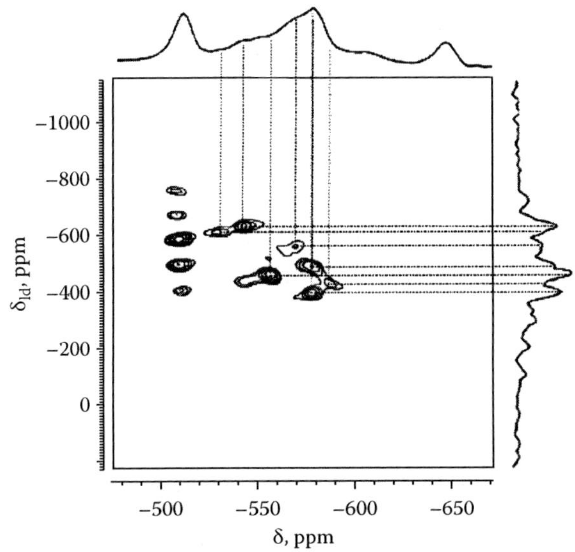

FIGURE 4.23 The ^{51}V 3Q MAS NMR spectrum recorded for a sample of a VO_x/ZrO_2 catalyst. The spectrum was obtained at the external magnetic field strength of 9.4 T. (Reproduced with permission from Lapina, O.B. et al., *Prog. Nucl. Magn. Reson. Spectrosc.*, 53, 128, 2008.)

Similarly to many quadrupolar nuclei, the multiquantum ^{51}V MAS NMR technique can be successfully applied for studies of materials. This is particularly productive for structural situations when magnetic shift anisotropies are relatively low and second-order quadrupolar effects are large. In these cases, the 3Q MAS NMR pulse sequence is most popular. Figure 4.23 depicts the ^{51}V 3Q MAS NMR spectrum recorded for a VO_x/ZrO_2 catalyst. As can be seen, the 1D projection obtained by this experiment is well resolved to show clearly the presence of 10 nonequivalent vanadium sites.

Applications of the heteronuclear correlation NMR spectroscopy, based on cross-polarization in ^1H–^{51}V pairs, the double resonance technique (SEDOR, REDOR, TRAPDOR, and REAPDOR) or the satellite transition spectroscopy are also available and can be found in Lapina et al. [27].

An analysis of ^{51}V NMR data, collected for amorphous or disordered systems, is particularly difficult. Generally the lines observed in the ^{51}V NMR spectra are strongly broadened and all the satellite transitions are present. The central transition lines, which are observed better, could be analyzed on the basis of a static ^{51}V NMR spectrum in combination with several MAS NMR spectra recorded at different spinning rates. Figure 4.24 shows the spectra obtained for a sample of catalyst VO_x/TiO_2, containing 20 wt% of V_2O_5, where vanadium is strongly bonded to its surface. As can be seen, even different spinning rates do not simplify interpretation of the MAS ^{51}V NMR spectra.

Only fitting procedures, performed for the static spectrum and the complex pattern of the MAS sidebands, taking into account the second-order quadrupolar effects, the components of ^{51}V quadrupolar and CSA tensors and their orientations, can help

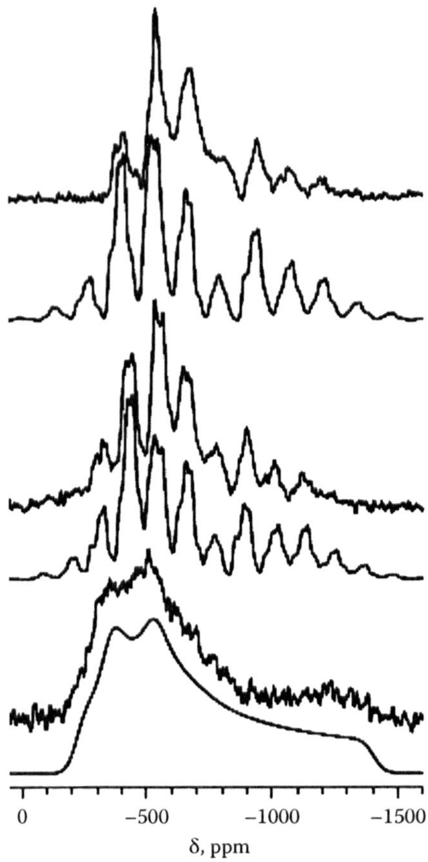

FIGURE 4.24 The ^{51}V NMR spectra recorded for a system VO_x/TiO_2 (20 wt% V_2O_5) at the external magnetic field of 9.4 T from top to bottom: the experimental spectrum at a spinning rate of 14.1 kHz; the simulated spectrum at 14.1 kHz, the experimental spectrum at 12.2 kHz, the simulated spectrum at 12.2 kHz, the experimental spectrum in a static sample, the simulated static spectrum. (Reproduced with permission from Lapina, O.B. et al., *Prog. Nucl. Magn. Reson. Spectrosc.*, 53, 128, 2008.)

in such cases. A good agreement should be reached between all the experimental and simulated NMR spectra fitted with *a single set* of parameters. The simulation procedures, performed for all the spectra in Figure 4.24, give a vanadium quadrupolar coupling constant of 14.7 MHz at an asymmetry parameter of 0.59, a chemical shift anisotropy of 650 ppm at an asymmetry parameter of 0.02, while the isotropic chemical shift δ(iso) is calculated as −611 ppm.

In spite of the difficulties discussed earlier, many vanadium oxide materials have been studied by solid-state NMR. These studies have resulted in identification and characterization of different VO_x sites in terms of Q_0 sites, symbolizing vanadium with tetrahedral oxygen coordination, containing isolated VO_4 units, and Q_1 sites, corresponding to the structural situation when V_2O_7 is formed, etc. It has been found that chemical shift anisotropy $\Delta\sigma$, asymmetry parameter η, the perpendicular

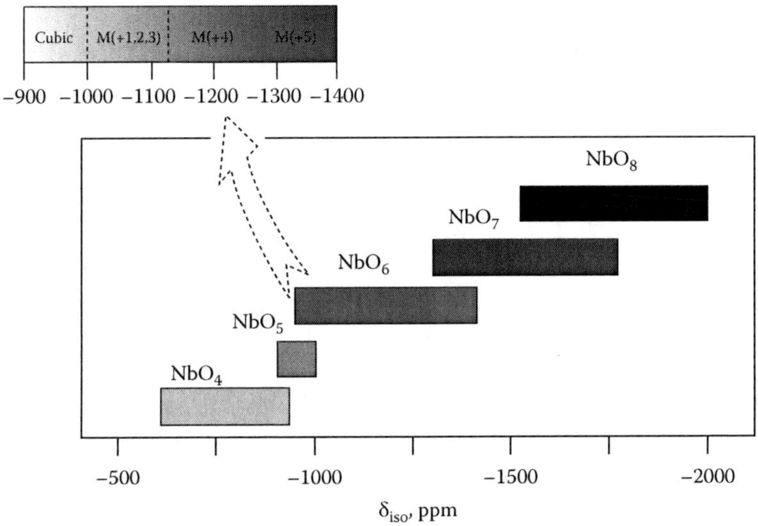

FIGURE 4.25 The ^{93}Nb NMR isotropic chemical shift scale for NbO_x structural units. (Reproduced with permission from Lapina, O.B. et al., *Prog. Nucl. Magn. Reson. Spectrosc.*, 53, 128, 2008.)

component of chemical shift tensor $\delta_\perp$, and quadrupolar coupling constant C_Q differ remarkably for the structural unites and can be used in their identification: tetrahedral Q_0 sites have generally $\Delta\sigma \leq 100$ ppm and $C_Q = 0$–6 MHz; tetrahedral Q_1 sites show $\Delta\sigma = 100$–200 ppm, $\eta = 0.1$–0.9, and $C_Q = 2.5$–10 MHz; tetrahedral Q_2 sites exhibit $\Delta\sigma = 200$–500 ppm, $\eta = 0.6$–0.8, $C_Q = 2$–7 MHz, associated with nonaxial VO_5 and VO_6 sites with $\Delta\sigma = 200$–400 ppm, $\eta = 0.3$–0.6, $C_Q = 3.5$–6 MHz; isolated octahedral VO_6 sites correspond to $\Delta\sigma = >700$ ppm, $\eta = 0$–0.1, $C_Q = 0$–2 MHz, and $\delta_\perp = 200$–350 ppm; trigonal pyramids VO_4 show $\Delta\sigma = 400$–500 ppm, $\eta = 0$–0.2, $C_Q = 1$–4 MHz, and $\delta_\perp = 400$-500 ppm.

Finally, the results of ^{93}Nb NMR measurements in different niobium oxide materials are shown in Figure 4.25. The figure illustrates how ^{93}Nb isotropic chemical shift changes as a function of the niobium oxidation state and environments.

4.6 SOLID-STATE ^{13}C NMR: WOOD AND WOOD PRODUCTS

Since the chemical shift range for nuclei ^{13}C is well known from solution ^{13}C NMR applied for organic/inorganic compounds, this section will describe specific applications of solid-state ^{13}C NMR, demonstrating techniques used in an analysis of wood and wood products important from the commercial and fundamental point of view.

Generally the following problems are constantly in focus of investigators: cellulose morphology in wood, the structure of lignin (see Chapter 2), strongly dependent of its origin, the carbohydrate lignin connectivity in wood, structural changes in wood due to pulping and bleaching processes, the studies of paper deteriorating due to aging and degradation effects, thermal modification of wood and also its bio- and photodegradation. All these aspects can be found in the recent review by Maunu [28].

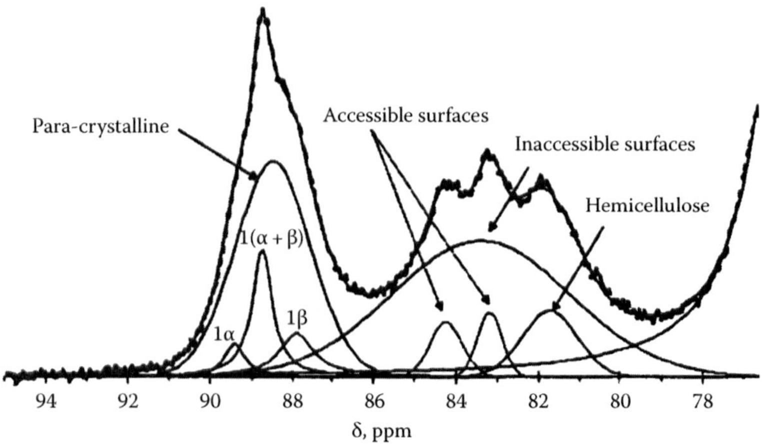

FIGURE 4.26 The ^{13}C CP {^{1}H} MAS NMR spectrum recorded in a sample of bleached pulp, showing the C4 region of the anhydroglucose rings. (Reproduced with permission from Maunu, S.L., *Prog. Nucl. Magn. Reson. Spectrosc.*, 40, 151, 2002.)

Among various NMR experiments applied for these studies, regular cross-polarization ^{13}C {^{1}H} MAS NMR spectroscopy plays a major role, giving a rapid analysis of samples. A typical example is shown in Figure 4.26, where C-4 signals of the anhydroglucose are broad and overlapped. However, they can be successfully "resolved" by deconvolution procedures available in software packages of NMR spectrometers, and assigned.

The protocols for determinations of the degree of crystallinity in cellulose samples, based on the modified cross-polarization pulse sequences, can be different. As noted in Chapter 3, the amplitudes for the Hartmann–Hahn matching can be adjusted by variable-amplitude cross-polarization experiments where the cross-polarization buildup data are collected with the incremented ^{1}H pulse duration from 5 μs to 11 ms. This variable-amplitude pulse technique has been successfully used to characterize the crystallinity of cellulose [29]. The protocol based on this experiment includes the subtraction procedures performed for the ^{13}C {^{1}H} CP solid-state NMR spectra obtained in an investigated sample and standard amorphous cellulose, as shown in Figure 4.27.

The second protocol is based on two pulse sequences shown in Figure 4.28. As can be seen, pulse sequence A includes the so-called spin-locking delay time, acting in the proton channel between the preparation and contact time pulses. Since the spin-lattice relaxation times of protons in the rotating coordinate systems differ remarkably in amorphous and crystal phases of cellulose, due to this delay time, the cross-polarization from ^{1}H to ^{13}C nuclei can be transferred only for the crystal fraction, relaxing slower. As a result, the final NMR spectrum will show only ^{13}C resonances characterizing the crystal phase. Then the spectrum can be compared with the regular ^{13}C CP MAS spectrum exhibiting the carbon resonances from both of these phases.

Approach B uses strong dipolar interactions between protons and carbons, which can be applied for a dephasing of the so-called protonated carbons in the CP

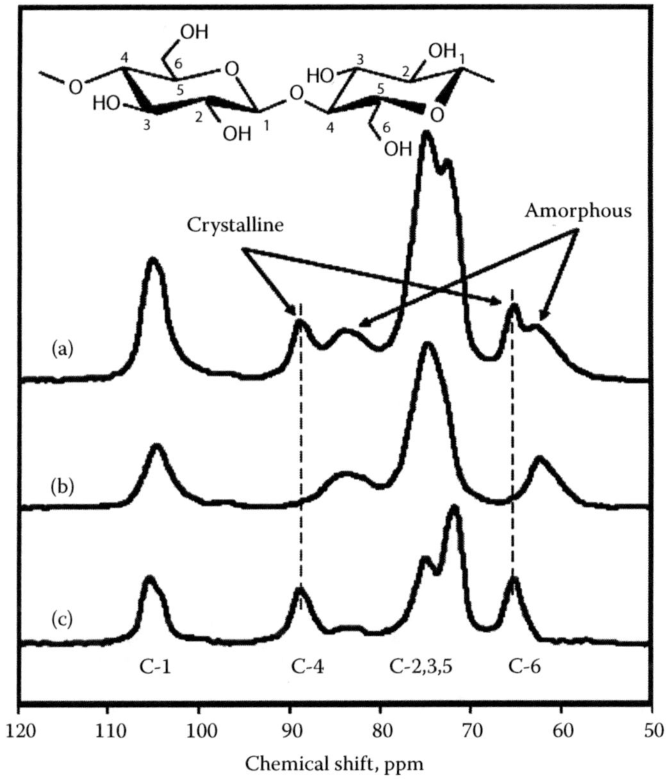

FIGURE 4.27 The subtraction procedure realized in order to determine the crystallinity index of cellulose by the solid-state ^{13}C CP NMR spectra: (a) Sigma a-cellulose, (b) amorphous cellulose, (c) "crystalline fraction" determined by subtracting (b) from (a). The conditions are ^{1}H and ^{13}C fields were matched at 53.6 kHz and a 1 dB ramp was applied to the proton rotating frame during the matching period. Acquisition time was 0.051 s and sweep-width was 20 kHz. Numbers of scans were around 10,000–20,000 at a contact time of 1 ms and a relaxation delay of 1.0 s. (Reproduced with permission from Park, S. et al., *Cellulose*, 16, 641, 2009.)

^{13}C MAS NMR spectra. In this pulse sequence, the dephasing delay time plays a major role, leading to effective relaxation of ^{13}C nuclei in carbons that are attached to protons. Then, the acquisition period will effective only for carbons that are remote from the protons and thus only these carbons will be observed in the final NMR spectra.

4.7 ^{2}H NMR IN SOLIDS AT DEUTERIUM NATURAL ABUNDANCE

Wide applications of solid-state ^{2}H NMR in materials science, particularly in probing large synthetic polymers or biological systems, are based on large sensitivity of deuterium line shape and deuterium relaxation to various motions of molecules or groups in molecules (see Chapter 5). Generally these studies can be successfully carried out for fully ^{2}H-enriched systems in static or spinning samples. In the structural context,

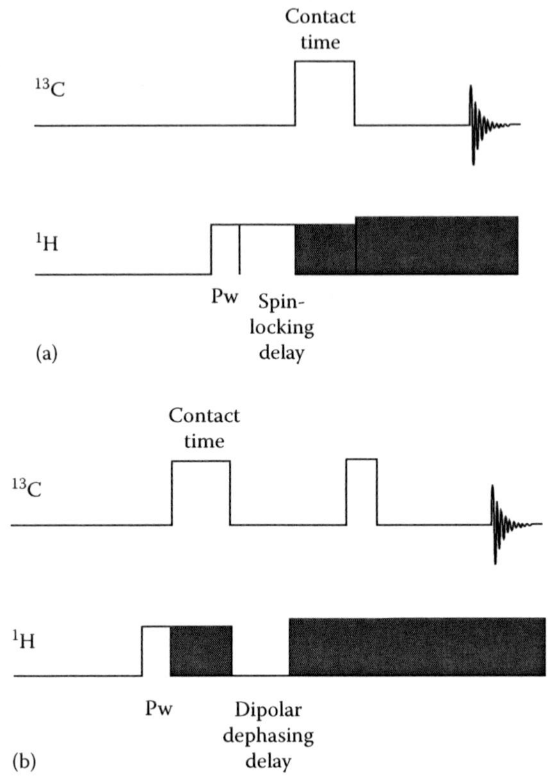

FIGURE 4.28 The spin-locking (a) and (b) dipolar dephasing pulse sequences applied for probing the crystallinity in samples containing amorphous and crystal phases. (Reproduced with permission from Maunu, S.L., *Prog. Nucl. Magn. Reson. Spectrosc.*, 40, 151, 2002.)

solid-state ^2H NMR is less attractive because the deuterium isotropic chemical shift range (expressed in Hz) is small (seven times less than that for protons).

Recording the solid-state NMR spectra at natural ^2H abundance becomes problematic due to strong decrease in overall sensitivity of NMR even in a static sample containing a single type of nuclei that show a classical Pake doublet. The situation does not change under MAS conditions: generally the spinning rates are small relative to the strength of the anisotropic quadrupolar and chemical shift anisotropy interactions. Under these conditions, an isotropic NMR signal should exhibit a very wide sideband pattern, again reducing the NMR sensitivity.

A remarkable progress in enhancement of the sensitivity can be achieved by magic angle spinning in combination with the cross-polarization technique and high-power proton decoupling considered in a recent review [30]. The cross-polarization and ^1H decoupling technique help to detect the ^2H NMR spectra, for example, in polycrystalline powders of plastic crystals and even to resolve different deuterium chemical shifts due to remarkable narrowing lines.

When proton spins are used as a reservoir for polarization transfer, theoretically it is clear that the cross-polarization effect can lead to increasing the signal-to-noise ratio in the spectra by a factor of 6.5. However in many cases, for example, in polymers, the theoretical sensitivity enhancement is not achieved. The reason is that due to an effective quadrupolar relaxation, the ^{2}H T_1 times are too short versus proton–deuteron cross-relaxation terms under the cross-polarization.

The approach recently developed to avoid this problem is based on indirect ^{2}H detection using the multiple cross-polarization periods [30]. Here, each of the periods is followed by a free evolution period, t_1, before the acquisition of the signal for the hydrogen spin. Then, the spectrum can be obtained by Fourier transformation of the amplitude for the hydrogen signal with respect to time t_1. Theoretically, the relative signal increase factor, R, for long cross-polarization periods in going from such an experiment to the standard cross-polarization experiment implying direct detection can be expressed as

$$R = \left[\frac{\gamma(^1H)}{\gamma(^2H)} \times n^{CP}\right] \times \exp\left(\frac{-(n^{CP}\tau^{CP})}{T_{1\rho}}\right) \quad (4.1)$$

where

n^{CP} is the number of CP periods

τ^{CP} is the contact time

$T_{1\rho}$ represents the ^{1}H relaxation time in the rotating coordinate system

This is a good method but as has been noted, such spectra can be distorted because of different hardware problems.

4.8 BETWEEN SOLUTION- AND SOLID-STATE NMR: NANOPARTICLES AND SUSPENSIONS

The field of NMR applications for studies of nanoparticles and nanomaterials is very large and various [31]. In many cases, nanoparticles, having a size from 10 to 1000 nm, can be dispersed within bulk samples in a gas, liquid, or solid phase, leading to heterogeneous systems. The nature of these systems defines physicochemical methods applied for their investigations. Dispersions in liquids, containing nanoparticles, nanospheres, and nanocapsules are of great interest in the context of this book, because they represent the heterogeneous systems situated between solutions and the solid state.

The solid particles dispersed in liquids are known to undergo two types of motions: self-diffusion and rotational diffusion in motion correlations times between 10^{-7} and 10^{-1} s as a function of their size. The smallest particles move rapidly resulting in completely averaging the anisotropy effects typical of solution-state NMR. In contrast, resonances of the largest and moving slowly particles can show shapes similar to solid-state patterns, as illustrated in Figure 4.29. In such a situation an analysis will require the methodologies generally used in solid-state NMR: from direct excitation of target nuclei to cross-polarization MAS excitations and spin-echo experiments.

Static ^{13}C NMR spectra can be obtained for suspension samples containing organic nanoparticles by ^{1}H–^{13}C cross-polarization experiments in combination with

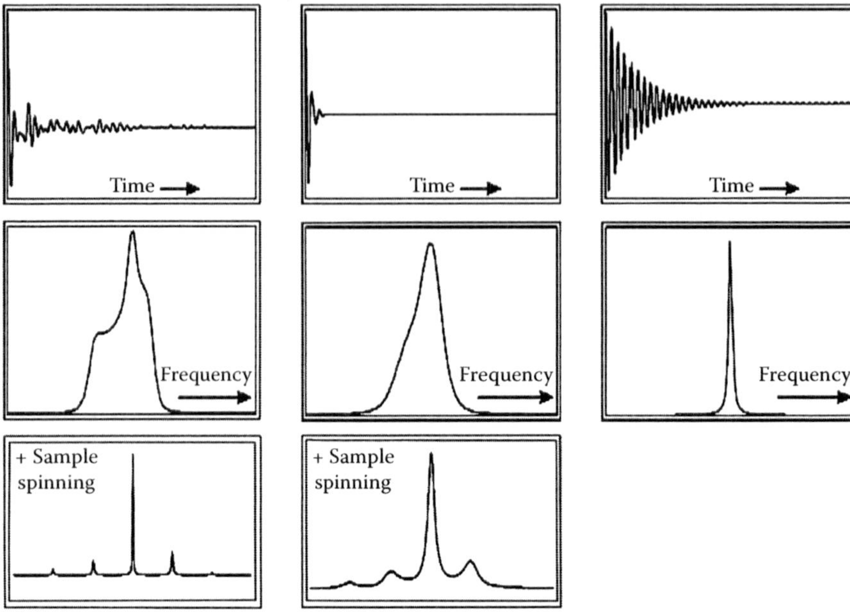

FIGURE 4.29 A presentation of how the rotational diffusion affects NMR time signal after a 90° pulse and the corresponding frequency spectrum: (left) slow reorientations result in a powder-like shape; (central) reorientations strongly affect the shape of signals; (right) fast reorientations produce a liquid-like signal. (Reproduced with permission from Mayer, C., *Prog. Nucl. Magn. Reson. Spectrosc.*, 40, 307, 2002.)

proton decoupling. Generally such ^{13}C NMR spectra show the resonances that are practically identical to those observed in solid-state NMR (see Figure 4.30). In the absence of the complete averaging, the spectra can be analyzed by computing the components of anisotropic ^{13}C chemical shift tensors. However, it has been emphasized that in contrast to rigid solids, simulations of the spectra and fitting procedures should include not only variations in magnetic tensor parameters but also the rotational diffusion tensor parameters and motion correlation times, as shown schematically in Figure 4.30. For example, the CH_n groups in cetyl palmitate nanoparticles can be well identified and characterized by different isotropic chemical shifts and tensor values σ_{11}, σ_{22}, and σ_{11}. The detailed theory of this analysis can be found in Mayer [31].

The C–H cross-polarization kinetics, measured at variations in the contact times, can also be used for characterization of nanoparticles. Usually the kinetic experiments result in the ^1H $T_1\rho$ relaxation times determined in the rotational coordinate system that in turn, characterize mobility of nanoparticles, for example, their segmental motions.

The poor spectral resolution in static samples obviously limits the spectral analysis. Principally the resolution can be improved by the MAS conditions. However, in contrast to regular solids, the samples with nanoparticles in liquid media create a serious technical problem that is well known from experiments with

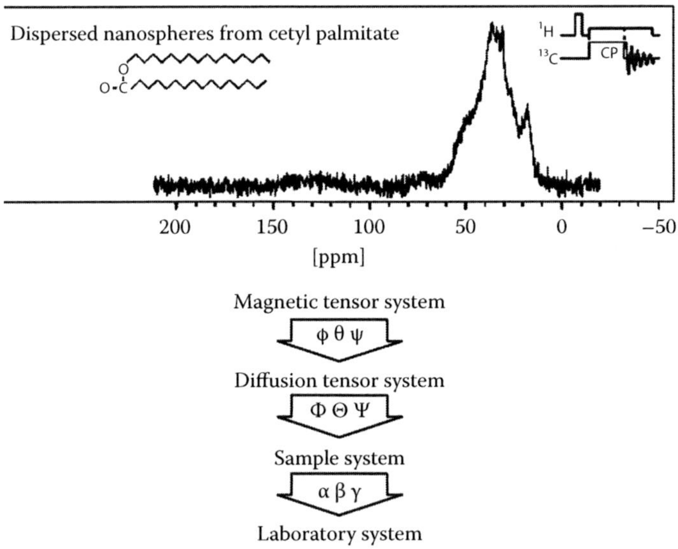

FIGURE 4.30 The cross-polarization ^{13}C NMR spectrum of water-dispersed cetyl palmitate nanospheres and the schema of spectral simulation procedures with varying parameters. (Reproduced with permission from Mayer, C., *Prog. Nucl. Magn. Reson. Spectrosc.*, 40, 307, 2002.)

ultracentrifugation. To avoid this problem one can use chemical stabilization of dispersions or adjusting the similar densities of liquid phase and the nanoparticles. Generally these manipulations provide only relatively low spinning rates typically between 1 and 3 kHz. Such MAS NMR spectra will show intense sideband patterns, which again can be simulated by variations in correlation times for slowly moving nanoparticles.

Finally, the off-angle spinning NMR experiments on nanoparticles are also available [31]. However it has been noted that the weakness of such experiments is notable loss in spectral resolution and signal-to-noise ratios in resulting NMR spectra.

NMR in suspensions, as an analytical instrument, attracts great interest from chemists working in the field of solid phase synthesis, chromatography, and catalysis. Among various immobilized species, catalytic systems, where one or two metal complexes are bound to oxide supports via bifunctional phosphine linkers (e.g., the bonding can be realized through diphenylphosphine groups), are described as most perspective. Different aspects of such systems, including applications of NMR have been recently represented by Blumel [32].

One of the important problems connected with understanding catalytic properties and mechanisms, is mobility of immobilized linkers in the presence of solvents. Since in these cases, the surface-bound fragments of catalysts contain the liquid and solid phase, a significant contribution to understanding can be provided by NMR. In fact, the bulk of the support can be studied by NMR, as well as the interface of the support with the linkers. As in the case of nanoparticles, NMR experiments are performed on samples representing *suspensions*. Blumel divides applications of NMR into two fields, when NMR data are collecting for a *static* or *spinning* sample.

Static samples of the suspensions imply usage of a regular solution NMR spectrometer, because mobility of bifunctional phosphine linkers increases significantly due to addition of solvents. It has been emphasized that this solvent effect does not connect with moving the whole support particles but is a result of the increasing mobility of a linker chain, itself. Quantitatively the solvent-addition effect is expressed in terms of nuclear relaxation dependent strongly on correlation times of motions, as is shown in Chapter 6. In this section one can be limited by the fact that spin-lattice relaxation times T_1 greatly shorten in the presence of solvents and are measured between 0.5 and 2 s. Thus, the shorter relaxation delays applied for pulse sequences offer rapid collection of the NMR data.

In static samples, containing suspensions, resonance lines of ^1H and ^{13}C nuclei remain still strongly broadened. This circumstance thus complicates their analysis. In contrast, ^{31}P resonances of bifunctional phosphine linkers look much better from this point of view and they can be well detected and analyzed. Figure 4.31 illustrates the situation when the ^{31}P NMR spectrum can discriminate between phosphorous linkers, which are covalently binding to the surface, and linkers that are simply adsorbed or situated in a solution. In fact, as clearly seen, the line-width, corresponding to covalently bound linkers, is measured by several kHz. At the same time, for adsorbed linkers, the line-width reduces strongly to 50–100 Hz only. Blumel shows how such NMR features can be used in studies of the interactions between phosphine moieties and the oxide support surface.

The solid-state multinuclear NMR equipment can be applied for characterization of the suspensions, when the MAS conditions are needed to reveal additional structural details of surface-bound linkers. The high-resolution MAS NMR spectroscopy uses two effects. First, the aforementioned influence of solvents increases mobility and reduces relaxation times. Second, the well-known line-narrowing effect under MAS conditions leads to observation of sharp resonances. This combination allows us to obtain the NMR spectra where signals from surface-bound linkers are highly resolved.

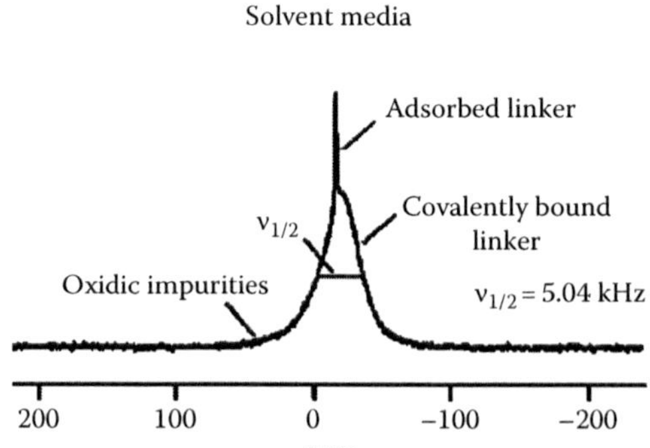

FIGURE 4.31 A surface oxide support in a solvent media (schematically). There are phosphine linkers that are chemically bounded to a surface of oxide support and absorbed phosphine linkers. The ^{31}P NMR spectrum of a $Ph_2P(CH_2)_3$-$[Al_2O_3]$ suspension in THF. (Reproduced with permission from Blumel, J., *Coord. Chem. Rev.*, 252, 2410, 2008.)

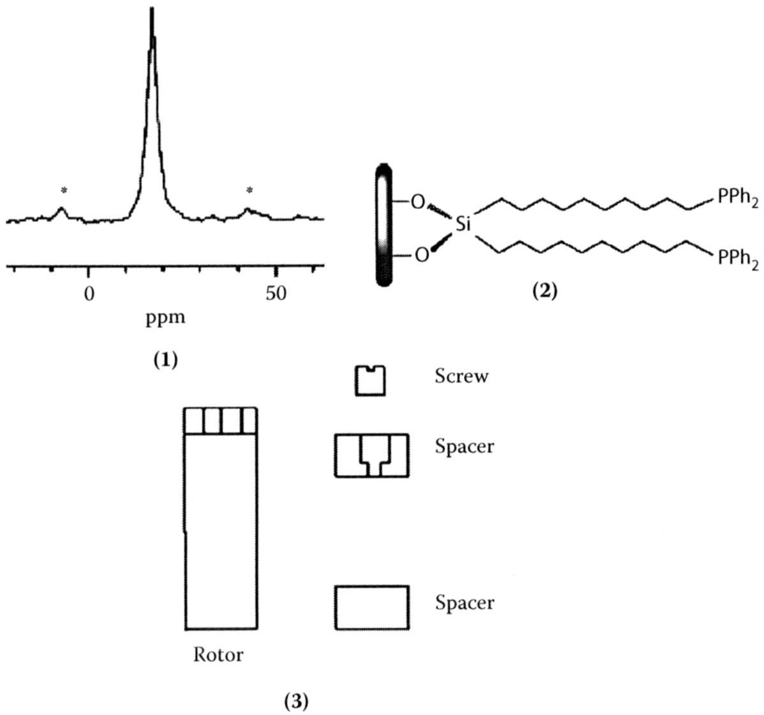

FIGURE 4.32 The ^{31}P {^{1}H} MAS NMR spectrum (**1**) obtained in a toluene suspension sample of linker (**2**) spinning at 2 kHz in the NMR rotor (**3**). (Reproduced with permission from Blumel, J., *Coord. Chem. Rev.*, 252, 2410, 2008.)

Figure 4.32 shows the ^{31}P MAS NMR spectrum recorded for one of the linkers in a suspension demonstrating the narrow phosphorous resonance.

Blumel has emphasized that the aforementioned experiments do not require specialized NMR equipment. Only NMR rotors should be modified. The modification is schematically shown in Figure 4.32, where a 4 mm rotor is equipped by a spacer and a screw with a tiny hole to remove surplus solvents. Generally a single pulse sequence in the presence of high-power proton decoupling can be used to collect such NMR spectra.

This technique can be successfully applied for other nuclei to characterize surface-bound linkers mobilized by the proper solvent. High-resolution ^{1}H and ^{13}C MAS NMR spectra can be obtained for such systems. Moreover, the high spectral resolution can be achieved in the 2D correlation NMR spectra providing the detailed structural information about surface-bound linkers and metal complexes that can be studied by variation in the nature of solvents. There is only one limitation of this method related to the requirement of a minimal mobility of the surface-bound species.

It has been noted that, for example, the ^{29}Si MAS NMR signals coming from the surface-bound ethoxysilane groups are invisible by this technique and only the bulk of the silica support can be seen. However, even in such cases, this fact can be used to distinguish surface-bound from unbound or cross-linked silane groups, or even silicon grease impurities.

NMR chromatography also focuses on measurements in suspensions and it is directed to separation of a simple mixture by diffusion coefficients that are supposed to be different for each component. This separation can be realized with pulsed magnetic gradients using self-diffusion NMR techniques. The data collected by diffusion-ordered spectroscopy (DOSY) are presented as a 2D contour map, where the chemical shift is plotted on the horizontal axis, while the diffusion rate is plotted on the vertical axis. As a result, each component in the mixture produces a separate 1D NMR spectrum, corresponding to its diffusion constant [33].

The main problem of this solution NMR technique is a very limited separation capability. It has been shown that this problem can be avoided by the DOSY NMR spectra, where structured media, first of all, silica gel is added to solutions for enhancement of the separation in the diffusion dimension. Under this condition, in the presence of solid silica, the signals coming from the liquid are broadened up to hundreds or even thousands of Hz, mainly due to inhomogeneous magnetic susceptibility. At the same time, signals coming from solids are broadened to tens of kilohertz because of chemical shift anisotropy. Thus broad signals can relax completely during the diffusion pulse sequence prior to acquisition. Therefore, solid-state NMR technique, and high-resolution magic angle spinning is needed to observe the spectrum. Caldarelli has developed this technique, details of which can be found in his review [34]. Figure 4.33 illustrates one of Caldarelli's experiments where on addition of octadecylsilyl (functionalized silica gel), the molecules in the mixture slow down proportionally to their affinity for the solid, leading thus to their good separation.

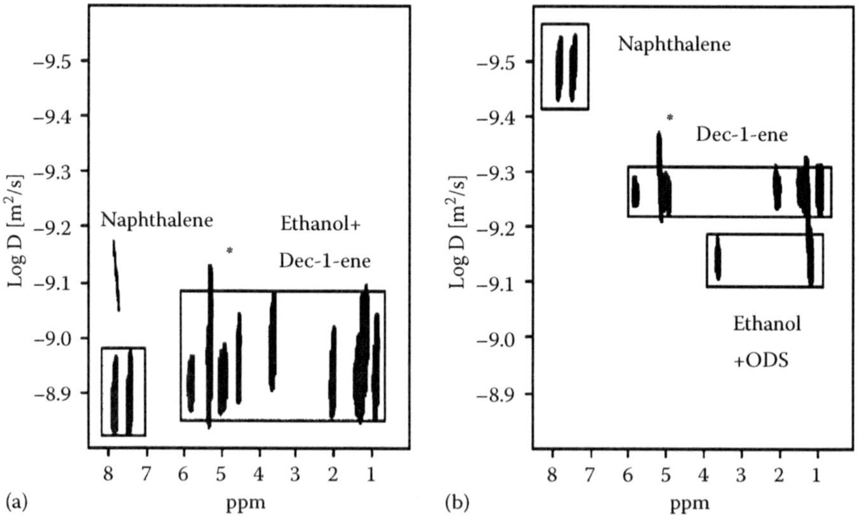

FIGURE 4.33 NMR Chromatography. (a) The regular DOSY NMR spectra of a mixture of molecules, dissolved in deuterated ethanol, which are poorly separated. (b) The addition of the functionalized silica gives excellent spectral separation. (Reproduced with permission from Caldarelli, S., *Magn. Reson. Chem.*, 45, S48, 2007.)

REFERENCES

1. Yang, Y., Beele, B., Blumel, J. 2008. *J. Am. Chem. Soc.* 130: 3771.
2. Lesage, A. 2009. *Phys. Chem. Chem. Phys.* 11: 6876.
3. Ashbrook, S. E. 2009. *Phys. Chem. Chem. Phys.* 11: 6892.
4. Gerstein, B. C. 1981. *Phil. Trans. Roy. Soc. Lond. Ser. A* 299: 521.
5. Schmitt, K. D., Haase, J., Oldfield, E. 1994. *Zeolites* 14: 89.
6. Cadars, A., Lesage, A., Trierweiler, M., Heux, L., Emsley, L. 2007. *Phys. Chem. Chem. Phys.* 9: 92.
7. Ooms, K. J., Feindel, K. W., Willans, M. J. et al. 2005. *Solid State NMR* 28: 125.
8. Lupulescu, A., Kolecha, M., Frydman, L. 2003. *J. Am. Chem. Soc.* 125: 3376.
9. Brown, S. P. 2007. *Prog. Nucl. Magn. Reson. Spectrosc.* 50: 199.
10. Pake, G. E. 1948. *J. Chem. Phys.* 16: 327.
11. Leskes, M., Vega, S. 2009. *J. Chem. Phys.* 130: 124506.
12. Ulrich, A. S. 2005. *Prog. Nucl. Magn. Reson. Spectrosc.* 46: 1.
13. Lee, J. E., Khitrin, A. K. 2008. *Concepts Magn. Reson. A* 32A: 56.
14. Fyfe, C. A., Lewis, A. R., Chezeau J. M., Grondey, H. 1997. *J. Am. Chem. Soc.* 119: 12210.
15. Spyros, A., Dais, P. 2009. *Prog. Nucl. Magn. Reson. Spectrosc.* 54: 195.
16. Hung, L. I., Wang, S. L., Kao, H. M., Lii, K. H. 2007. *Inorg. Chem.* 46: 3301.
17. Massiot, D., Drumel, S., Janvier, P., Bujoli-Doeuff, M., Bujoli, B. 1997. *Chem. Mater.* 9: 6.
18. Bakhmutova-Albert, E. V., Bestaoui, N., Bakhmutov, V. I., Clearfield, A., Rodriguez, A. V., Llavona, R. 2004. *Inorg. Chem.* 43: 1264.
19. Schnell, I. 2004. *Prog. Nucl. Magn. Reson. Spectrosc.* 45: 145.
20. Hanna, J. V., Smith, M. E. 2010. *Solid State NMR* 38: 1.
21. Chappell, H., Duer, M., Groom, N., Pickardc, C., Bristowe, P. 2008. *Phys. Chem. Chem. Phys.* 10: 600.
22. Chapman, R. P., Widdifield, C. M., Bryce, D. L. 2009. *Prog. Nucl. Magn. Reson. Spectrosc.* 55: 215.
23. Siegel, R. S., Nakashima, T. T., Wasylishen, R. E. 2005. *Concepts Magn. Reson. Part A* 26A: 62.
24. Sandland, T. O., Du, L. S., Stebbins, J. F., Webster, J. D. 2004. *Geochim. Cosmochim. Acta* 68: 5059.
25. Gerothanassis, I. P. 2010. *Prog. Nucl. Magn. Reson. Spectrosc.* 57: 1.
26. Madhu, P. K., Pike, K. J., Dupree, R., Levitt, M. H., Smith, M. E. 2003. *Chem. Phys. Lett.* 367: 150.
27. Lapina, O. B., Khabibulin, D. F., Shubin, A. A., Terskikh, V. V. 2008. *Prog. Nucl. Magn. Reson. Spectrosc.* 53: 128.
28. Maunu, S. L. 2002. *Prog. Nucl. Magn. Reson. Spectrosc.* 40: 151.
29. Park, S., Johnson, D. K., Claudia, I., Ishizawa, C. I., Parilla, P. A., Davis, M. F. 2009. *Cellulose* 16: 641.
30. Lessot, P., Courtieu, J. 2009. *Prog. Nucl. Magn. Reson. Spectrosc.* 55: 128.
31. Mayer, C. 2002. *Prog. Nucl. Magn. Reson. Spectrosc.* 40: 307.
32. Blumel, J. 2008. *Coord. Chem. Rev.* 252: 2410.
33. Pemberton, C., Hoffman, R., Aserin, A., Garti, N. 2011. *J. Magn. Reson.* 208: 262.
34. Caldarelli, S. 2007. *Magn. Reson. Chem.* 45: S48.

5 Dynamics in Diamagnetic Materials from Solid-State NMR: Methods, Measurements, and Analysis

Many chemical and physical properties of materials depend on their dynamics. Besides isotropic chemical shift, magnetic shielding anisotropy, and dipolar and quadrupolar coupling, the dynamics directly affects number and shapes of observed resonances. Therefore, minimal representations about dynamic behavior of objects and its influence are needed in interpretation of NMR spectra even in the structural context, when the number of detected resonances and their shapes are a primary source for description of crystalline, amorphous, and strongly disordered systems.

Another aspect in chemistry and physics of materials is quantitative investigation of their dynamics, including types of motions, their frequency spectra, and energies. The types of motions in solids can be varied, from high-frequency librations and extremely low energy internal rotations, sometimes tunneling, as in the case of CH_3-rotations, to slower rotations and segmental reorientations or translation diffusion. In addition, many motions can become cooperative, further contributing to their frequencies and energetics. Understanding the dynamics can help in the design of new molecular systems with specific mechanical, optical, or transport properties. For example, in practice, it is important to describe the behavior of amorphous objects near glass transitions.

Dynamic behavior of porous and dynamically heterogeneous materials represents a special case. Generally such systems can be formulated as systems with an open solid matrix, which is filled with a liquid. This category includes biological tissues, gels, plasticized polymers, zeolites, etc. The main feature of these systems is the small sizes of the liquid spaces varying between tens of Angstroms to a few microns. In this circumstance, the dynamics of liquids can be very complex and dependent on structural parameters of the matrix. Similarly, translation diffusion in immobilized catalysts in the presence of solvents is also of great interest.

The influence of dynamics on NMR can be easily demonstrated by the following statement. The static ^{1}H NMR spectra of rigid solids are well known to show very broad resonances due to the strong dipolar proton–proton coupling. In contrast, in solutions, the lines are sharp due to rapid isotropic molecular motions. Thus, motions can average the anisotropic interactions in solids, reducing the line width. Then, monitoring the line widths at different temperatures is the direct way to obtain information on correlation times and activation energies of motions in solids.

A priori solids can show slow dynamics within the frequency diapason between 1 Hz and 1 kHz, moderate motions, characterized by kHz frequencies, and finally, fast dynamics, when molecular or group motions reach frequencies up to ~100 MHz [1]. From the theory of NMR, fast dynamics mainly contributes to spin–lattice relaxation (T_1) behavior of target nuclei. Correlation times of the fast motions can be determined directly by the spin–lattice relaxation time measurements. Motions connected with reorientations of chemical shift anisotropy tensors or quadrupolar coupling tensors occur on the same timescale as the FID is collected (or faster). Thus, they formulate the condition $\tau_c \leq T_2$ and contribute strongly to line shapes [2]. Then, a computer analysis of line shapes will provide characterizations of these motions.

Slow dynamics, for example, conformational transitions or segmental reorientations in solid polymers, can be observed from changes of orientation-dependent

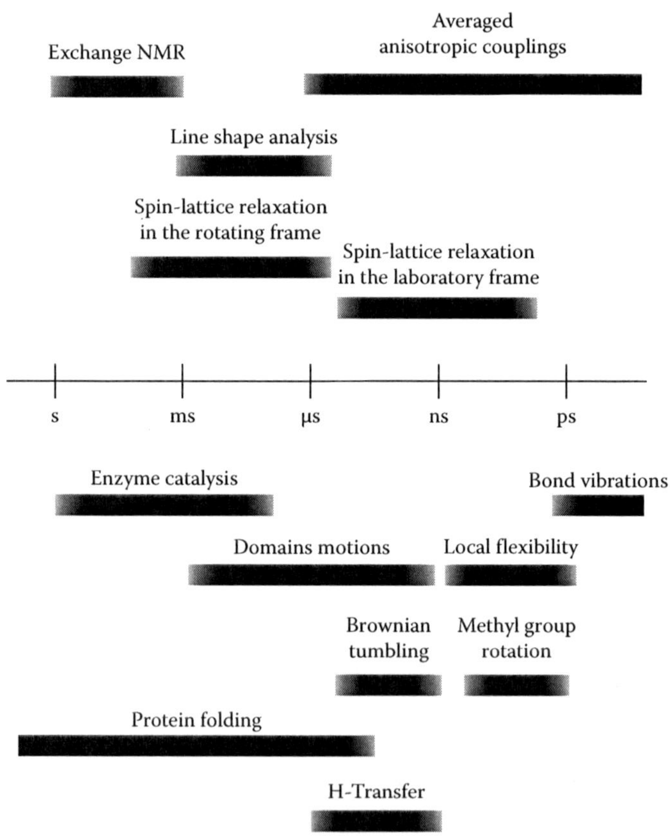

FIGURE 5.1 Dynamic NMR experiments and timescale of dynamic processes associated with proteins. (Reproduced with permission from Krushelnitsky, A. and Reichert, D., *Prog. Nucl. Magn. Reson. Spectrosc.*, 47, 1, 2005.)

NMR frequencies. They can be probed by exchange NMR spectroscopy. Finally, intensity of resonances, detected by cross-polarization NMR experiments, is known to be dependent on molecular mobility. Therefore, monitoring the CP intensity can be also used to recognize and characterize the molecular motions.

This chapter illustrates how to apply different methods of solid-state NMR in practice to describe molecular mobility in materials in terms of motional types and their frequencies. The focus is on nuclear relaxation phenomena and relaxation measurements requiring a more detailed understanding.

Krushelnitsky and Reichert [2] give an excellent illustration of relationships between dynamic NMR and various motions and processes associated with proteins and shown in Figure 5.1. As we can see, the timescale is very large covering the diapason between seconds and picoseconds.

5.1 VARIABLE-TEMPERATURE SOLID-STATE NMR EXPERIMENTS AND A LINE-SHAPE ANALYSIS

Dynamic processes occurring on the same timescale as the FIDs affect line shapes in NMR spectra that can be documented by variable-temperature NMR experiments. Methodically and technically, this is the simplest version of dynamic NMR. There are only two general requirements: the NMR spectra should be carefully recorded with good phasing and baseline correction; the temperature should be well controlled. Sometimes even these simplest NMR experiments will be sufficient to reveal the presence of molecular mobility at a qualitative level. For example, mesoporous materials of the MCM-41 family, synthesized in the presence of cationic surfactants, show clearly the temperature dependence in the ^{13}C MAS NMR spectra. This dependence illustrates, thus, intense motions of the surfactant aggregated into a gel-like state [3].

Quantitative interpretations of the variable-temperature NMR spectra obviously require precise temperature control in the sample. In fact, the disparities between the temperature setting in a variable-temperature unit and the true temperature in NMR samples are well known. Generally, temperature calibrations in solid-state NMR are carried out on the basis of the well-established temperature dependence of the ^{207}Pb chemical shift determined in lead nitrate. This technique is used as an internal NMR thermometer particularly under magic angle spinning (MAS) conditions providing precise measurements of the isotropic chemical shift values. On the other hand, on increasing the MAS rate, the frictional heating effects will change the local temperature and the calibration should also include the MAS rate.

There is a protocol suggested by Guan and Stark [4] to determine the temperature in MAS NMR probes. The authors have obtained the empirical equations (5.1) and (5.2) applicable for the BioMAS ($T_S(b)$) and FastMAS ($T_S(f)$) NMR probes with spinning rates changing typically between 8 and 35 kHz:

$$T_S(b) = 0.97T_0 + 1.34°C\exp\left(\frac{\omega_r}{7.53\,kHz}\right) - 0.77°C \tag{5.1}$$

$$T_S(f) = 0.98T_0 + 3.79°C \exp\left(\frac{\omega_r}{19.6\,\text{kHz}}\right) - 3.49°C \quad (5.2)$$

where

T_0 is given set temperature

ω_r is an MAS rate

Other calibration procedures are also available. Among them, the method based on measuring the frequency difference between the CH_3 and OH groups in ^{1}H NMR of methanol is most popular. Generally the ^{1}H NMR spectra are recorded for a standard sample of tetrakis (trimethylsilyl) silane soaked in liquid methanol.

When the correlation time of a motion is well compared with time T_2 (or is shorter than T_2), then the motions will average anisotropic interactions leading to a line-shape evolution in static or MAS NMR spectra. The typical temperature evolution of line shapes observed in static NMR spectra of deuterium is shown in Figure 5.2 [2]. As can be seen, on freezing the motion, from bottom to top, the line transforms showing different widths and quadrupolar splitting.

A great advantage of NMR is that shapes of resonances are mathematically known and can be calculated using the semiclassical or quantum mechanical theory. The algorithm of such calculations or, in other words, a line-shape analysis, applied for static samples, is generally based on the Bloch–McConnell equation describing the vectors of the transverse magnetizations, M(t), which undergo an exchange

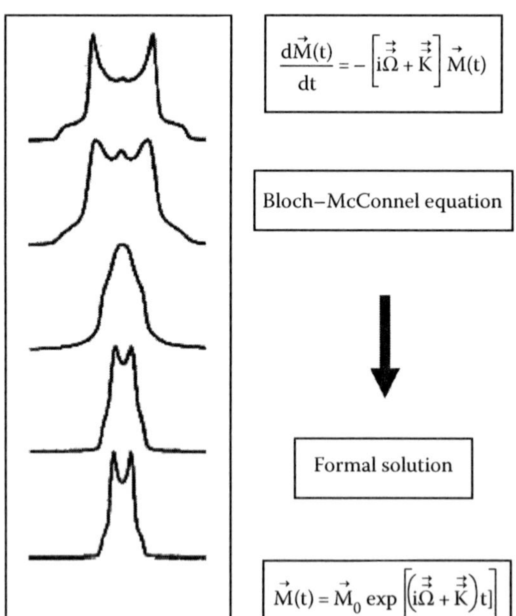

FIGURE 5.2 The temperature evolution of ^{2}H NMR resonance and the Bloch–McConnell formalism.

between sites N in Figure 5.2. Here and also in the formal solution as an exponential function, Ω is the [N × N] diagonal matrix that represents the anisotropic frequencies for the exchanging nuclei and K is again the [N × N] matrix that describes the exchange [2]. In other words, the elements in the Ω matrix depend on the spectral parameters, for example, the CSA, while the K matrix includes the exchange rates. Then, the computer simulation and fittings to the experimental line shapes will result in determinations of motion correlation times.

The situation changes under MAS conditions: the Ω matrix can become time dependent because of the mechanical (macroscopic) sample rotation. This effect is expressed as characteristic additional broadenings observed for resonance lines in the MAS spectra. In such cases, there are more complicate computer treatments, for example, the Floquet procedures, details of which can be found in Krushelnitsky and Reichert [2].

It should be emphasized again that the types of motions (i.e., their geometry), and the frequencies (i.e., correlation times) and activation energies can be principally extracted from the experimental NMR data by a line-shape analysis. However, in this method there are two important problems. First, often different motional models can equally describe the experimentally observed line shapes. It is obvious that in such a situation the calculation by itself cannot distinguish between the different types of motions, thus requiring independent data. Second, a description of a molecular motion in terms of a single correlation time, can be incorrect in the solid state (see the following correlation time distributions).

5.1.1 Dynamics in Polymeric Materials from a Line-Shape Analysis

For nuclei that are spectroscopically convenient for experimental observation, there are no principal limitations in their choice as targets to elucidate details of molecular dynamics in different materials by the line-shape-analysis method. However, in practice, ^{1}H, ^{13}C, and ^{15}N nuclei, among nuclei with spins of 1/2, and ^{2}H, among quadrupolar nuclei, are most convenient and popular, particularly for dynamic description of polymeric systems. Protons are generally used due to high NMR sensitivity. The ^{13}C and ^{15}N NMR technique works better at previous labeling procedures. A deuterium line-shape is extremely sensitive to the translational, librational, and rotational motions with the activation energies associated with thermally activated jumps between structurally different sites. This ^{2}H NMR approach is probably most powerful for characterizations of polymers and many other materials. Even the coking phenomena in zeolites [5] can be monitored by ^{2}H NMR.

Besides improved NMR sensitivity, materials enriched with ^{13}C, ^{15}N, or ^{2}H have a great advantage because the experiments are aimed at selective observation of labeled groups and thus, a line-shape analysis applied for static samples can be remarkably simplified. In turn, more simple schemes of calculations will provide more reliable data. Therefore, the labeling is often used in studies of polymeric systems.

Figure 5.3 illustrates dynamic effects expected for nuclei with spins of 1/2 and deuterium nuclei. Line shapes of nuclei with spins of 1/2 are dominated by isotropic chemical shifts (noted here as σ_{iso}), the chemical shift anisotropy (noted as δ), and the asymmetry parameter, η. As is seen, a regular NMR powder pattern can be

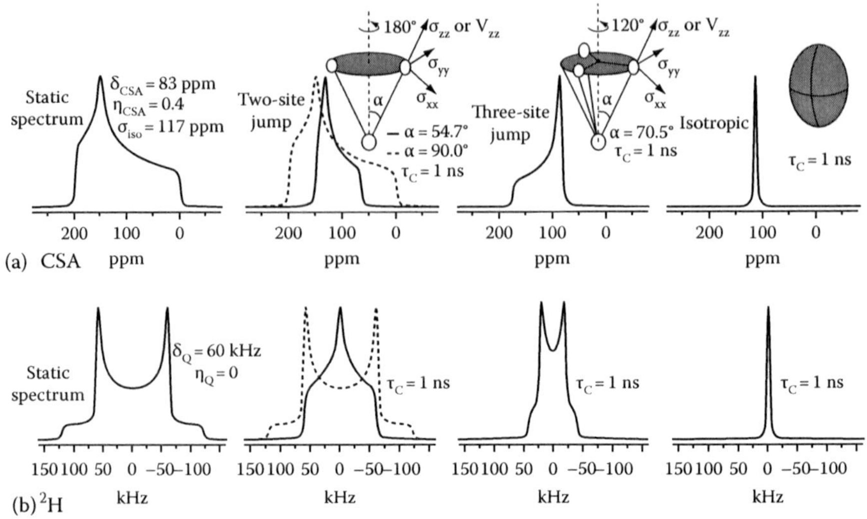

FIGURE 5.3 The powder line shapes calculated for nuclei with spins of 1/2 (a) and ^2H nuclei (b) at different kinds of molecular motions: (a) the shapes governed by the chemical shift anisotropy parameters; (b) the shapes dominated by quadrupolar interactions. (Reproduced with permission from Azevedoa, E.R. et al., *Prog. Nucl. Magn. Reson. Spectrosc.*, 47, 137, 2005.)

observed when the correlation times τ_C (for processes leading to reorientations of molecules or molecular groups with respect to the external magnetic field) is significantly longer than the inverse of the anisotropy factor, $\tau_C \gg 2\pi/\delta$. This case is generally termed as a slow exchange [1]. Rapid completely isotropic reorientations obviously will lead to an isotropic resonance, observed in the spectra as a sharp line. The spectral parameters and geometry and correlation times of the motion obtained by the simulations are also seen in this figure.

In the intermediate exchange region, that is, at $\tau_C \sim 2\pi/\delta$, the line shape will depend not only on the motional correlation time value but also on the type of the molecular motion and its geometry. The spectra show how dramatically the line shape changes in going from two-site jumps at different geometries to the three-site jump motion.

Similarly to the influence of the motional geometry on chemical shift anisotropy parameters, δ and η, observed for nuclei with spins of 1/2, the deuterium spectra, represented in Figure 5.3, show relationships between line-shapes of quadrupolar nuclei and parameters of motions. Two important conclusions can be made. First, for any nonaxial motions, for example, the three-site jumps in Figure 5.3, the resulting average tensor is axially symmetric due to the equal rotationally averaged "x" and "y" principal component. Second, in the case of isotropic motions, the three principal components are completely averaged and a single resonance will be observed, when the motions are very fast.

The ^2H and ^{15}N static NMR spectra of a-tactic poly(acrylonitril) $[C_3H_2N]$n-aPAN [1] can be a practical example of the line-shape analysis applied for characterization of motions in solid polymers. The NMR spectra recorded between 10°C and 140°C

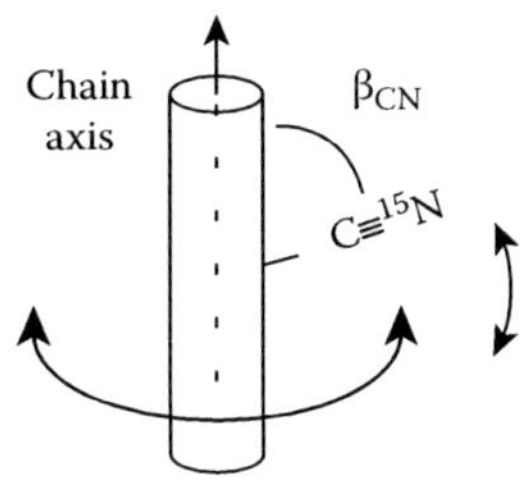

SCHEME 5.1 Rotational diffusion around the main chain in a-tactic poly(acrylonitril).

have shown a pronounced line-shape evolution corresponding to rotational diffusion around the main chain with a restricted distribution of amplitudes as Scheme 5.1 shows.

Besides polymer solids, the variable-temperature solid-state ^2H NMR spectra are often applied for probing molecular mobility in porous systems again due to the high sensitivity of deuterium line-shapes to the translational, librational, and rotational motions. Generally, molecular mobility of small molecules absorbed by porous systems, for example, benzene, ethanol, etc., are the focus of these studies [5,6]. Finally, the detailed theoretical description of a deuterium line-shape analysis, performed for solid dimethyl sulfone-d_6 by Brown and coworkers, can be found in Brown et al. [7].

5.2 MOLECULAR MOBILITY FROM THE SOLID-STATE NMR RELAXATION

As noted earlier, the line-shape approach does not require a special NMR experience on the experimental level, while the approach based on nuclear relaxation does. Therefore, this section will consider the general elements of the relaxation theory, features of relaxation time measurements, and calculations of the relaxation curves and the interpretations of the relaxation data in terms of molecular mobility in solids.

5.2.1 Measurements of Relaxation Times and Errors of the Measurements

The hard- and software of modern solid-state NMR spectrometers provide simple and convenient procedures for accurate measurements of relaxation times T_1, T_2, $T_{1\rho}$, and T_{1d} and also selective relaxation times.

Spin–lattice T_1 time can be determined by the *inversion recovery*, *saturation recovery*, or *progressive saturation* NMR experiments. Other methods of direct T_1 determinations are also available but less popular. These measurements are based on applications of radio frequency pulses, which, as we know, can be of different power and length. When the macroscopic magnetization is considered in a coordinate system, shown in Figure 1.6, the equilibrium state corresponds to the magnetization vector located along the OZ direction. On the application of an irradiating radio frequency field, as a short radio frequency pulse, along the OY axis (i.e., perpendicular to the plane of Figure 1.6), the magnetization vector converts to be situated along the OX axis. Since NMR signals are technically registered along this direction, a 90° pulse will generate a maximal magnetization along the OX axis, corresponding to maximal line intensities in the NMR spectra. It is obvious that a 180° pulse converts the magnetization vector from OZ to −OZ and its projection on the OX axis will be equal to zero. Thus after Fourier transformation intensities of resonances will also be close to zero.

Both pulses operate in the standard inversion recovery experiments:

$$(\mathrm{RD} - 180° - \tau - \mathrm{AT})_n \tag{5.3}$$

which provide determinations of the *nonselective* T_1 times. Here n is the number of scans needed to reach a good signal-to-noise ratio and pulses should be *hard* and *short*.

Typically in the inversion recovery experiments, the hard pulses have durations of ~5–10 μs and the power should be sufficient to excite a range of frequencies of order 10^5 Hz. The delay time, τ, between the inverting and registering pulses is typically varying from values of $\tau \ll T_1$ to $\tau \approx 3T_1$. Then, in a very short τ delay ($\tau \ll T_1$), the first 180° pulse inverts the magnetization vector to the −OZ axis and the registering 90° pulse converts this vector from −OZ to −OX, resulting in a signal with a "negative" intensity as shown in Scheme 5.2.

The NMR data, obtained at τ variations in the inversion recovery experiments, show their evolution from the "negative to positive" resonances that can be treated by a standard nonlinear three-parameter fitting routine of NMR spectrometers to calculate T_1 values.

It should be emphasized that the nuclear relaxation in solids can be an exponential or nonexponential process, as is discussed later. Here it should be emphasized that applications of the standard three-parameter fitting programs, implying only a mono-exponential function (see, e.g., Figure 5.4), is not permissible in the absence of accurate independent information on the relaxation nature. It is obvious that such treatments will be the source of large principal errors. Finally, in order to improve the quality of spin inversion, the 180° pulse in the pulse sequence can be replaced with a composite pulse cluster 90°ϕ 240°ϕ 90°ϕ. Pulse sequences, used for T_1 time determinations by saturation recovery experiments, are similar to those in the inversion recovery experiments. However, here the 90° saturating pulse replaces the inverting 180° pulse.

Experiments in the rotating coordinate system are based on the spin-locking pulse sequence and directed to determinations of the $T_{1\rho}$ times. Generally when a

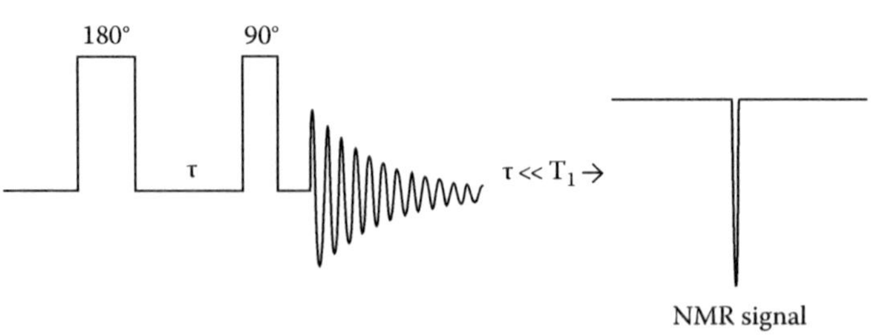

SCHEME 5.2 A NMR signal appears with "negative" phase when shortest times, τ, are applied.

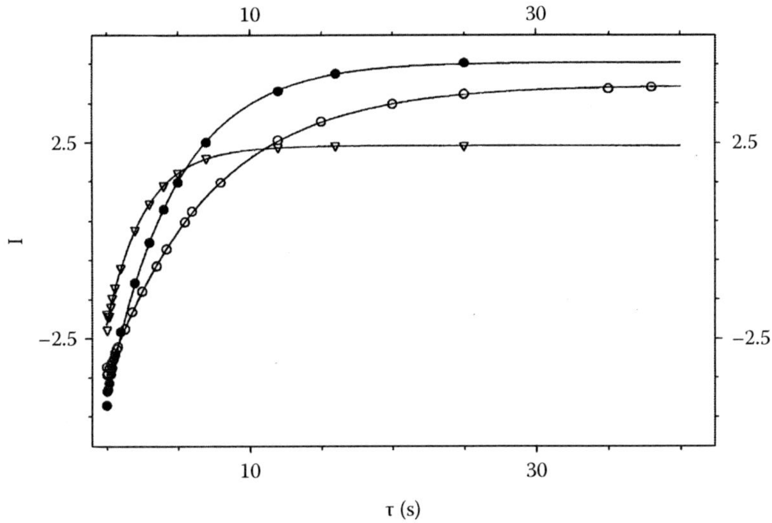

FIGURE 5.4 Examples of treatments performed for exponential relaxation curves by a standard nonlinear three-parameter fitting routine.

90° pulse reorients the magnetization vector (see Figure 1.6) and then the irradiating field is switching off, nuclear spins relax due to motions occurring in the lattice at frequencies close to Larmor's frequency. However, if the irradiating radio frequency field will operate as long as the τ time by 90° changes in its phase, then, the nuclear magnetization will be aligned along the direction of the applied radio frequency field. Because this effective radio frequency field is weak relative to the external magnetic field, B_0 (generally, strength of the locking field is less by several orders), the created magnetization will decay with a specific time constant, $T_{1\rho}$, governed by molecular motions characterized by significantly lower frequencies. By analogy with the inversion recovery experiments, variation in the τ values leads to determination of the spin–lattice relaxation time in the rotating coordinate system, $M(\tau) = M_0 \exp(-\tau/T_{1\rho})$. In rigid solids with restricted molecular motions, the $T_{1\rho}$ values differ remarkably from T_1 times providing characterizations of corresponding low-frequency molecular motions.

By definition, the spin–spin relaxation times, T_2, describe the transverse component recovery. The T_2 times are usually measured by the Carr–Purcell pulse train experiments directed to observations of echo:

$$(90°x' - \tau - 180°y' - \tau \text{ (first echo)} - \tau - 180°y' - \tau \text{ (second echo), etc.)} \quad (5.4)$$

The NMR decays, collected at τ variations, are treated by the standard fitting routine of NMR spectrometers to calculate T_2 times. Again, the spin–spin relaxation in solids can be exponential or nonexponential.

Quantitative interpretations of relaxation times will be reliable at minimal errors in the experimental T_1, $T_{1\rho}$, and T_2 determinations. One can accept the following instrumental factors having a significant influence on accuracy in relaxation time measurements: (a) calibrations of radio frequency pulses (the practical effects of bad pulse calibration is considered later), (b) homogeneity of radio frequency pulses, (c) a control of temperature in a sample, and (d) signal to noise ratios in the resulting NMR spectra. The latter factor can strongly influence the final result.

Noise effects in relaxation experiments have been quantitatively analyzed by Daragan et al. [8], using the exponential relaxation process as an example. Since line intensity (I) in the inversion recovery data is a function of the time τ, as $I(\tau) = \exp(-\tau/T_1)$, the semilogarithmic coordinates will obviously lead to straight lines that, in turn, should show slopes equal to $1/T_1$. It has been demonstrated that due to the presence of a spectral noise, errors in determinations of the ln(I) values depend on the τ durations. This effect is shown in Figure 5.5. As one can see, the points, obtained at the *tails* of relaxation curves (i.e., when the τ/T_1 ratio is larger than 1), give growing errors. This means that in practice such points should be weighted less in calculations of final results.

Bad choice of the parameters varied in relaxation time measurements can also have strong effects. Obviously relaxation delays in the pulse sequences should provide a *complete* relaxation of nuclei in each cycle of the measurements. When they are too short, the peak intensities in the collected NMR spectra will be distorted. In fact, as is well known, the delays close to T_1 values, will recover only ~63% of an equilibrium nuclear magnetization while a 99% recovery of the equilibrium magnetization will require a delay equal to five T_1. Therefore it is reasonable to perform the preliminary experiments for rough estimations of relaxation times.

Based on experience, one can formulate a best set for the τ values in T_1, T_2 experiments. In the case of a simple mono-exponential relaxation, this set should be between $0.1 * T_1$ and $1 * T_1$ with a number of the τ values more than 16–20. However in practice, since the type of nuclear relaxation is often unknown, the process should be properly characterized, for example, in the inversion recovery experiments, by

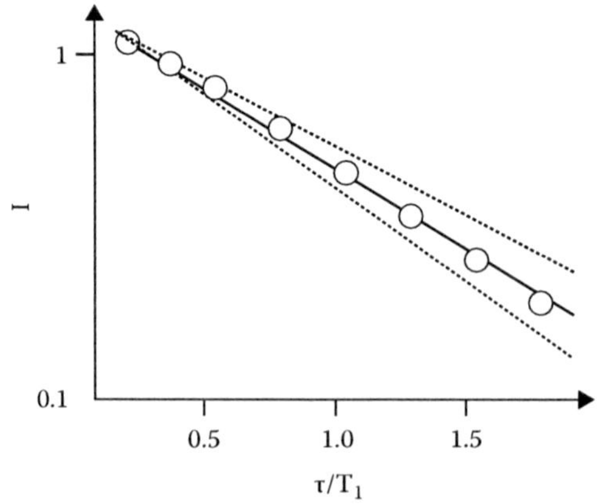

FIGURE 5.5 The plots of the line intensity versus the τ/T_1 values in semilogarithmic coordinates (the points and the solid line). The dashed lines illustrate the increasing errors due to the presence of noise in the NMR spectra.

the τ values covering a large range from $0.1 * T_1$ to $3 * T_1$. Under these conditions, the exponential and nonexponential relaxation can be discriminated. For example, appearance of "overshoots" in the inversion recovery curves, particularly at τ values close to $T_1/2T_1$, is a good test for the presence of an unusual nuclear relaxation.

Finally it should be emphasized that the standard NMR adjustments and some experience in relaxation experiments generally lead to results, relaxation times in which are determined with errors less than 3%–5%. Such errors are not significant in terms of relaxation time interpretations.

5.2.2 How Molecular Motions Affect Nuclear Relaxation

The relaxation mechanisms in solids are different. As summary one can accept the following:

1. In the absence of paramagnetic centers nuclear relaxation is dominated by dipole–dipole coupling and/or quadrupolar interaction.
2. Relaxation by chemical shift anisotropy can appear in experiments at high resonance frequencies (i.e., at highest magnetic fields).
3. Heavy nuclei with spins of 1/2, such as ^{207}Pb [9], can relax by a spin–phonon Raman scattering mechanism, where coupling is caused by phonons. The phonons create a local magnetic field by modulating the valence electron shell motion relative to the nuclear/electron core. Phenomenologically, in such cases, the T_1 times are independent of the magnetic field strength and inversely proportional to the square of the temperature [9].

The fluctuating magnetic fields appearing due to dipole–dipole or quadrupolar coupling lead to nuclear relaxation with rates $1/T_1$, $1/T_2$, or $1/T_{1\rho}$ that can be expressed via the strength of internuclear coupling C and the spectral density function $J(\omega_0, \tau_C)$ (see Chapter 1). Here however, we stress again that the $J(\omega_0, \tau_C)$ is the factor illustrating how much the influence of the C on relaxation weakens, when frequencies of the field fluctuations move off Larmor's frequency, ω_0 ($\omega_0 = 2\pi\nu_0$).

When molecular motion is isotropic and can be characterized by a single correlation time at a given temperature, the spectral density function takes a simple form. According to the theory of Bloembergen, Purcell, and Pound [10], nuclear T_1, T_2 relaxation for two identical nuclei due to their dipolar coupling is given by

$$\frac{1}{T_1^{DD}} = \left(\frac{2}{5}\right)\gamma^4\hbar^2 r^{-6} I(I+1)\left\{\frac{\tau}{(1+\omega_I^2\tau^2)} + \frac{4\tau}{(1+4\omega_I^2\tau^2)}\right\} \tag{5.5}$$

$$\frac{1}{T_2^{DD}} = \left(\frac{1}{5}\right)\gamma^4\hbar^2 r^{-6} I(I+1)\left\{3\tau + \frac{5\tau}{(1+\omega_I^2\tau^2)} + \frac{2\tau}{(1+4\omega_I^2\tau^2)}\right\} \tag{5.6}$$

where ω_I is the resonance frequency. This mechanism is particularly powerful in strong dipolar coupling, for example, for two protons, particularly at short internuclear distances r between them and at a molecular motional frequency close to ω_H. Similarly relaxation of nucleus A via heteronuclear dipolar coupling by nucleus B is described by equations

$$\frac{1}{T_1} = \left(\frac{4}{30}\right)\left(\frac{\mu_0}{4\pi}\right)^2 r(A-B)^{-6}\gamma_A^2\gamma_B^2\hbar^2 I_B(I_B+1) \times \left\{\frac{3\tau}{(1+\omega_A^2\tau^2)} + \frac{6\tau}{(1+(\omega_A+\omega_B)^2\tau^2)} + \frac{\tau}{(1+(\omega_A-\omega_B)^2\tau^2)}\right\} \tag{5.7}$$

$$\frac{1}{T_2} = \left(\frac{4}{30}\right)\left(\frac{\mu_0}{4\pi}\right)^2 r(A-B)^{-6}\gamma_A^2\gamma_B^2\hbar^2 I_B(I_B+1) \times \left\{4\tau + \frac{3\tau}{(1+\omega_A^2\tau^2)} + \frac{6\tau}{(1+\omega_B^2\tau^2)} + \frac{\tau}{(1+(\omega_A-\omega_B)^2\tau^2)} + \frac{6\tau}{(1+(\omega_A+\omega_B)^2\tau^2)}\right\} \tag{5.8}$$

By analogy, for quadrupolar nuclei, interacting with the electric field gradients, eq_{ZZ}, T_1 relaxation is described as

$$\frac{1}{T_1(Q)} = \left(\frac{3}{50}\right)\pi^2(2I+3)(I^2(2I-1))^{-1}\left(\frac{e^2q_{ZZ}Q}{h}\right)^2\left(1+\frac{\eta^2}{3}\right)\left(\frac{\tau}{(1+\omega_Q^2\tau^2)} + \frac{4\tau}{(1+4\omega_Q^2\tau^2)}\right) \tag{5.9}$$

where η is the asymmetry parameter (the η transforms to zero at an axial symmetry). Generally, the asymmetry parameter cannot be found by relaxation experiments, requiring, thus, independent data. For example, the η can be found from line shapes.

Since the molecular motion correlation times τ_C are temperature dependent via $\tau = \tau_0 \exp(E_a/RT)$, the T_1 plots in semilogarithmic coordinates are symmetrical, V-shaped, and go through minima as shown in Figure 5.6. The slopes of the T_1 curves correspond to the E_a values and the τ-independent minima can be applied for determining the internuclear distances or quadrupolar parameters. It should be emphasized that the left wing of the T_1 curve shows field-independent relaxation times, while the right wing corresponds to field-dependent values. The corresponding conditions are represented in the figure. Since molecular (or group) motions are generally slow in the solid state, the right wing is typical of solids.

A similar expression for spin–lattice relaxation time, measured in the rotating coordinate frame, and given for homonuclear coupling

$$\frac{1}{T_{1\rho}} = \left(\frac{1}{5}\right)\gamma^4\hbar^2 r^{-6} I(I+1)(3J^D(2\omega_1) + 5J^D(\omega_1) + 2J^D(2\omega_1)) \tag{5.10}$$

also shows a minimum. As seen, the $T_{1\rho}$ (min) is shifted toward lower temperatures relatively to the $T_{1\,min}$.

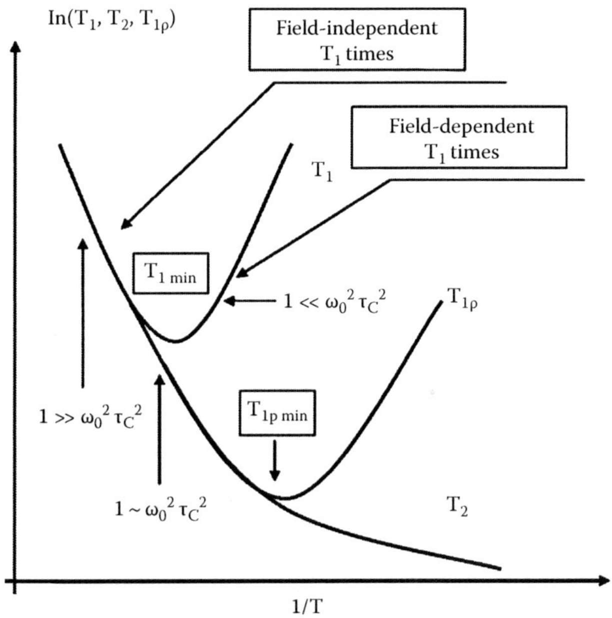

FIGURE 5.6 The influence of temperature on relaxation times T_1, T_2, and $T_{1\rho}$ (schematically).

Thus, in principle, the variable-temperature measurements help to characterize molecular mobility in terms of motional correlation times and activation energies. However, the main problem in such characterizations is the type of the correlation function, which is generally unknown. To solve this problem, the correlation function is usually approximated by a simple model with a limited number of dynamic parameters and then these parameters are determined from the relaxation times.

5.2.2.1 Correlation Time Distribution versus Single Correlation Time

Reorientations of molecules or functional groups in amorphous systems and polymeric materials, for example, can be very complex and cannot be represented by a single molecular correlation time at a given temperature. In such cases, they use the concept of a distribution of correlation times τ_C, when the τ_C becomes a center of the distribution while its width is represented by a distribution parameter. There are a number of correlation time distributions: Cole–Cole, Davidson–Cole [11], Fuoss [12], or Havriliak–Negami's model [13]. These functions can be incorporated into the spectral density functions for quantitative interpretations of relaxation data.

It is important to discriminate between symmetric and asymmetric distributions as the most typical cases describing molecular motions in solids. While the latter is known to have insignificant influence on the $T_{1\,min}$ values, the symmetric distribution can lead to remarkable underestimations of activation energies, E_a, and overestimations of $T_{1\,min}$ values, when the temperature-dependent T_1 data are

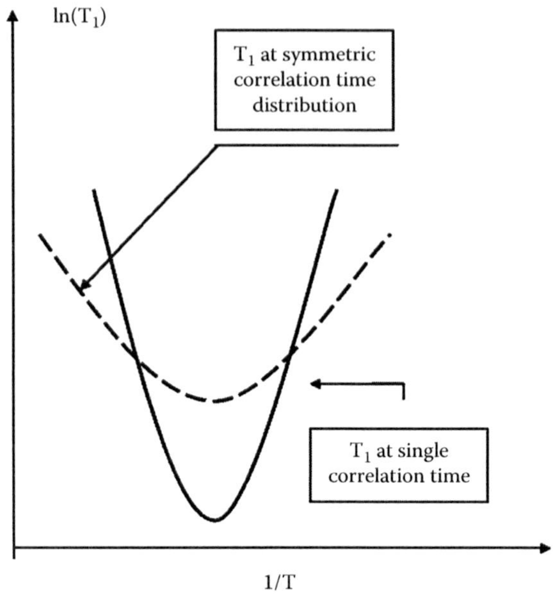

FIGURE 5.7 Schematic illustration of how a symmetric correlation time distribution affects the relaxation times in the presence of a single correlation time.

analyzed in terms of isotropic reorientations. This effect is represented schematically in Figure 5.7. It becomes clear that if distribution of correlation times is not taken into account, there is a large chance to overestimate, for example, internuclear distances.

One remark is also important in the framework of qualitative T_1 interpretations: when the symmetric correlation time distributions are present, the limiting magnitudes for the slopes of the $\ln(T_1)$ versus 1/T curves, that is, $E_a(obs)/R$, are proportional to the distribution parameters β:

$$E_a(obs) = \beta E_a \tag{5.11}$$

where E_a and $E_a(obs)$ are the real and experimentally observed activation energies, respectively, and $0 < \beta \leq 1$. Thus, the real energies can be obtained at observations of the $T_{1\,min}$ values by the relaxation experiments performed at various magnetic fields, which, in turn, result in determination of the $\tau_{C\,min}$ values at different temperatures. Then, the dependence $\tau_{C\,min}$ versus 1/T produces the real energy value. If this value does not correspond to an E_a value obtained from the slopes of the $\ln(T_1)$ versus 1/T curves, then a correlation time distribution is present. Probably this is a direct way to recognize the presence of a correlation time distribution.

Details of Havriliak–Negami's model applied for probing dynamics in amorphous polymer systems can be found in the work of Forte et al. [13]. The authors, by fittings of the variable temperature ^1H and ^{13}C T_1 curves obtained at different frequencies, have shown how to recognize and characterize segmental main chain motions, rotations of the methyl groups, and even librations of C–H bonds.

Finally, readers can be referred to the reviews [14–16] for a more detailed understanding of the theory of nuclear spin–lattice/spin–spin relaxation in solids, the analysis of second moments in solid state NMR spectra, basic relaxation experiments, and their interpretations, including solids with very mobile fragments like the quantum tunneling methyl groups.

5.2.3 Relaxation Approaches: Methodology and Examples

The dependence of dipole–dipole relaxation times on the τ_C and internuclear distances (or quadrupolar coupling) is the basis of their potential application for probing materials in the structural and dynamic contexts. Even relaxation times themselves, carefully measured, for example, at room temperature for different nuclei in materials, are very useful for their structural analysis and planning advanced NMR experiments.

Good examples in the aforementioned context are diamagnetic porous silica-based materials and zeolites. Klinowski [17] gives a summary of the ^{29}Si and ^{27}Al relaxation data, characterizing such systems:

1. ^{29}Si spin–lattice relaxation times, T_1, are *relatively* short and typically measured as 5–30 s (however in some systems they can reach 100 s or more).
2. ^{29}Si T_1 relaxation is not practically affected by the presence of ^{27}Al nuclei in the matrix.
3. Nuclear spin–lattice relaxation is not controlled by spin diffusion via paramagnetic centers (see Chapters 7 and 8) due to the large distance between neighboring silicon nuclei and the relatively low natural abundance of ^{29}Si nuclei.
4. The ^{29}Si T_1 effects of molecular oxygen absorbed by the silica matrix are dramatic, while other paramagnetic impurities or water play a secondary relaxation role.
5. T_1 ^{29}Si NMR relaxation times of crystallographically different silicon atoms are different.
6. The ^{29}Si T_1 values increase in going from crystalline to amorphous zeolites because many Si atoms in the amorphous samples are inaccessible to oxygen.
7. Spin–lattice relaxation times of ^{27}Al nuclei in zeolites are measured between 0.3 and 70 ms and depend on the temperature. Their relaxation is governed by the electric quadrupole interactions with the crystal electric field gradients that are modulated by translational motions of polar sorbate molecules.
8. The silicon T_1 times elongate remarkably when organic components are added to zeolites.
9. ^{27}Al T_1 times dramatically increase upon dehydration of the samples. It should be emphasized that T_1 times are very sensitive to the chemical prehistory of silica-based materials.

All these features can obviously be used as important structural criteria.

Commonly the methodology for characterization of molecular dynamics in diamagnetic materials [2] is based on the following:

1. Investigating the temperature and/or resonance frequency dependencies of T_1 times, where the number of minima in the temperature dependence or the number of steps in the frequency dependencies will correspond to the number of dynamic processes characterized by strongly different molecular motion correlation times. Then, the types of motions can be identified using an appropriate mathematical formalism.
2. Measurements of the dipolar line widths, T_2(DD), that, in terms of a two spin system (spin of 1/2), show the values of the residual dipolar interaction averaged by molecular motions having correlation times greatly shorter than the T_2 (generally in the order of 10^{-5} s).
3. Measurements of the temperature-dependent $T_{1\rho}$ times, which are sensitive to motions in the microsecond diapason.
4. Measurements of T_{1D} and $T_{1\rho\rho}$ (a time in the doubly rotating frame) times sensitive to slow motions in the submillisecond–millisecond diapason. Reichert et al. [2] consider the measurements of the T_{1D} and $T_{1\rho\rho}$ times and their quantitative analysis. Generally they are more complex and not always obvious.

A priori, as mentioned earlier, the problem in quantitative interpretations of relaxation times in terms of molecular mobility is related, first of all, to uncertainty in the form of motional correlation function [2]. Here besides the concept of correlation time distribution, the correlation function can be approximated by a simple model, giving a model-free approach

$$C(t) = S^2 + (1 - S^2)f(t) \tag{5.12}$$

where

S^2 is taken as an order parameter

f(t) is a decay function that satisfies the conditions f(0) = 1 and f(∞) = 0

The simplest form of the f(t) is

$$f(t) = \exp\left(\frac{-t}{\tau_C}\right) \tag{5.13}$$

In turn, the S^2 parameter characterizes the averaging of dipolar coupling or quadrupolar interaction related to the motional amplitude. Finally, the order parameter S^2 can be expressed by

$$S^2 = \langle P_2(\cos\theta(0,\infty))\rangle \tag{5.14}$$

where P_2 represents the coordinate part in the function of spectral density, for example, $P_2(x) = (3x^2 - 1)/2$.

Thus, the S^2 depends on the type of molecular motions. Generally the order parameter is known as a function of angular amplitudes for such motional models, such as jumps between two equally populated states and continuous rotational diffusion or wobbling with a cone [2].

The careful ^{13}C relaxation NMR study of a polylysine sample [18] is an excellent example of probing dynamics in solid polymers. The authors have analyzed the T_1 $T_{1\rho}$ (off-resonance) and ^1H-decoupled $T_{1\rho}{}^d$ times, measured at different resonance frequencies and temperatures. The variable-temperature data are shown in Figure 5.8, where the minimal relaxation time values are clearly observed at temperatures between 280–300 K. Totally, the data have resulted in 15–25 relaxation parameters characterizing the dynamics. The main conclusion was as follows: the polymer exhibits two motional processes with correlation times on the scale 10^{-9} and 10^{-4} s.

Other approaches for characterization of dynamics by solid-state NMR are also available. For example, the REDOR experiments for determination of dipolar-coupling in microcrystalline ubiquitin can be found in the recent work by Ernst et al. [19]. These experiments have been combined by the authors with the autocorrelated and cross-correlated ^{15}N relaxation rates to accurately describe the amplitudes and correlation times of backbone dynamics on picosecond and nanosecond timescales.

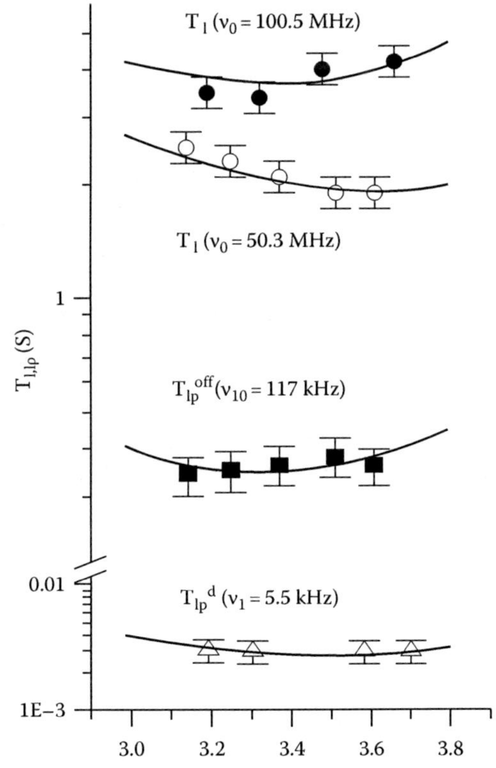

FIGURE 5.8 The temperature dependencies (in 10^3 K/T) of NMR relaxation times: T_1, $T_{1\rho}$ (off-resonance), and $T_{1\rho}{}^d$ times measured for Cα in a sample of solid polylysine. Solid lines show the results of fitting procedures. (Reproduced with permission from Krushelnitsy, A. et al., *Biopolymers*, 73, 1, 2004.)

5.3 ONE- AND TWO-DIMENSIONAL EXCHANGE SOLID-STATE NMR SPECTROSCOPY

The phenomenon of chemical exchange in NMR can be formulated as an exchange between two nonequivalent spin states, A and X,

$$\begin{array}{ccc} A & \Leftrightarrow & X \\ \nu^A;\tau^A & & \nu^X;\tau^X \end{array} \tag{5.15}$$

which averages the spectral characteristics of these states. This process can represent a physical migration of target nuclei, for example, between different molecular groups (^{1}H migration along hydrogen bonds, etc.) or consists of changing the magnetic environments, for example, nuclei participating in conformation motions. The chemical exchange directly affects line-shapes of the A, X resonances, when the exchange frequencies are comparable with chemical shift differences, $(\nu^A - \nu^X)$. Finally, when the exchange frequencies $<\nu^A - \nu^X$, chemical exchanges are *slow*. The latter is typical of solids.

Generally speaking, the exchange NMR experiments in solids can detect relatively slow motions on the frequency scale between 1 Hz and 1 kHz, observed, for example, in polymeric materials typically undergoing segmental reorientations or changing conformations.

The one-dimensional (1D) exchange NMR approach is based on the so-called stimulated echo pulse sequence [1] shown in Figure 5.9. This sequence helps to evaluate two- and multiple-time correlation functions that describe molecular (or group) rotational process. It is often applied for organic and inorganic materials. As seen, the experiment includes the following sections. The first exciting 90° radio frequency pulse is applied that is followed by time τ for the evolution of magnetization due to chemical shift anisotropy or quadrupolar interactions. Then, the storage pulse is used and single quantum coherence converts into the term that does not evolve during the long mixing time, t_m. Then, the third pulse provides collection of the FID data. Avoiding the theoretical details [1], it would be emphasized that the decrease in the echo intensity is dependent on the mixing time in order to determine the two-time correlation function for an exchange process. It has been also noted that the number of sites participating in the exchange will also effect on the echo, when the mixing time is long.

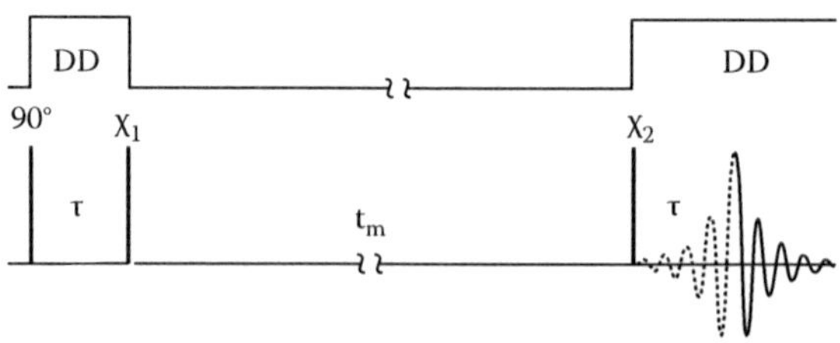

FIGURE 5.9 Pulse sequence for stimulated-echo experiments. (Reproduced with permission from Azevedoa, E.R. et al., *Prog. Nucl. Magn. Reson. Spectrosc.*, 47, 137, 2005.)

Different versions of stimulated-echo experiments are available and widely applied for organic polymers to recognize the timescales and amplitudes of slow motions. Among them, the 1D pure exchange (PUREX) NMR technique is most popular. In contrast to other pulse sequences, the PUREX helps to suppress the resonances that belong to *immobile* molecules or groups. In fact, the 1D PUREX NMR spectra will selectively exhibit signals coming from the nuclei located in segments, which reorient during the mixing time. The 1D one-dimensional exchange spectroscopy by sideband alternation (ODESSA) pulse sequence is more specific. It is aimed at slow molecular reorientations, analyzed on the basis of spinning sidebands. These experiments are very successful, for example, in studies of molecular motions that are responsible for α- and β-relaxation in the glassy polymers, such as poly(n-hexyl methacrylate). For example, it has been established that motions of the alkyl group govern the β-relaxation of the polymer, while motions of the main chain and carboxyl groups participate in both relaxation processes.

Finally, slow motions in polymer materials or biopolymers and proteins can be characterized by the center band-only detection of exchange (CODEX) methods, described in the reviews [1,2]. These approaches are aimed at analysis of the resonance intensities that correspond to the isotropic lines in the high-resolution MAS NMR spectra.

Figure 5.10 depicts the radio frequency pulse sequences applied, for example, in the two-dimensional (2D) exchange NMR experiments on static samples. Here the first 90° pulse excites the transverse magnetization that develops during the t_1 time due to the chemical shift anisotropy interactions. Then, a second acting 90° pulse stores one component of the magnetization situated along the Z-direction, which remains during the long mixing time t_m. Then, after the t_m time, the registering pulse is applied and the magnetization develops again under the chemical shift anisotropy interactions during the t_2 time. As in the case of any 2D NMR, the signal observed will depend on times t_1 and t_2 and the 2D Fourier transformation will give a 2D NMR pattern. As expected, the final spectrum will contain the intensity distribution as a function of the NMR frequencies and in addition, as a function of molecular motions occurring during the mixing time. Thus, in the absence of the motions, the resulting 2D NMR spectra will be purely diagonal. In contrast, in the presence of such

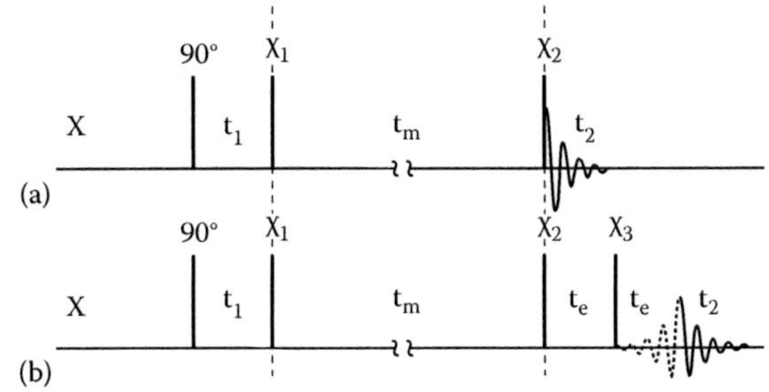

FIGURE 5.10 Two-pulse sequence generally used for the 2D-exchange NMR experiments. (a) The so-called basic scheme applying a single 90° excitation pulse; (b) a modified sequence with a single 90° excitation pulse and echo acquisition. (Reproduced with permission from Azevedoa, E.R. et al., *Prog. Nucl. Magn. Reson. Spectrosc.*, 47, 137, 2005.)

motions, the orientationally dependent frequencies change and lead to appearance of nondiagonal resonances.

Similar principles are used for echo-pulse sequences. Spiess and coworkers [20] have demonstrated the advantages of 2D exchange NMR by probing molecular reorientations in poly-oxymethylene. It has been found that this polymer undergoes helical jumps.

Generally, spectroscopists name the 2D exchange spectrum as a map of reorientational probabilities. Because these probabilities depend on the molecular motion correlation times and geometry of motions, the 2D exchange patterns are sensitive to both of these factors. This circumstance explains why the different versions of the 2D exchange NMR techniques, for example, the 2D PUREX NMR spectra, or the rotor-synchronized 2D MAS exchange NMR technique, pioneered by Veeman et al. [21], are widely used for probing molecular motions in polymeric solids [1,2]. For example, the review of Bohmer et al. [22] describes in detail dynamics in glassy systems including distributions of small jump angles, mapping rotational correlation times, and jump geometries.

5.4 MOLECULAR MOBILITY FROM THE CROSS-POLARIZATION NMR EXPERIMENTS

The cross-polarization (CP), first of all, from protons to carbons, can be incorporated into the different 2D exchange NMR experiments, considered in the previous section, to detect the ^{13}C nuclei in the spectra at better signal-to-noise ratios. On the other hand, as shown in Chapter 3, the cross-polarization parameters, the T_{XH} constants, and the ^1H $T_{1\rho}$ relaxation times in the rotating frame, extracted from the CP NMR experiments, depend on molecular (or molecular group) mobility. Thus, potentially the CP kinetics opens the way to probing dynamic behavior of solids.

Note that that the ^1H $T_{1\rho}$ times in a sample can be averaged due to the pronounced proton spin-diffusion relaxation over few nm distances. Thus the ^1H $T_{1\rho}$ times can characterize large proton networks up to several hundred of nm, providing information on the homogeneity of a solid within the volume in the subnanometer range.

When a material investigated is actually homogeneous, all ^1H $T_{1\rho}$ values measured for different protons should be very similar and thus, a single proton spin reservoir can be considered for the whole sample. In contrast, if different domains take place in a nonhomogenous material, then the ^1H $T_{1\rho}$ magnitudes will be separated in these domains, because their sizes are larger than the spatial averaging by spin-diffusion. All these aspects are important at probing various composite materials, for example, woods [23], where the chemical interactions at a molecular level and molecular (or group) mobility correlate with the macroscopic properties.

The ^1H $T_{1\rho}$ and T_{CH} magnitudes can be easily obtained by the CP kinetic experiments, where the ^{13}C signal intensities are measured as a function of the proton–carbon contact times. Under MAS conditions, the NMR spectra are well resolved and thus different ^{13}C sites can be characterized separately in terms of corresponding motions, mainly on a semiquantitative level. The following summary of the cross-polarization in numerous experiments, performed for crystalline, amorphous or semicrystalline polymer systems, may be useful for investigators planning studies of molecular mobility by CP:

1. ^{1}H $T_{1\rho}$ times increase, when crystallinity or the degree of ordering increases.
2. The ^{1}H $T_{1\rho}$ time shortening can indicate either an amorphous state in materials or decrease in their degree of ordering.
3. A strong increase in the T_{CH} constants, connected with dipolar C–H coupling, is generally induced by molecular motions.

As an example, the CP NMR data collected for some polyethylene-glycol-treated woods by the ^{13}C MAS experiments have been fitted to the experimental CP kinetic curves to show the presence of two components of the polyethylene glycol moieties with different ordering and mobility. The moieties have been characterized by T_{CH} = 105.9 μs and ^{1}H $T_{1\rho}$ = 0.80 ms and by T_{CH} = 11.0 μs and ^{1}H $T_{1\rho}$ = 3.65 ms, respectively [23].

The wide-line-separation (WISE) pulse sequence (or similar versions of the separate local field [SFL] experiments) represents a regular cross-polarization experiment modified by incorporation of an extra time period t_1 [1]. As shown in Figure 5.11, after the regular ^{1}H–^{13}C polarization transfer, the ^{13}C magnetization will depend on the evolution of the ^{1}H magnetization during time t_1. Then, at incrementing time t_1,

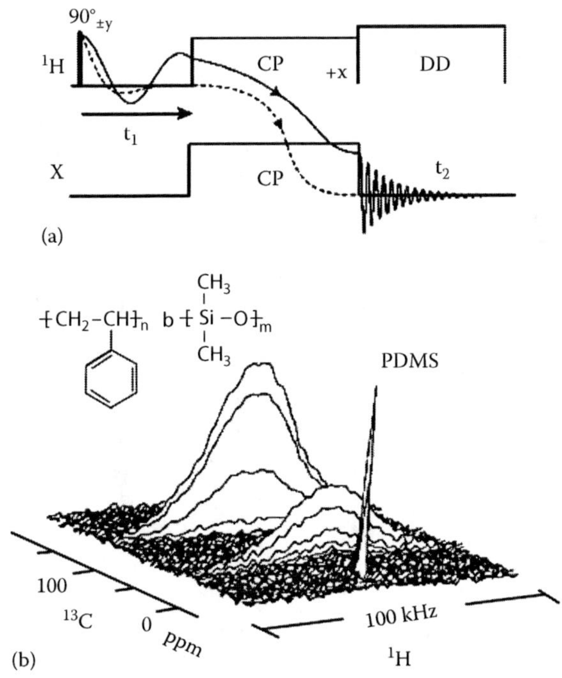

FIGURE 5.11 (a) The pulse sequence operating in the 2D WISE experiments. (b) Application of the aforementioned experiment to recognize the rigid and mobile phases in the block copolymer PS-b-PDMS (see the text). (Reproduced with permission from Azevedoa, E.R. et al., *Prog. Nucl. Magn. Reson. Spectrosc.*, 47, 137, 2005.)

usually applied for performing 2D experiments, a 2D time matrix can be obtained to give finally a 2D NMR spectrum after a Fourier transformation. Figure 5.11 shows that in practice such 2D NMR data will include the high resolution MAS spectra observed for ^{13}C nuclei and the wide-line spectra corresponding to ^{1}H nuclei.

The WISE experiments are relatively simple and attractive from the practical point of view. They can distinguish between regions in investigated materials if these regions have different mobility. For example, the WISE experiment, applied for a sample of the block copolymer polystyrene-*b*-poly(methyl siloxane), named in Figure 5.11 as PS-PDMS, clearly shows that the polystyrene fraction is relatively rigid and corresponds to broad ^{1}H lines in proton axis. At the same time, more narrow ^{1}H resonances are observed for the PDMS fraction, which is therefore rather mobile.

5.5 MOLECULAR MOBILITY FROM MAGNETIC RELAXATION DISPERSION EXPERIMENTS

Molecular mobility in disordered solids, for example, in porous materials and other heterogeneous molecular systems is of great interest particularly when the solid matrix of the materials is filled with a liquid of a gas, named here as small molecules. In fact, dynamic behavior of the matrix and small molecules, and particularly their mutual connection, play a major role for understanding the mechanisms of catalysis.

It is well known that liquids located at matrix surfaces or interfaces are quite mobile. Generally their translational diffusion coefficients are similar to those of the bulk liquids and thus, the mobility of liquid molecules is not solid-like. The local rotational and translational reorientations of liquids at interfaces in porous materials are quite fast and therefore the potentially anisotropic magnetic interactions in the "liquid spin system" are significantly averaged. For this reason, dynamics of small molecules in heterogeneous materials can be probed by regular NMR techniques. This statement can be well illustrated by the ^{2}H NMR spectrum recorded for a sample of the supermicroporous silica-based material SiO_2–Al_2O_3 (2 wt% of Al) with pores filled with $^{2}H_2O$ [24]. As Figure 5.12 shows, the ^{2}H NMR spectrum does not reveal the presence of a rigid powder shape even at 220 K. Only at 200 K the line transforms toward a rigid deuterium powder pattern that is resolved at 150 K showing quadrupole splitting of 148 kHz, typical of solid water.

The variable-temperature NMR experiments can be used to characterize behavior of small molecules absorbed by porous materials and situated in pores. Since liquids within small pore spaces freeze or melt at lower temperatures than the bulk liquids, the Gibbs–Thompson temperature depression of the melting point can be applied for studies of pore-size distributions [25,26]. This liquid-state ^{1}H NMR technique can characterize, for example, pore-size distribution and pore-wall thickness in the mesoporous materials of the FSM-16 family after their saturation with water or benzene or other liquids. This approach is carefully considered in Chapter 6. Dynamic behavior of olefin molecules, such as 1-butene or 1-pentene, absorbed by zeolite materials, can be successfully described by the regular (solution-like) high-resolution ^{1}H and ^{1}H nuclear overhauser effect spectroscopy (NOESY) MAS NMR techniques on the basis of ^{1}H chemical shifts measured for the adsorbed olefins as

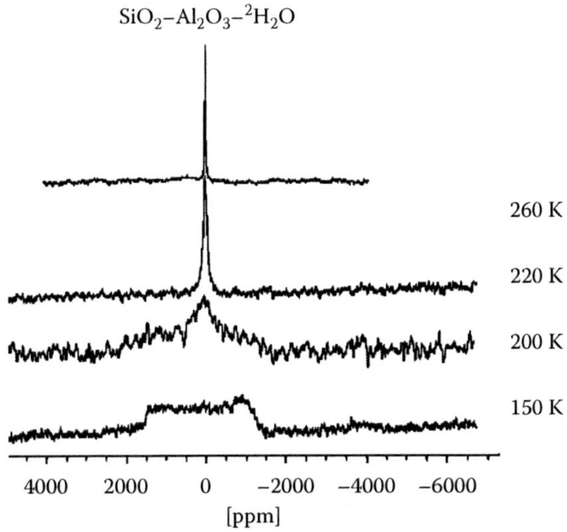

FIGURE 5.12 The variable-temperature ^2H NMR spectra recorded for the supermicroporous silica-based material SiO_2–Al_2O_3 (2 wt% of Al) containing ^2H_2O. (Reproduced with permission from Bakhmutov, V.I. et al., *Micropor. Mesopor. Mater.*, 118, 78, 2009.)

a function of pore fillings [27]. From these studies it is now known that the chemical shifts of olefins at lower pore fillings are very close to those measured in the gas phase, while at higher pore fillings, the chemical shift values approach those in the bulk liquids. More detailed descriptions of the adsorbed molecules in terms of their fast librations and overall reorientation–translation motions are also available.

Again the regular solid-state ^2H NMR technique is widely applied for probing mobility of absorbed molecules in porous systems due to the extremely high sensitivity of deuterium line shapes to translational, librational, and rotational motions. In general, this ^2H NMR method consists of recording the variable-temperature ^2H NMR spectra of deuterated liquids (e.g., benzene, ethanol, etc.) to document, upon cooling, their temperature evolution from an isotropic resonance at fast isotropic reorientations to the powder shape pattern showing quadruple splitting at slow motions [5–7].

The regular NMR relaxation measurements, selectively focused on absorbed small molecules, can describe their dynamics. These studies can be based on the ^1H T_1 and $T_{1\rho}$ [28] or ^1H–^{13}C cross-relaxation time [27] measurements performed for organic template molecules or for liquids in porous media by ^1H relaxation imaging [29]. It is also remarkable that the application of the ^1H T_1 technique is very important for chemical and food industry. These experiments help to recognize the structure and dynamics of water or oil at the surface, to determine dynamic parameters, such as diffusion coefficients, activation energies, time of residence, and coefficients of surface affinity, and to show the microporosity of ultra high performance concretes.

Sometimes the molecular dynamics in strongly heterogeneous systems with pore spaces filled with liquids becomes too complex [30] to be investigated by regular

NMR relaxation approaches. Generally such dynamics can be characterized by magnetic relaxation dispersion, which is based on the magnetic field dependence of the spin–lattice relaxation rates $1/T_1$. This approach obviously entails the NMR relaxation experiments performed at different magnetic field strengths. Technically this can be done, for example, by mechanically moving a sample from one magnetic field to another or by rapidly changing the electric current in an available magnet. The latter approach is obviously preferable and more convenient.

Since the NMR sensitivity strongly depends on the applied magnetic field, the relaxation experiments would be performed with the high sensitivity even at low magnetic field strengths. The modern NMR techniques using magnetic field switching help to achieve this result and the relaxation dispersion profiles can be obtained at different field strengths practically with the same signal-to-noise ratios. In spite of this technical solution of NMR sensitivity, the experiments on highly sensitive nuclei ^1H still remain most popular. It would be emphasized that there is one important limitation in application of this method: a rapidly decaying component can be lost during the magnetic field switching time. Therefore, generally the modern technique is capable of measuring the $1/T_1$ magnitudes up to $2000\,s^{-1}$.

As mentioned earlier, the major attention of relaxation dispersion studies is concentrated on protons. In this connection, Korb and Bryant [30] distinguish between proton-poor systems (inorganic solids) and proton-rich systems (generally organic solids). In inorganic systems, target protons of a liquid component and nuclei situated in the matrix of a material, for example, ^{29}Si or ^{27}Al, are generally not coupled. The absence of the coupling remarkably simplifies an analysis of the relaxation dispersion profiles. Generally such experiments are directed to understanding the interactions of liquid to surface in porous systems. Recently Korb [29] has revealed a significant difference between water and other solvents when they come into contact with high surface area in microporous chromatographic silica glasses. Figure 5.13 illustrates the dependences of proton spin–lattice relaxation rates, $1/T_1$, on magnetic field strengths, expressed in frequency units (MHz). The experiments have been carried out for a calibrated porous glass sample with a field cycling NMR spectrometer. The solid lines show the best fittings to the relaxation theory, details of which can be found in Korb [29]. Then, the corresponding variable-temperature experiments have resulted in determination of the effective activation energy for the protons moving on the surface.

Protons of a liquid component situated on the surface of a proton-rich system and protons from the solid component can be magnetically coupled and due to this coupling, the relaxation process can become very complex and nonexponential. However it has been emphasized that under these circumstances, the relaxation dispersion experiments and the resulting dispersion profiles better characterize the complex mobility of liquids than the regular relaxation approaches. Moreover, the dispersion profiles are preferable because they do not require variable-temperature conditions.

Following the classification of Korb and Bryant [30], proton-rich solids, such as polymer systems and/or proteins, are of greatest interest as objects of the ^1H relaxation dispersion experiments because the experiments can provide additional information on their mobility. The advances in the speed of current-switched field-cycling NMR instruments provide a direct measurement of the dispersion profiles for the

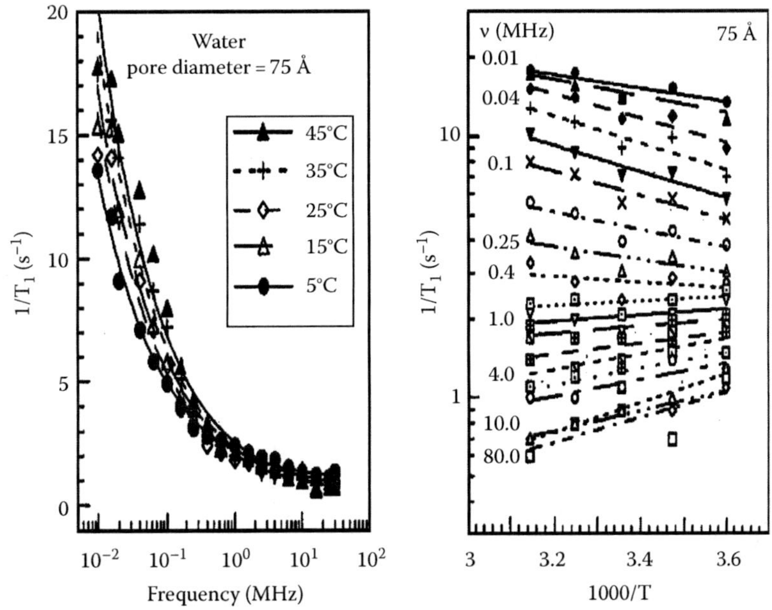

FIGURE 5.13 The ^{1}H spin–lattice relaxation dispersion measured in a sample of calibrated porous glass at different temperatures. (Reproduced with permission from Korb, J.P. and Bryant, R.G., *Adv. Inorg. Chem.*, 57, 293, 2005.)

protein protons, when the proteins are dry or especially treated and confined in gel. The profiles look very similar to those in Figure 5.13.

Different relaxation models are applied for quantitative treatments of the relaxation data. The details of these models can be found in the reviews [29,30]. For example, the relaxation dispersion, observed for protein molecules experimentally, can be described by a power law

$$\frac{1}{T_1(\omega)} = A\omega^{-b} + C \tag{5.16}$$

where b is 0.78 determined for dry proteins. It has been found that the field dependence of the proton relaxation rates does not depend on the dynamics of the side chains in the polypeptide molecules, but is dictated by the dynamics of the main chain itself. In other words, the motions of the main chain modulate the time-dependent proton–proton dipolar couplings, which in turn cause proton relaxation.

REFERENCES

1. Azevedoa, E. R., Bonagambaa, T. J., Reichertb, D. 2005. *Prog. Nucl. Magn. Reson. Spectrosc.* 47: 137.
2. Krushelnitsky, A., Reichert, D. 2005. *Prog. Nucl. Magn. Reson. Spectrosc.* 47: 1.
3. Simonutti, R., Comotti, A., Bracco, S., Sozzani, P. 2001. *Chem. Mater.* 13: 771.
4. Guan, X., Stark, R. E. 2010. *Solid State NMR* 38: 74.

5. Cheah, K. Y., Alexander, P., Gladden, A. P. 1997. *Appl. Catal. Gen.* 148: 387.
6. Benesi, A. J., Grutzeck, M. W., O'Hare, B., Phair, J. W. 2004. *J. Phys. Chem. B* 108: 17783.
7. Brown, M. J., Void, R. L., Hoatson, G. L. 1996. *Solid State NMR* 6: 167.
8. Daragan, V. A., Loczewiak, M. A., Mayo, K. H. 1993. *Biochemistry* 32: 10580.
9. Grutzner, J. B., Stewart, K. W., Wasylishen, R. E., Lumsden, M. D., Dybowski, C., Beckmann, P. A. 2001. *J. Am. Chem. Soc.* 123: 7094.
10. Bloembergen, N., Purcell, E. M., Pound, R. V. 1948. *Phys. Rev.* 73: 679.
11. Cole, K., Cole, R. J. 1941. *Chem. Phys.* 9: 341.
12. Fuoss, K. 1941. *J. Am. Chem. Soc.* 63: 385.
13. Forte, C., Geppi, F. C., Malvaldi, M., Mattoli, V. 2004. *J. Phys. Chem. B* 108: 10832.
14. Beckmann, P. A. 1998. *Phys. Rep.* 171: 85.
15. Klein, P. G., Ries, M. E. 2003. *Prog. Nucl. Magn. Reson. Spectrosc.* 42: 31.
16. Horsewill, A. J. 1999. *Prog. Nucl. Magn. Reson. Spectrosc.* 35: 359.
17. Klinowski, J. 1991. *Chem. Rev.* 91: 1459.
18. Krushelnitsy, A., Faizullin, D., Reichert, D. 2004. *Biopolymers* 73: 1.
19. Schanda, P., Meier, B. H., Ernst, M. 2010. *J. Am. Chem. Soc.* 132: 15957.
20. Hagemeyer, A., Schmidt, K., Spiess, H. W. 1989. *Adv. Magn. Reson.* 13: 85.
21. Dejong, A. F., Kentgens, A. P. M., Veeman, W. S. 1984. *Chem. Phys. Lett.* 109: 337.
22. Bohmer, R., Diezemann, G., Hinze, G., Rosser, E. 2001. *Prog. Nucl. Magn. Reson. Spectrosc.* 39: 191.
23. Bardet, M., Gerbaud, G., Giffard, M., Doan, C., Hediger, S., Le Pape, L. 2009. *Prog. Nucl. Magn. Reson. Spectrosc.* 55: 199.
24. Bakhmutov, V. I., Shpeizer, B. G., Prosvirin, A. V., Dunbar, K. R., Clearfield, A. 2009. *Micropor. Mesopor. Mater.* 118: 78.
25. Jackson, C. L., McKenna, G. B. 1990. *J. Chem. Phys.* 93: 9002.
26. Xie, X., Satozawa, M., Kunimori, K., Hayashi, S. 2000. *Micropor. Mesopor. Mater.* 39: 25.
27. Bohlmann, W., Michel, D., Roland, J. 1999. *Magn. Reson. Chem.* 37: S126.
28. Gougeon, R., Chezeau, J. M., Meuer, B. 1995. *Solid State NMR* 4: 281.
29. Korb, J. P. 2001. *Magn. Reson. Imag.* 19: 363.
30. Korb, J. P., Bryant, R. G. 2005. *Adv. Inorg. Chem.* 57: 293.

6 Strategies in Solid-State Multinuclear NMR: Studies of Diamagnetic Porous Materials

Porous molecular systems with a developed surface, susceptible to chemical modifications of the matrix by displacement reactions in the lattice, to regulations of surface properties by immobilization, and to addition of acidic/basic or metallic centers, are of great interest in modern materials science. The design of such (often strongly amorphous) systems, aimed at controlling micro- and macrostructures and fitting pore sizes and channels, requires reliable methods of investigation, such as solid-state NMR. Similar problems exist with materials of other nature requiring, thus, a common strategy for their solution.

This chapter formulates the NMR strategy, illustrated by a number of typical examples, which can be used in practice by researchers, experienced or beginners.

Structural investigations of porous molecular systems would be done by *multinuclear* NMR, magic angle spinning (MAS), and/or static, providing, thus, larger sets of experimental data. The NMR experiments entailing spectral and relaxation methods should be aimed at

1. Characterization of the matrix and a surface of materials and behavior of surfactant molecules
2. Description of molecules or groups of molecules situated in pore spaces, including their dynamics
3. Characterization of porosity
4. Establishment of the nature, locations, and distributions of doping (metal) ions and determination of their coordination numbers and oxidation states

All these key questions are considered here on the basis of silica-based materials, thus opening a way to understanding the problems in solid-state NMR applied to diamagnetic solids.

6.1 POROSITY OF MATERIALS FROM NMR

Porosity is one of the most fundamental characteristics of various materials from the well-known zeolites, important in catalytic chemistry, to wood and wood products widely applied in chemical laboratories and industry. Recently, the porous or rather microporous self-assembled molecular systems became very attractive as materials for the storage of gases, again based on using the pore spaces. Here the selective uptake and retention of small gaseous molecules, first of all, such as dihydrogen and gaseous alkanes, are most attractive in the context of their storage, separation, and transport.

A good example of such potentially storage systems is the two component mercury-containing complex shown in Scheme 1.1. This complex forms the solid-state three-dimensional (3D) network with a large pore space, depicted in Figure 6.1. Due to the pores and channels, the self-assembled system is capable of retention of small alkane molecules that can be directly observed by solid-state NMR. Thus, this method provides a reliable proof for the gas storage [1].

The corresponding ^{1}H MAS NMR spectra are shown in Figure 6.1. As we can see, the ^{1}H MAS NMR spectrum, prior to exposure to an alkane molecule, consists

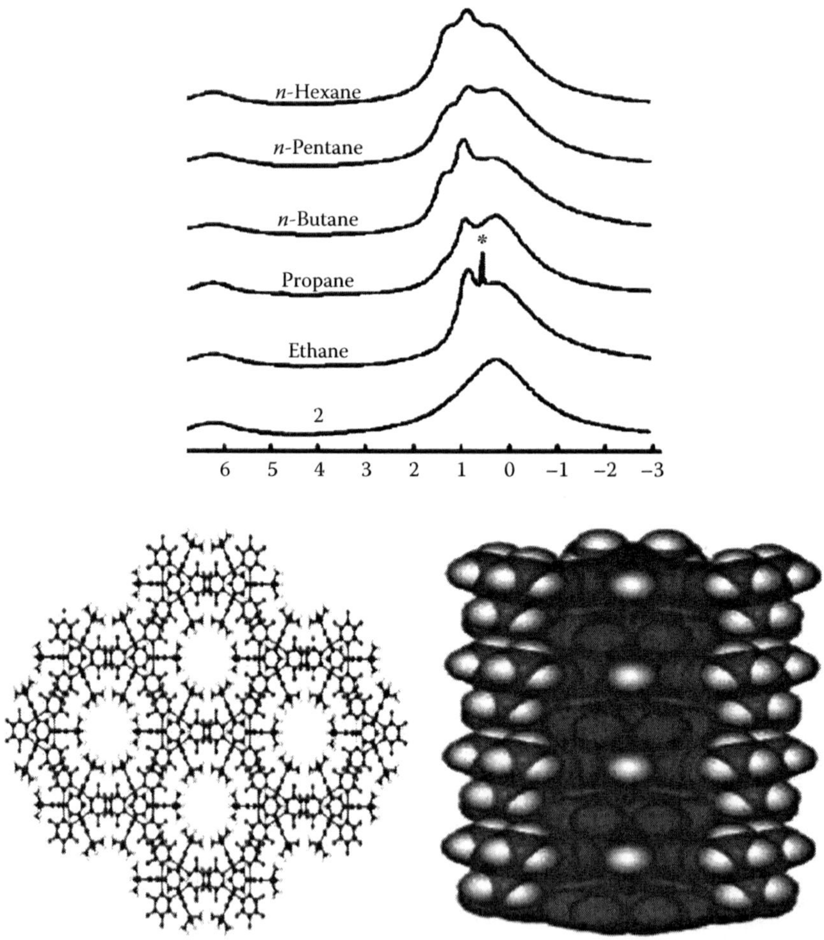

FIGURE 6.1 Bottom: the solid-state 3D network formed by the self-assembled two-component mercury-containing complex in Scheme 1.1. Top: the ^{1}H MAS NMR spectra of the system before and after alkane adsorption. The peak assigned to free ethane gas is indicated (*). (Reproduced with permission from Taylor, T.J. et al., *Angew. Chem. Int. Ed.*, 45, 7030, 2006.)

of two broad resonances assigned to the methyl groups (δ = 0.18 ppm) and the aromatic ring moiety (δ = 6.29 ppm) in accordance with the chemical formula in Scheme 1.1. After exposure to ethane, for example, an additional resonance appears at δ = 0.90 ppm that can be well attributed to adsorbed ethane molecules versus free ethane molecules, located in the rotor headspace, detected as a sharp resonance with δ = 0.59 ppm. It should be emphasized that the resonance due to the entrapped ethane molecules is only slightly shifted with respect to that of the free gas. It may be observed that the walls of the channels do not generate significant magnetic anisotropy within the internal pore space.

The ^{1}H MAS NMR spectrum after exposure to propane contains two new signals observed at δ of 1.15 and 0.75 ppm, which can be assigned to the methylene and methyl groups of the entrapped hydrocarbon, respectively. In the spectra of the samples exposed to *n*-butane, *n*-pentane, and *n*-hexane, the intensity of the methylene signal reasonably increases with respect to that of the methyl signal. For *n*-pentane and *n*-hexane, the two different types of methylene groups are not resolved, but they are detected as a single resonance.

Finally, again in the context of signal assignments, the following features play a very important role: (a) the ^{1}H MAS NMR resonances coming from the methyl groups (δ = 0.18 ppm) and the aromatic rings of the matrix are accompanied by an intense sideband pattern (in Figure 6.1 this effect is not demonstrated) due to strong dipolar coupling, while the gases show only isotropic resonances due to fast molecular reorientations in the pore space; (b) the circumstance can be used to observe the isotropic ^{2}H resonance even at a natural abundance.

Returning to the porosity, it should be noted that at the present time, there are many physicochemical methods for probing different porous media and to determine pore sizes and pore distributions though the volume of materials: gas adsorption, mercury porosimetry, neutron scattering, imbibition, and differential scanning calorimetry (DSC). Among them, NMR spectroscopy occupies a deserved place, providing significant (sometimes unique) contributions to this field.

Since all these methods have advantages and disadvantages, a range of experiments would be done using the different techniques to fully characterize porous media. Complementary experiments can give greater information on the pore structure and surface properties that is more detailed than that obtained by a single technique alone.

In many cases a good agreement can be reached within different methods applied for measuring porosity. The examples are shown in Figure 6.2 illustrating the pore size distributions obtained by three independent methods: NMR, gas adsorption, and DSC [2]. As can be seen, the solid and dashed curves corresponding to different methods demonstrate a satisfactory coincidence though certainly some deviations occur due to the specifics of each method. Nevertheless the agreement clearly proves the reliability of NMR measurements. It should be added that NMR can be applied for unusual samples with partially filled pores and/or for dual phase systems. This is a unique feature of NMR.

NMR, collecting the time- or frequency dependent data applied for porosity measurements, will be obviously focused on the behavior of mobile small molecules (in liquid or gas phase) that are situated within pores. *A priori* effects expected in measurements will be related to changes in physical properties of these molecules and their dynamics due to various surface interactions. Depending on the characteristics

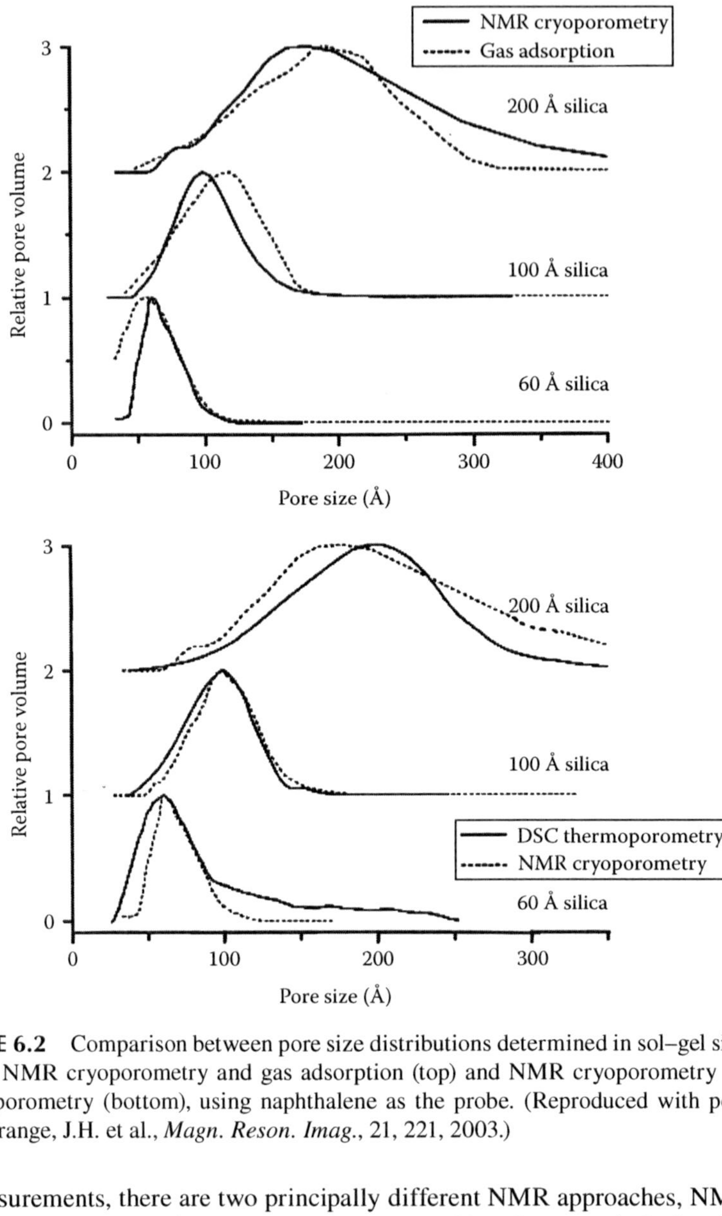

FIGURE 6.2 Comparison between pore size distributions determined in sol–gel silica samples by NMR cryoporometry and gas adsorption (top) and NMR cryoporometry and DSC thermoporometry (bottom), using naphthalene as the probe. (Reproduced with permission from Strange, J.H. et al., *Magn. Reson. Imag.*, 21, 221, 2003.)

of measurements, there are two principally different NMR approaches, NMR cryoporometry and NMR relaxometry, which are considered separately in Sections 6.1.1 and 6.1.2.

6.1.1 NMR Cryoporometry

NMR cryoporometry is a spectroscopic approach aimed at regular registration of relatively narrow NMR signals from a liquid phase situated in pore space. Ideologically this method is close to DSC thermoporometry, described by Ishikiriyama and

Todoki [3]. In fact, both of the methods are based on the so-called melting-point suppression effect, therefore generally leading to comparable pore size distributions as demonstrated in Figure 6.2. At the same time, NMR cryoporometry has remarkable advantages. First, the signal amplitude measured by NMR is directly proportional to the volume of a liquid located only in the *filled* pores. Second, NMR cryoporometry deals with static measurements, when the temperature of a sample does not change during the experiments. Thus no correction is needed for dynamic processes that occur in measurements by the DSC method.

Pore spaces and/or the pore size distributions in glasses, silica-based materials, wood and wood products, and other porous media [4] can be satisfactorily studied by the experiments on proton and deuterium nuclei. Both of them are the most spectroscopically convenient targets. Generally the ^{1}H and/or ^{2}H signal amplitudes are measured in the nonfrozen fraction of confined liquids, ^{2}H-enriched in the case of deuterium NMR, as a function of temperature. In such experiments variation in the temperature depends on the pore sizes. For example, the temperature varies in steps of ca. 0.1–1 K, when the size of pores corresponds to the mesoporous range [5].

Interpretation of the variable-temperature NMR measurements is commonly based on the well-known theory of Gibbs–Thomson, the elements of which are shown in Scheme 6.1.

The theory describes the melting-point suppression effect that can be written by Equation 6.1, where constants T_0 and T represent the melting points of the bulk and confined liquid, respectively. The intensity of an NMR signal, I, observed in the experiment as the temperature dependent parameter, is generally plotted as the pattern I versus X = 10^3/T (noted later as 1/T). Then, the I value is expressed by Equation 6.2, where N is the number of phases in a sample investigated, and magnitudes I_{0i}, X_{ci}, and σ_i express the intensity, the inverse transition temperature and the width of the temperature distribution curve of the phase, noted as i, respectively.

$$\Delta T = T_0 - T = T_0 - \frac{10^3}{X} = kp/R \tag{6.1}$$

$$I(X) = \sum_{i=1}^{N} \frac{I_{0i}}{2}\left[1 - \mathrm{erf}\{(X - X_{ci})/\sqrt{2}\sigma_i\}\right] \tag{6.2}$$

$$\mathrm{erf}\,(Z) = \frac{2}{\sqrt{\pi}} \int_0^z \exp\,(-u^2)du \tag{6.3}$$

$$\frac{dI}{dR} = \frac{10^3 k_p}{\sqrt{2\pi}(RT_0 - k_p)^2} \sum_{i=1}^{N} \frac{I_{0i}}{\sigma_i} \exp\left[-\left\{\frac{10^3 R - X_{ci}(RT_0 - k_p)}{\sqrt{2}\sigma_i(RT_0 - k_p)}\right\}\right]^2 \tag{6.4}$$

SCHEME 6.1 Elements of the theory of the Gibbs-Thomson applied for interpretations of NMR data.

The term erf(Z) in this equation represents the error function. Finally, differentiating Equations 6.1 and 6.2 relative to the R and the X parameters, respectively, leads to expression (6.4) that describes the pore-size distribution as a curve dI/dR versus R.

Aksnes and Kimtys [5] have demonstrated practical application of this theory by probing the porosity in some glasses and silica gels prepared with controlled pore sizes. Benzene and perdeuterated benzene molecules situated in pore spaces have been taken as NMR targets. Figure 6.3 shows the experimental curves, I versus 1/T, obtained by the variable-temperature ^{2}H NMR spectra of different silica gels and glass treated with C_6D_6. The dashed curves in the figure show the best fittings to Equation 6.2 illustrating, thus, an excellent agreement between the theory and the experimental data. It should be noted that errors in such measurements come from the well-known inaccuracy in signal integration.

One of the typical methodical problems appearing in NMR cryoporometry measurements performed for ^{1}H nuclei in liquids, for example, in porous silica-based materials, is connected with possible contributions to the detected intensity from absorbed water molecules and/or silanol groups in the silica surface. It has been recommended that the aforementioned effects can be negligible, when measurements performed at ^{1}H and ^{2}H frequencies in samples with the normal and deuterated probe molecules and their treatments give results that agree well [5].

The second problem can be connected with a possible simultaneous detection of the liquid and solid components in the recorded NMR spectra. It has been established that the components can be well distinguished by the spin–spin relaxation times T_2 measured by the spin-echo experiments [5].

Methodologically similar NMR investigations can be found in the work of Xie et al. [6]. Here the pore size and a pore wall thickness in materials of the FSM-16

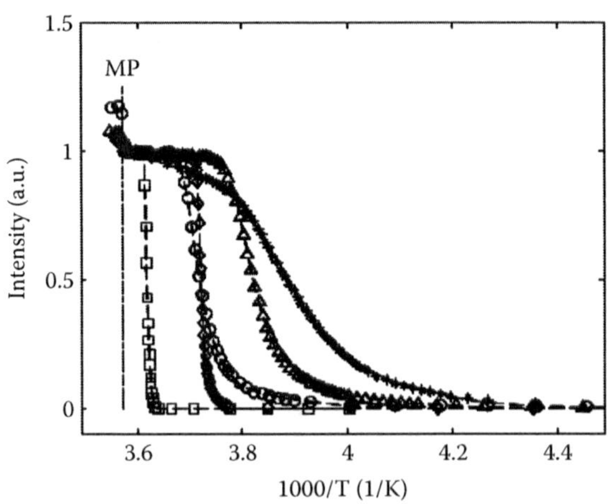

FIGURE 6.3 The curves "intensity of ^{2}H NMR signals (arbitrary units) versus temperature (1/T)" obtained for C_6D_6 molecules located in pore spaces of 4 nm silica gel (noted as *), 6 nm silica gel (noted as △), 10 nm silica gel (noted as ○), 7.9 nm glass (noted as ◇), and 23.9 nm glass (noted as □). (Reproduced with permission from Aksnes, D.W. and Kimtys, L., *Solid State Nucl. Magn. Reson.*, 25, 146, 2004.)

family have been determined by ^{1}H NMR measurements performed for water-saturated samples. The authors have emphasized that the cryoporometry NMR results have demonstrated a good agreement with parameters obtained by N_2 adsorption isotherms and transmission electron micrographs. This is again adequate proof of the reliability of NMR, applied in this context.

6.1.2 NMR Relaxometry

Since target objects in studies of porosity are liquids, the regular NMR relaxation measurements and/or the various pulsed field-gradient methods generally applied in solutions can be successfully used at probing porous media [4]. The main idea used in such NMR studies is that a porous structure should notably reduce the molecular motions and diffusion of the pore-filling fluids.

The theoretical background of this phenomenon is shown in Scheme 6.2 [7]. In the framework of this theory, the dynamic behavior of the pore-filling fluid in porous media is characterized by a reduced diffusion constant, D, relative to that measured in bulk fluid. In fact, this is reasonable because the molecules situated closely to the matrix surface have a smaller free space and thus the net displacement is reduced. Then, this effect is directly proportional to the surface (S)-to-volume (V) ratio (SVR) in the porous materials, and corresponds to a reduced diffusion coefficient, expressed by Equation 6.5. Here, D_0 characterizes the bulk diffusion coefficient, the S parameter expresses the total surface area, the V_P magnitude is the total pore volume, and D(t) represents the diffusion constant at the diffusion time, t. Finally, A is a constant that depends on dimensionality.

It has been established that this theory works well, when the distance $(D_0t)^{1/2}$ is greatly smaller than the pore size. Under this condition, the diffusion coefficient is defined as $\langle \Delta X(t)^2\rangle/2t$, where the X(t) parameter is taken as the net displacement along one direction at time t, and the $\Delta X(t)^2$ represents the square displacement.

$$\frac{D(t)}{D_0} \approx 1 - A\sqrt{D_0 t}\,\frac{S}{V_P} \tag{6.5}$$

$$\Delta r\!: D(t) = \langle \Delta r(t)^2\rangle/6t \tag{6.6}$$

$$\frac{D(t \to \infty)}{D_0} = \frac{\Phi}{T} \tag{6.7}$$

$$\begin{aligned}\langle \Delta r^2\rangle &= \iint |r_1 - r_2|^2 P(r_1, r_2, t)\,dr_1, dr_2\\ &= \frac{1}{V_P}\iint |r_1 - r_2|^2\,dr_1, dr_2\\ &= 2(\langle r^2\rangle - \langle r\rangle^2),\end{aligned} \tag{6.8}$$

$$\langle \Delta r^2\rangle = 3d^2/5 \tag{6.9}$$

SCHEME 6.2 The theory describing the dynamic behavior of the pore-filling fluid in porous systems.

For an isotropic molecular diffusion in 3D spaces, the D(t) coefficient can be defined from the vector displacement in Equation 6.6. When the $(D_0t)^{1/2}$ distance is larger than the pore size, then the second effect appears to be expressed via the long time diffusion. Under these conditions, reducing the long time diffusion coefficient relative to D_0 will be connected with factor T in expression (6.7). This factor represents a geometrical parameter determined by the pore structure and the Φ parameter is the porosity.

If the system investigated can be classified as a closed system, the long time displacement will be limited by the size of the enclosing structure. Then at the limit of long diffusion time, or in other words, when the initial and final positions are not correlated, the mean square displacement can be expressed as shown in Equation 6.8. In this equation, the symbol P represents the diffusion propagator, $P = 1/V_P$ at long time and the V_P symbol corresponds to the pore volume. Finally, it should be added that in the case of a sphere characterized by diameter d, the (Δr^2) factor can be given by Equation 6.9. Thus, since the diffusion displacement in NMR experiments is measured along one direction, this displacement is defined as $(\Delta r^2)/3$.

In practice, diffusion measurements in porous media can be performed by various NMR techniques using one- (1D) and/or two-dimensional (2D) radio frequency pulse sequences. The task of these experiments is determination of spin–spin relaxation times T_2. Among different pulse sequences, the 1D CPMG pulse sequence can be accepted as the most popular pulse train resulting in the T_2 magnitudes. At CPMG applications, the echo signal, M, is assumed to be a sum of multiple exponential decays with a distribution of the times T_2, noted as function $F(T_2)$:

$$M(\tau_2) = \int e^{-\tau_2/T_2} F(T_2) dT_2 \tag{6.10}$$

It has been emphasized that the $F(T_2)$ function can be described by the signal amplitude $M(\tau_2)$ measured at different times (τ_2).

Typically, 2D NMR experiments are aimed at the creation of a 2D matrix with two independently varied times: τ_2 and τ_1. The pulse sequences and time sections applied in the sequences are shown in Figure 6.4, where all the radio frequency pulses should be well calibrated. As we can see, these sequences are based on application of the pulsed-field gradients.

The details of mathematical treatments of the collected NMR data to determine T_2 times can be found in the work of Song [7]. The author also demonstrates a large number of diffusion measurements performed in different porous media and the corresponding results.

Limitations in molecular motions in pore spaces of investigated media can affect both the relaxation processes, T_1 and T_2, as expected from general relaxation theory. Potentially, these effects of enhancing nuclear relaxation rates for molecules located at a pore surface can allow calculations of the pore diameter by *direct* T_1 and T_2 time measurements. Generally, a fast exchange is assumed between molecules at the

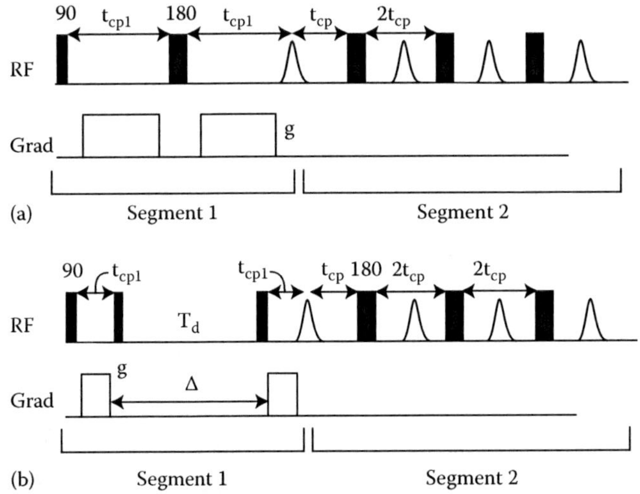

FIGURE 6.4 2D Diffusion–relaxation correlation sequences, based on applying the pulsed-field gradients. (a) The first segment is a *spin-echo*, when the echo signal appears at a time $2t_{cpl}$ after the first pulse. (b) The first segment is a *stimulated echo*, which appears at a time t_{cpl} after the third pulse. The detection segment for both measurements is a standard CPMG pulse train. (Reproduced with permission from Song, Y.Q., *Prog. Nucl. Magn. Reson. Spectrosc.*, 55, 324, 2009.)

surface and in pores [2]. Then the relaxation rates, $1/T_1$ and $1/T_2$, will be proportional to the S–V ratio in the porous media according to Equation 6.11:

$$\frac{1}{T_1} = \frac{\rho_1 S}{V}$$
$$\frac{1}{T_2} = \frac{\rho_2 S}{V} \tag{6.11}$$

Then, fitting a distribution of relaxation times to the experimental T_1/T_2 relaxation data will result in estimations of a final pore size distribution. This approach is simple but very sensitive to signal-to-noise ratios in the collected NMR spectra. As a result, these measurements can lead to distributions that seem to be very broadened compared to real ones. It is obvious that under this circumstance, such distributions are difficult to compare to pore size distributions obtained by other independent methods. Moreover, the relaxivity constants are also liable to changing the pore surface composition.

6.1.3 NMR Behavior of Gases in Pore Spaces

A special place in NMR porosity measurements belongs to probing the nuclear relaxation of gaseous molecules in pore spaces of porous media [8–10].

The conditions of collecting NMR data and their interpretations depend significantly on the nature of the absorbed gas molecules and relaxation mechanisms operating in samples. For example, Stupic et al. [9] have investigated quadrupolar relaxation of hyperpolarized ^{83}Kr nuclei to apply them as probes located close to surfaces of porous polymeric and glass systems. In fact, it has been shown that the ^{83}Kr T_1 times (measured in practice between 2 and 7 s) depend strongly on the chemical composition of the surfaces in the vicinity of the gas, and also on chemical treatments of the glass surfaces.

A very intriguing idea has recently been suggested to apply the imaging NMR experiments for porosity determinations in porous media focusing on behavior of C_2F_6 gas [11]. Again, the ^{19}F relaxation time of C_2F_6 gas situated in pores is remarkably longer than that of bulk gas. At the same time the time increases on decreasing the pore sizes.

The strategy in the experiments on gas molecules and interpretation of data collected for the porosity measurements can be considered here by the example where ^{19}F relaxation has been carefully investigated in fumed porous silica samples treated with CF_4 gas [8].

Small molecules in *nonviscous* media are well known to experience fast rotations leading to nuclear spin–lattice relaxation via the spin-rotation (SR) mechanism. It is remarkable that electrons in the molecules create magnetic fields even in the absence of the external magnetic field. Normally, the fast rotational molecular reorientations result in the spin–rotation relaxation rate $1/T_1(SR)$ expressed via

$$\frac{1}{T_1(SR)} = \frac{I_r^2 C^2}{9\hbar^2 \tau_c} \tag{6.12}$$

where

I_r is the molecular inertia moment
C is the spin-rotational constant, measured in frequency units
τ_c is the molecular motion correlation time

It should be emphasized that this relatively simple relationship is valid only for spherical molecules. Following from Equation 6.12, effectiveness of the SR mechanism depends on the C values that are significant for heavy nuclei, for example, ^{19}F or ^{31}P nuclei. In the case of ^{19}F nuclei, the C constants can reach up to 2000 Hz. Finally Equation 6.12 demonstrates a unique feature of the spin-rotational mechanism: $T_1(SR)$ times decrease with increasing the temperature. This effect is a good test for the presence of SR relaxation.

Going back to gases situated in pore spaces of media, for example, CF_4 gas in silica materials, the nuclear relaxation becomes more complex. In fact, the NMR ^{19}F spin–lattice relaxation times (T_1) are dominated by spin–rotation interactions that are mediated by the molecular collision frequency. Under these conditions, due the presence of pores, additional collisions of molecules with the pore walls significantly affect the T_1 times. This complexity obviously requires adequate mathematical models for treatment of relaxation data. In spite of their variety, the elements of a basic mathematical model can be represented in Scheme 6.3.

$$T_{1,b} = \frac{1}{aT}\left(\frac{f_{g,b}}{b_g} + \frac{\omega^2}{(f_{g,b}/b_g) + c\omega^2\sqrt{T}}\right) \tag{6.13}$$

$$T_{1,s} = \frac{1}{aT}\left(\frac{f_{g,b}}{b_g} + \frac{f_{w,s}}{b_{w,s}} + \frac{\omega^2}{(f_{g,b}/b_g)+(f_{w,s}/b_{w,s})+c\omega^2\sqrt{T}}\right) \tag{6.14}$$

$$f_{w,s} = \frac{v_{g,b}S}{4V} \tag{6.15}$$

SCHEME 6.3 A basic mathematical model used for interpretations of T_1 NMR relaxation dominated by spin-rotation mechanism where T_1 times mediated by collisions.

In accordance with the classical Bloembergen's relaxation theory, applied for fluorinated gases, the relationship between spin–rotation interaction and molecular collision frequency provides calculations of the ^{19}F T_1 times for a bulk gas (noted as $T_{1,b}$), for given pressure and temperature from the curve-fitting data collected with variations in temperatures and pressures. In turn, to calculate the collision frequency of gas molecules in bulk, $f_{g,b}$, one can use Clausius's state equation, the Lennard–Jones collision cross-sections, and the term showing the molecular velocity that increases slightly with density. If the molecular collision frequency is known, then the bulk gas $T_{1,b}$ is expressed via Equation 6.13 in Scheme 6.3. Here "a" is the coupling constant, "ω" is Larmor's frequency, "T" represents absolute temperature, and "c" corresponds to thermally mediated intramolecular relaxation mechanisms in the absence of collisions. The factor "b_g" represents the average number of the collisions necessary for spin–rotation exchange. This exchange depends on the temperature as $b_g = 1/(s\Omega + d)$, where "s" and "d" are the slope and intercept relating the cross-section, written for change in molecular angular momentum to the Lennard–Jones cross-section for collisions X.

Kuethe et al. [8] emphasizes that a simple extension of this expression, applied for collisions of gas molecules with pore walls, produces Equation 6.14, where "$f_{w,s}$" is the frequency of collisions with the wall and the "$b_{w,s}$" symbolizes the average number of wall collisions needed for spin–rotation interactions. Following from simple gas dynamics, the average wall collision frequency for molecules located in a pore space can be expressed via Equation 6.15. Here S is surface area, the "$\upsilon_{g,b}$" is the average velocity, V is the pore volume and "$f_{w,s}$" is the factor proportional to the S–V ratio.

In practice, pressure and temperature are parameters to adjust the "$f_{w,s}$" factor. This basic model, after minor additions, has been applied for fumed silica samples with the already known S–V ratios to show experimentally how the ^{19}F T_1 times change with confinement. Then the authors [8] measured the T_1 times for CF_4 gas in these samples by varying pressures and used developed mathematical models reflecting the change in T_1 values to finally fit the experimental data. It is interesting that the CF_4 gas density in pores is greater than that of the bulk gas. Thus it has been demonstrated that this approach represents a relatively rapid and precise way to measure S–V ratios from the increasing T_1 due to confinement in pores.

6.2 STRUCTURAL FEATURES OF SILICA LATTICE AND SURFACE BY SOLID-STATE ^{1}H ^{29}SI, ^{27}AL, AND ^{17}O NMR SPECTRA

Generally in probing the matrix of silica-based materials, the main focus is on identification of different structural units forming the matrix, and determinations their ratios. Simonutti et al. [12] have described multinuclear MAS NMR experiments performed on mesoporous silica materials of the MCM-41 family prepared in the presence of cetyltrimethylammonium chloride as a surfactant. This work can be used as a good example to demonstrate in practice the strategy in such studies. Figure 6.5 shows the ^{29}Si {^{1}H} MAS NMR spectra recorded by a single pulse sequence for noncalcined and calcined MCM-41 systems with very long relaxation delays (up to 100 s) to ensure complete relaxation of the silicon nuclei. Because ^{29}Si–^{29}Si dipolar coupling is relatively weak, the MAS NMR spectra *did not show spinning sideband patterns*. The spectrum of the noncalcined material exhibits two intense peaks at −99 and −109 ppm, which are generally attributed to group (Si–O)$_3$*Si*–OH, named as Q^3, and group *Si*(Si–O)$_4$ named as Q^4. A weaker peak at −89 ppm is from geminal silanol groups (Si–O)$_2$*Si*–(OH)$_2$, named usually as Q^2.

On calcination, the resonances significantly broaden, partially overlapping, and the Q^4 signal becomes more intense. The Q^3 and Q^2 species are present on the surface of the silica matrix and the intensity ratio (Q^3 + Q^2)/Q^4 is associated with the surface area. Then a decrease in this ratio, as observed on calcination, corresponds to partial condensation of the silanolic groups and shrinkage of the channels. In addition, the line-broadening effect in the calcined material can also be related to the shrinkage due to a distortion of the siloxanic tetrahedrons and an inhomogeneous broadening. Finally, quantitative information can be obtained by deconvolution performed for the signals in the aforementioned spectra. It should be added, however, that the results

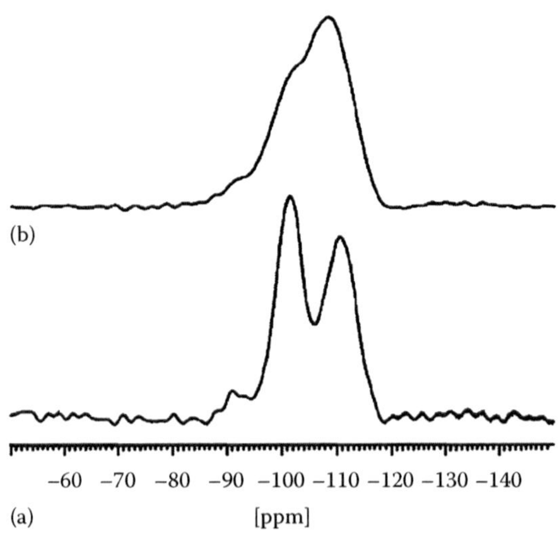

FIGURE 6.5 The single pulse ^{29}Si {^{1}H} MAS NMR spectra of noncalcined (a) and calcined (b) material MCM-41. (Reproduced with permission from Bakhmutov, V.I., *Chem. Rev.*, 111, 530, 2011.)

of such deconvolution procedures, as shown in previous chapters, will depend on the line shapes applied at treatments.

When silica-based materials are very amorphous and the ^{29}Si NMR signals can be completely overlapped, the simple one-pulse ^{29}Si {^1H} MAS NMR spectra are not sufficient. Recently Davis et al. [13] have suggested a technique for analysis of Si(Q) distributions based on 2D magic angle flipping experiments applied for potassium disilicates enriched by ^{29}Si. It has been emphasized that regular 1D ^{29}Si NMR spectra were completely unresolved. Figure 6.6 shows the pulse sequence applied in

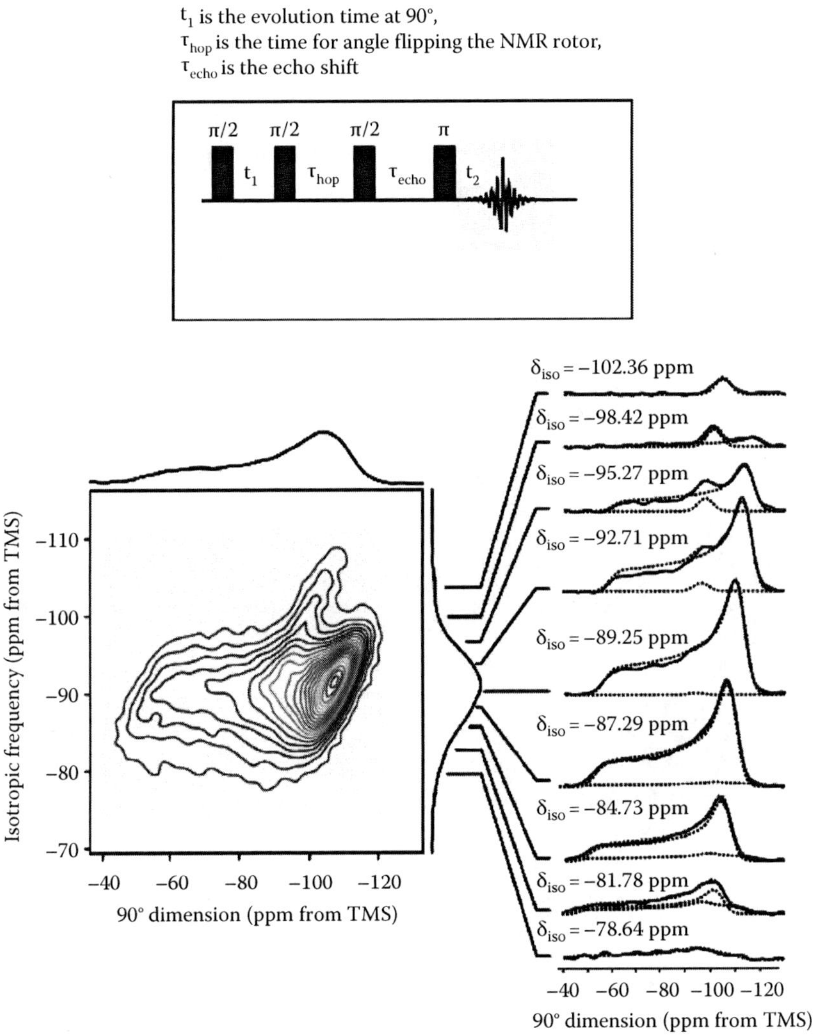

FIGURE 6.6 The pulse sequence applied for collection of the 2D ^{29}Si MAS NMR spectrum in a sample of the $K_2O \cdot 2SiO_2$ glass. The right part in the spectrum shows the selected experimental cross-sections, presented as solid lines. The dashed lines show the corresponding fittings. (Reproduced with permission from Davis, M.A. et al., *J. Phys. Chem. A*, 114, 5503, 2010.)

this work and the 2D contour plot of the ^{29}Si magic angle flipping spectrum recorded for an enriched $K_2O \cdot 2SiO_2$ glass sample. From the data, by using differences in ^{29}Si anisotropic line shapes, which are characteristic for each Q(n) units, one can obtain accurate and quantitative Q(n) populations. Disadvantage of the technique is related to enriching procedures.

A similar problem of low resolution can appear in the ^{29}Si MAS NMR spectra of aluminosilicates and zeolites containing Al atoms, where line assignments and/or deconvolution procedures can be very problematic because the resonances of ^{29}Si nuclei in the (3Si, 1Al) and (3Si,OH) structural units often overlap. Moreover it is difficult to distinguish one group from another. The latter is particularly important because the Si[3Si, 1Al] units are generally associated with the acidic SiOHAl groups, while the Si[3Si, 1OH] units correspond to nonacidic silanol groups in zeolite defects.

Luo and coworkers [14] have shown how to apply the triple-resonance techniques based on the $^{1}H/^{27}Al/^{29}Si$ TRAPDOR–CP and $^{1}H/^{27}Al$ TRAPDOR pulse sequences. For example, the different peaks observed in the ^{29}Si CP NMR spectra of materials MCM can be distinguished and assigned by this triple-resonance NMR approach. Combination of the $^{1}H/^{27}Al$ double-resonance NMR experiments with the $^{1}H/^{27}Al/^{29}Si$ triple-resonance spectra finally reveals the presence of two different acidic sites.

Since oxygen is one of the key chemical elements forming the matrix of materials, the solid-state ^{17}O MAS NMR spectra can be used as an excellent tool to probe and characterize the different oxygen bridges [15]. However, due to poor NMR sensitivity at the low natural abundance of the ^{17}O isotope (0.037%) the applications of the NMR technique are not simple and require ^{17}O-enriched samples. Examples of such applications are shown in the following. The ^{17}O MAS NMR spectra of several polydimethylsiloxane (PDMS) oxide nanocomposites prepared from $Me_2Si(OEt)_2$ (dimethyldiethoxysilane) and metal alkoxides $M(OR)_n$, where M = Ti, Zr, Nb, and Ta are presented in Figure 6.7. The spectra show clearly the presence of the self-condensed Si–O–Si (0–100 ppm) and M–O–M (>250 ppm) bonds. Moreover, several ^{17}O NMR signals are clearly observed for these M–O–M species, which correspond to the multiple coordination modes of oxygen atoms with transition metal elements. Finally, in the intermediate chemical shift range, intense resonances can be observed for co-condensed Si–O–M bridges. This feature proves that good interconnection within the oxide network is attained.

On the quantitative level, the structure of the matrix can be accurately solved by a combination of solid-state NMR with powder x-ray diffraction (XRD) data when XRD is quite informative. This approach has been recently applied for the determination of zeolite structures, where the important step of structure solution includes the solid-state ^{29}Si double-quantum dipolar recoupling NMR experiments, sensitive to the distance dependent dipolar interactions between naturally abundant ^{29}Si nuclei in the zeolite framework [16].

As we see, solid-state NMR potentially provides the precise characterization of amorphous porous materials, where the key experiments are based on applications of high-resolution MAS techniques. In practice, however, structural conclusions can be strongly dependent on spectral assignments, when they play a major role.

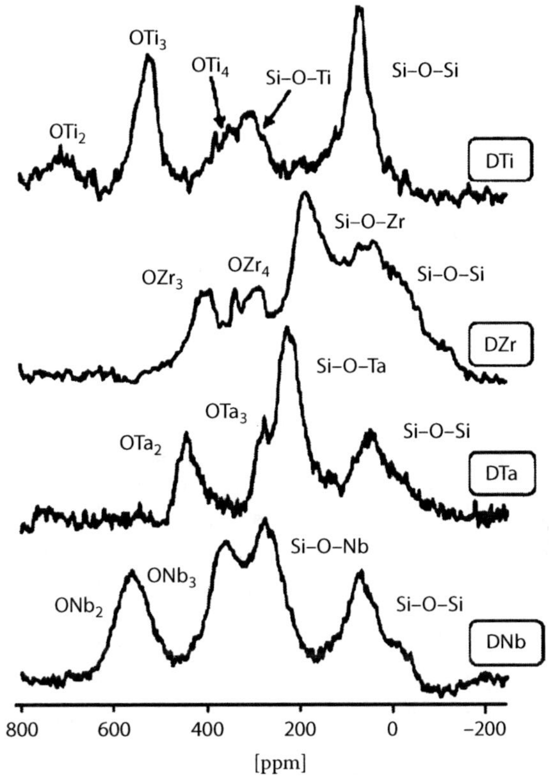

FIGURE 6.7 The ^{17}O isotropic chemical shifts obtained for different M–^{17}O–M oxo bridges, where M = Si, Ti, Zr, Nb, and Ta. (Reproduced with permission from Bonhomme, C. et al., *Acc. Chem. Res.*, 40, 738, 2007.)

In this connection, Bonhomme has emphasized that the first-principles calculations of NMR parameters can help to assign the experimentally observed resonances [15].

The quantum mechanical calculations of isotropic chemical shifts, magnetic shielding anisotropy, or quadrupolar interaction are based on detailed descriptions of all the electrons and nuclei in the presence of external magnetic field B_0. Generally they can be realized by using available program packages. However these many-body problems are not simple and require some approximations. One of the most reliable approaches is density functional theory (DFT) considering the energy as a function of the total charge density. Generally DFT calculations of real objects imply their approximation by modeled clusters that are terminated with target nuclei such as ^1H, for example. Besides the DFT, there are other theoretical approaches. For example, Pickard and Mauri have suggested a plane wave pseudopotential method [15]. This method has shown excellent agreement between the calculated and experimental NMR parameters obtained for numerous silica polymorphs.

Properties of silica-based (and other) materials and sphere of their applications strongly depend on two factors: the nature of their matrix and structural features of surface spaces. The factor of surface can include, for example, three important aspects.

First, hard conditions in preparation of the materials (high-temperature calcinations) can significantly affect their surface and these effects would be controlled. Second, the silica-based materials are often prepared in the presence of surfactant molecules. Therefore their proximity to surface is of great interest. Third, the surface of silica can be modified chemically, for example, by immobilized ligands. It is obvious that the reactions used for these treatments would be analyzed by a convenient method.

Acidity or protons situated in the silica surface can be well investigated by the highly resolved ^1H NMR spectra in MAS experiments [12]. These spectra are sufficient to distinguish the different OH groups located in the surface: the *isolated* Si–OH groups show ^1H signals at 1.3–2.2 ppm, whereas resonances from the *hydrogen-bonded* SiOH groups are shifted to 3.9–4.0 ppm. Generally these signals are broadened due to a distribution of the hydrogen bond strength. Resonances of acidic protons are shifted toward lower fields.

Detailed characterizations of the surface are constantly a focus of researchers working in the field of catalysis, where different ligands and linkers can be immobilized on the surface. A good example of solid-state NMR applications in this context is given by Blumel [17], describing different bifunctional phosphine linkers used for immobilizing the metal complexes. Phosphine parts in these linkers coordinate with metal centers, while their alkoxysilane groups are binding to the silica matrix. Since the number of alkoxy groups per silane unit can be one, two, or three, its influence on the binding performance is of significant interest. In other words, the reactions of silanes with silica surfaces should be controlled. Blumel shows that this control can be reached by ^{13}C {^1H} CP MAS NMR. For example, the NMR spectra of silica, treated with Me_3SiOEt, show clearly that the ethoxy group is retained on the silica surface quantitatively. However, with more than one ethoxy group, the reaction can become more complex. Figure 6.8 illustrates the ^{29}Si CP/MAS NMR spectrum obtained for silica with an immobilized vinyltriethoxysilane. The main unexpected feature of this spectrum is a 1:2:1 ^{29}Si pattern, observed between −60 and −80 ppm, which corresponds to the different numbers of ethoxy groups. The same patterns are

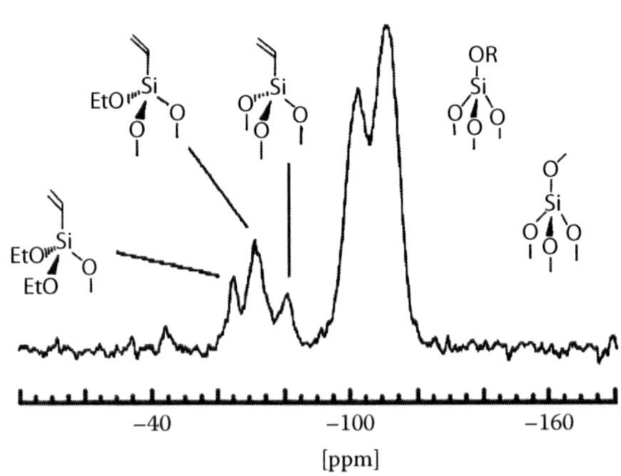

FIGURE 6.8 The ^{29}Si CP {^1H} MAS spectrum recorded for a silica sample with immobilized vinyltriethoxysilane. (Reproduced with permission from Blumel, J., *Coord. Chem. Rev.*, 252, 2410, 2008.)

generally detected for phosphine or other linkers containing triethoxysilane groups after their immobilization on silica. It is obvious that without the ^{29}Si and ^{13}C CP/MAS NMR spectra, the amount of ethoxy groups is unknown, and thus, the nature of the surface coverage cannot be determined by other methods than solid-state NMR.

In addition, the ^{29}Si CP/MAS signals of the silica matrix, observed between −103 and −112 ppm, can give valuable information about dryness of the silica surface after modification. Methodically it can be done by measuring these signal intensities at variations in the contact time [17].

Solid-state NMR is a powerful and unique tool to probe the proximity of macromolecular chains to silica or silicate surfaces in materials synthesized in the presence of surfactant molecules. While the surfactant molecules can be directly observed in ^{13}C CP MAS NMR spectra [12], their proximity to silica surface can be determined by ^{1}H–^{29}Si cross-polarization kinetics. In fact, the transfer of magnetization from the surfactant protons to the silicon nuclei in the nonprotonated silicon atoms can be significant, when these molecules are situated on the pore surface. Figure 6.9 demonstrates the ^{1}H–^{29}Si cross-polarization kinetics for noncalcined and calcined mesoporous silica materials of the MCM-41 family. The curves can be treated by fittings to give the cross-polarization time constants for silicon units SiQ(3) and SiQ(4). In a noncalcined MCM-41 sample, these constants were calculated as 1.0 and 7.1 ms for SiQ(3) and SiQ(4) units, respectively. The very short cross-polarization time constant obtained for SiQ(3) species proves close locations of ^{1}H nuclei in surfactant molecules. In contrast, the SiQ(4) unit shows the greatly elongated time. Thus, the close contacts are absent and magnetization is transferred only from the protons of geminal silanol groups. It is notable that the cross-polarization kinetic curve changes quantitatively in going from the noncalcined sample to calcined MCM-41. Now the ^{1}H–^{29}Si cross-polarization constants are determined as 4.2 and 22.3 ms for units Si(Q3) and Si(Q4), respectively. However in spite of this, relatively, they exhibit the same tendency.

6.2.1 Distribution of Diamagnetic Metal Ions from Solid-State NMR

Solid-state NMR implies two approaches to localization of metal atoms in porous systems: direct and indirect. The former is preferable; the latter is simpler on the experimental level while both of them are limited by NMR sensitivity particularly when the absolute amount of metal ions is low. Indirect localization consists of observation matrix nuclei, for example, ^{29}Si nuclei in silica matrix. Generally this approach is based on comparison of the spectra recorded for metal ion free systems and systems containing the metal ions. As we know, the diapason of ^{29}Si isotropic chemical shifts in the silica matrix, including silicon species Q(2)–Q(4), is not great. Nevertheless it is often sufficient to detect metal–O–Si species. Figure 6.10 shows the ^{29}Si {^{1}H} single-pulse MAS NMR spectrum of a calcined SiO_2–Al_2O_3 sample with three strongly overlapped peaks centered at −92, −100, and −108 ppm, belonging to units SiQ(2), SiQ(3), and SiQ(4), respectively [18]. These resonances are much better resolved in the ^{29}Si CP MAS NMR spectrum obtained when a ^{1}H spin-lock section has been incorporated into the standard CP pulse sequence. The ^{29}Si {^{1}H} single-pulse MAS NMR spectrum of a SiO_2–Al_2O_3–ZnO material shows very broadened lines with an intensity maximum observed at −108 ppm. The resolution

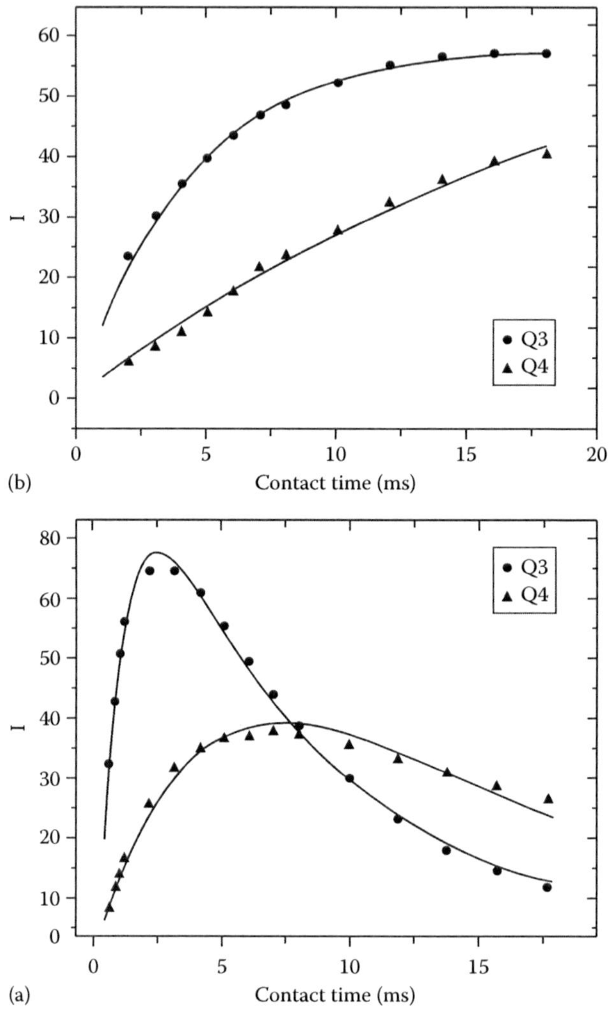

FIGURE 6.9 The room-temperature ^{1}H–^{29}Si cross-polarization kinetics measured as evolution of ^{29}Si magnetization at a variable contact time. The solid lines are nonlinear least-squares fittings. (a) Noncalcined MCM-41 and (b) calcined MCM-41. (Reproduced with permission from Simonutti, R. et al., *Chem. Mater.*, 13, 771, 2001.)

does not change in the CP MAS NMR experiment. Since the Al concentration in the Zn containing samples is very small, the peak can be well assigned to Si–O–Zn sites in the silica matrix.

Resonance frequencies of heavy nuclei, such as ^{27}Al, ^{51}V, $^{47,49}Ti$, and $^{117,119}Sn$, etc., strongly depend on atomic coordination numbers, oxidation states, and types of chemical bonding. Therefore direct observation of the nuclei and determination of their isotropic chemical shifts will result in a more reliable localization of the corresponding atoms inside or outside the matrix of materials.

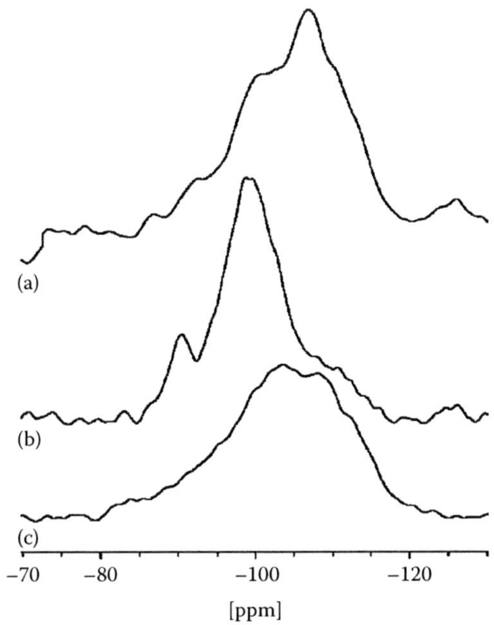

FIGURE 6.10 The ^{29}Si {^{1}H} solid-state MAS NMR spectrum of a calcined sample Al_2O_3–SiO_2 (5 mol% alumina) (a); the ^{29}Si CP spin-lock MAS NMR spectrum of the same sample (b); the ^{29}Si {^{1}H} MAS NMR spectrum of a calcined sample Al_2O_3–SiO_2–ZnO (5 mol% alumina, 20 mol% ZnO) (c). (Reproduced with permission from Shpeizer, B.G. et al., *Micropor. Mesopor. Mater.* 90, 81, 2006.)

For Al ions in silica-based materials, this key structural problem can be solved relatively easily. When the ^{27}Al {^{1}H} MAS NMR spectra show resonances with isotropic chemical shifts around 54–62 ppm, aluminum atoms are incorporated into the silica matrix as four-coordinated species [19]. The presence of lines between 0 and −6 ppm, points to extra-framework aluminum. More accurate and complete characterizations of different Al-moieties require a better spectral resolution by reducing the second order quadrupolar broadening. The effect can be reached by the 2D ^{27}Al 3Q-MAS approach [20], as applied for some titanium-substituted zeolites. Here the Al species represent four-, five-, and six-coordinated Al atoms clearly observed in the 2D ^{27}Al MAS NMR spectra in Figure 6.11, which can be quantitatively analyzed by deconvolution. Moreover, as it follows this figure, reducing the second order quadrupolar broadening helps to observe two different tetrahedral and octahedral Al environments. In fact, these atoms cannot be detected by application of the regular one pulse sequence. It should be emphasized that ^{27}Al nuclei are very convenient targets for the aforementioned studies even at low Al concentrations due to the 100% natural abundance and short spin–lattice relaxation times, caused by quadrupolar interactions.

In contrast to the fast-relaxing ^{27}Al nuclei, $^{117,119}Sn$-T_1 relaxation times in diamagnetic solids are very long and the natural $^{117,119}Sn$ abundance is relatively low. These circumstances obviously complicate collection of ^{119}Sn NMR data particularly

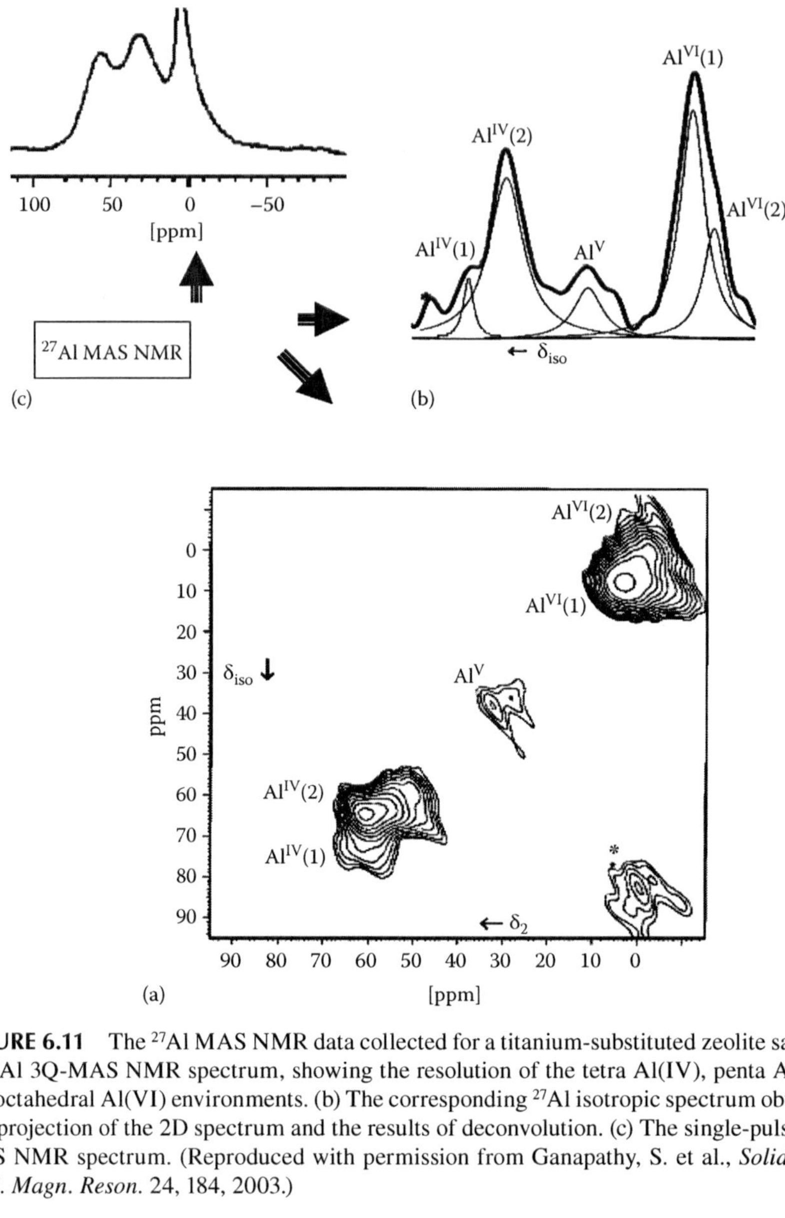

FIGURE 6.11 The ^{27}Al MAS NMR data collected for a titanium-substituted zeolite sample: (a) ^{27}Al 3Q-MAS NMR spectrum, showing the resolution of the tetra Al(IV), penta Al(IV), and octahedral Al(VI) environments. (b) The corresponding ^{27}Al isotropic spectrum obtained as a projection of the 2D spectrum and the results of deconvolution. (c) The single-pulse ^{27}Al MAS NMR spectrum. (Reproduced with permission from Ganapathy, S. et al., *Solid State Nucl. Magn. Reson.* 24, 184, 2003.)

at low tin concentrations. Such materials can be conventionally characterized by experiments with ^{119}Sn-*enriched* samples. For example, Renz et al. [21] have characterized catalytic sites in some tin-containing amorphous materials by directly measuring isotropic ^{119}Sn chemical shifts. It is well established that ^{119}Sn chemical shifts between −670 and −740 ppm can be attributed to framework Sn^{IV} sites with octahedral or tetrahedral coordination. For example, the ^{119}Sn resonance at $\delta = -688$ ppm, observed in the MAS NMR spectra of dehydrated zeolite Sn-MFI, can be assigned to octahedral Sn^{IV} units while the shoulder at −439 ppm can be attributed to framework tetrahedral tin sites. In addition, the authors have revealed that this shoulder moves to −579 ppm, due to the formation of *five-coordinate Sn species upon hydration*. Similarly, the ^{119}Sn MAS NMR spectrum of hydrated zeolite Sn-Beta-2 exhibits two signals at −690 and −740 ppm, which transform to a resonance at −445 ppm in the dehydrated sample. This signal was attributed to tetrahedral Sn^{IV} framework sites, where the coordination number increases by bonding to extra water molecules.

Figure 6.12 illustrates the series of the ^{119}Sn {^1H} MAS NMR spectra, recorded at the ^{119}Sn natural abundance for supermicroporous materials SiO_2–SnO_2, synthesized at different Sn contents [22].

As can be seen, the spectrum of the material with 14 wt% of tin exhibits a sideband pattern with an isotropic resonance observed at −600 ppm, typical of bulk SnO_2 with six-coordinated tin(IV) atoms. In addition to this signal, the spectrum of the material with 3 wt% of tin displays another line that is up-field shifted. Furthermore, the material with 2 wt% of tin exhibits only this up-field resonance (−697 ppm). This evolution of the NMR spectra was suggested to indicate that only ~2 wt% of tin atoms can be incorporated into the silica surface, likely forming catalytic centers. In full agreement with these assignments, the ^{119}Sn MAS NMR spectrum of the *dehydrated* material doped with 3 wt% of tin exhibits a low-field sideband pattern with an isotropic chemical shift of −460 ppm, corresponding to four-coordinated Sn sites situated in the silica matrix surface. Again, this transformation is reversible.

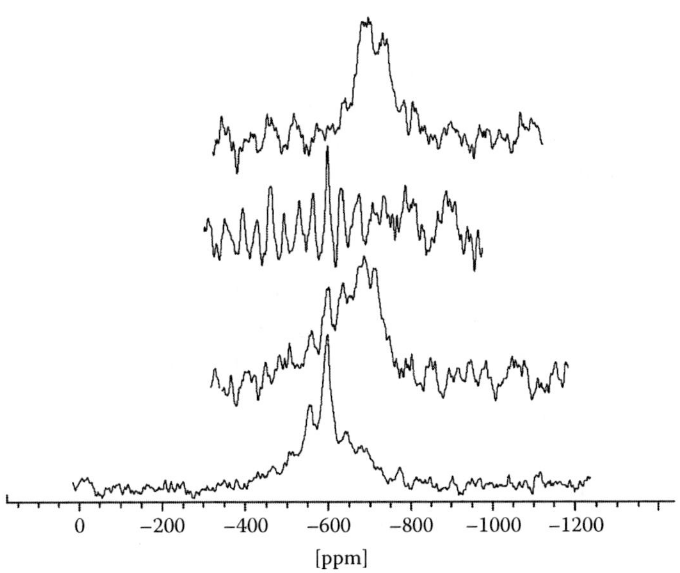

FIGURE 6.12 The ^{119}Sn Hahn-echo MAS NMR spectra (synchronized) recorded at a spinning rate of 6 kHz, from top to bottom: material SiO_2–SnO_2, containing 2 wt% of tin; dehydrated material SiO_2–SnO_2 containing 3 wt% of tin; the same material obtained in an air/water atmosphere; and material SiO_2–SnO_2, containing 14 wt% of tin. (Reproduced with permission from Bakhmutov, V.I., *Chem. Rev.*, 111, 530, 2011.)

Finally it should be noted again that such spectra require a large number of accumulations and a long experimental time.

Ganapathy et al. [20] have demonstrated applications of solid-state NMR techniques for studies of titanium-substituted zeolites. In contrast to the ^{27}Al nuclei, the ^{47,49}Ti nuclei show very strong quadrupolar coupling. Therefore the ^{47,49}Ti NMR experiments in *static* samples were preferable. The experimental and computer-simulated NMR spectra revealed a ^{49}Ti resonance with δ(iso) of −850 ppm and a quadrupole coupling constant of 5 MHz, which can be used as a good test for the octahedral Ti atoms that are incorporated into the matrix of zeolites. Other quadrupolar nuclei, such as ^{71}Ga [23], and ^{51}V [24,25], can be also used as targets. Different Ga- and V-containing materials have been investigated in static and spinning samples to determine the characteristic magnitudes of the quadrupole and shielding anisotropy tensors as a function of the nuclear environments.

6.3 MOLECULAR MOBILITY IN DIAMAGNETIC POROUS MATERIALS

The whole arsenal of methods of dynamic solid-state NMR, described in Chapter 5, can be applied for probing molecular mobility in diamagnetic porous systems. In practice, however, a line-shape analysis is more popular. Ashbrook [26] has emphasized that the dynamic effects are more evident for line-shapes of quadrupolar nuclei and first of all, deuterons. Therefore, quadrupolar target nuclei are more preferable in such studies than nuclei with spins of 1/2. When lines are broadened by the first-order quadrupolar interactions, then molecular motions affecting NMR are situated on a time scale of 100 ns–10 μs. In turn, when central-transition line-shapes are affected by the second-order quadrupolar broadenings, the molecular reorientations can be between 100 μs and 1 ms. In general, one can say that MAS NMR spectra are sensitive to molecular mobility on a timescale of 10–100 μs.

Studies of molecular mobility in porous materials are especially important in the context of understanding their catalytic properties and catalytic mechanisms operating on the surface. Among various systems potentially active in catalysis, mesoporous silica–based materials with different functional groups on surface and particularly their local dynamics are of significant interest. Shantz et al. [27] have illustrated a typical application of ^{2}H solid-state NMR techniques to probe mobility of functional groups immobilized on the surface in materials of the MCM-41 and SBA-15 families. Figure 6.13 shows the functionalized SBA-15 material and the static ^{2}H NMR spectra recorded in this sample at different temperatures. As clearly seen, the quadrupolar splitting in the powder pattern is small versus the static quadrupolar coupling constant (QCC) of 150–200 kHz. Just this result without calculations shows clearly the high mobility of the functional groups. The fitting procedures applied for the spectra gave QCC value of 19.3 kHz at both temperatures with an asymmetry parameter <0.05. The line shape can be well described by two types of rotations, shown in the Figure 6.13.

The second example is connected with nonmobile ligands on the surface. Figure 6.14 shows the room-temperature ^{2}H MAS NMR spectrum, recorded for a sample

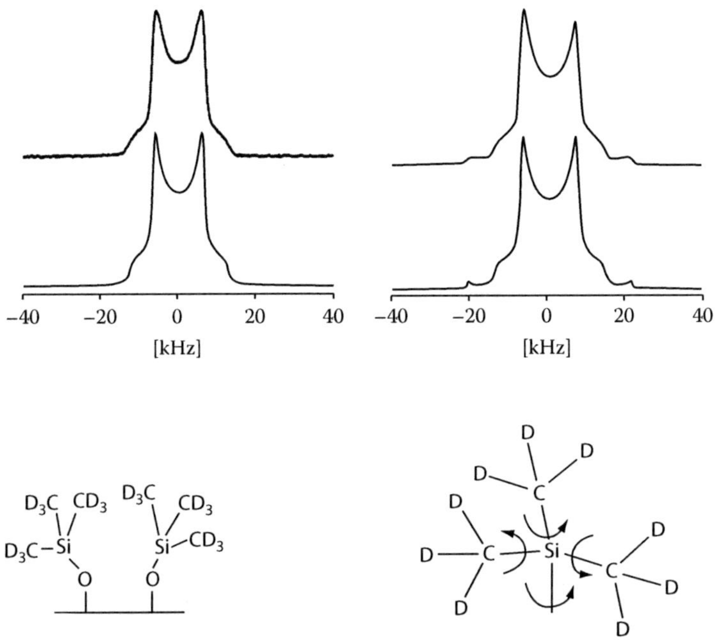

FIGURE 6.13 The ^2H NMR spectra recorded at 20°C (left) and −100°C (right) in a static sample of the functionalized SBA-15 material (top). The theoretical spectra (bottom) simulated on the basis of two rotational processes affecting the line shape. (Reproduced with permission from Wang, Q. et al., *J. Phys. Chem. C*, 113, 18142, 2009.)

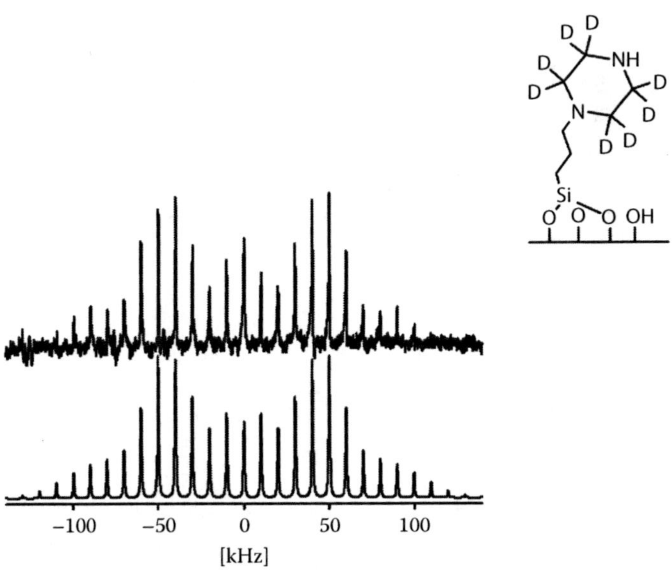

FIGURE 6.14 The ^2H MAS NMR spectrum of material 1-propylpiperazine- MCM-41 recorded at a spinning rate of 10 kHz (top) and the best fitting result (bottom). (Reproduced with permission from Wang, Q. et al., *J. Phys. Chem. C*, 113, 18142, 2009.)

of the material 1-propylpiperazine-MCM-41 at a spinning rate of 10kHz. The best fitting result is also represented. In contrast to the aforementioned case, the NMR spectrum corresponds to rigid deuterons. In fact, the QCC value is calculated as 160kHz with a zero asymmetry parameter that is very similar to the completely rigid deuterons, exhibiting QCC ≈ 170kHz. Thus the piperazine group on the surface is rigid on the microsecond time scale.

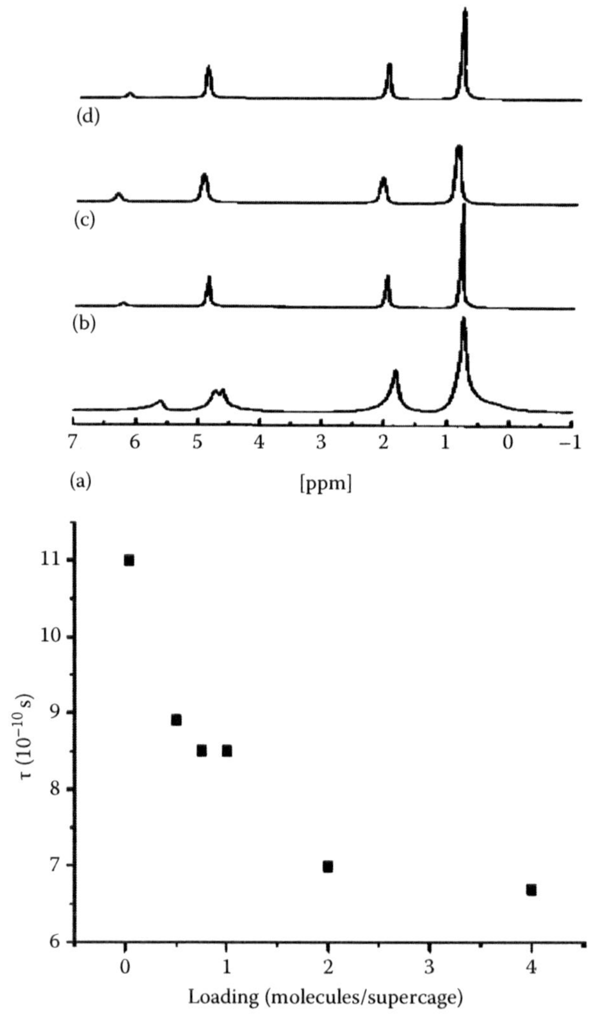

FIGURE 6.15 Top: the ^{1}H MAS NMR spectra of 1-butene adsorbed on the surface of zeolite NaX (b–d) compared with the spectrum obtained in the gas phase (a); (b) −2 butene molecules per supercage; (c) 0.5 molecules per supercage; (d) −0.1 molecules per supercage. Bottom: the dependence of the correlation time of the olefinic part of 1-butene absorbed on NaX on loadings. (Reproduced with permission from Bohlmann, W. et al., *Magn. Reson. Chem.*, 37, S126, 1999.)

To end this chapter, let us again dwell on the dynamic behavior of the olefin molecules, like 1-butene or 1-pentene, absorbed by zeolites and manifestation of the dynamics in solid-state NMR. The reorientation mobility of these olefins in zeolite NaX can be found in Bohlmann et al. [28]. The main feature of such systems is the molecular mobility of adsorbed small molecules is characterized by correlation times that are situated between the typical values measured in solids and liquids [29], and the correlation times, themselves, are very close to the inverse values of the resonance frequencies. Thus, the different NMR techniques can be applied for determination of the spectral density functions at different resonance frequencies. Then they can be applied for checking the models for correlation functions rather than the application of the single-correlation time models used generally in practice.

According to Bohlmann et al. [28], 1-butene, absorbed in zeolite NaX, exhibits the highly resolved ^{1}H MAS NMR spectra recorded at different pore fillings as shown in Figure 6.15. At the same time, the molecules in the gas phase show strongly broadened resonances. This circumstance provides applications of NMR technique generally used in liquids, for example, ^{1}H NOESY NMR technique and cross-relaxation time measurements. For example, the comparison of the different *homonuclear* and *heteronuclear* cross-relaxation times helps to describe in detail the motions of the guest molecules that, in contrast, cannot be described by a single correlation time. Butene mobility has been characterized by fast libration-type motions of the molecules and the overall reorientation–translation motions with correlation times that are comparable to the lifetimes of the molecules in a cavity with respect to their jumps into a neighboring cavity.

One of the important results in studies of the rotation–translation motions characterizing the adsorbed molecules in the large zeolite cavities (plying an important role in catalysis) is the fact that their correlation times change on decreasing the pore-filling factor. This dependence, illustrated in Figure 6.15, has been rationalized by increasing the interaction strength between the gas molecules and the adsorption sites in the NaX zeolites. It has been noted that the model corresponds to the adsorption effected by the cations that are situated on the different crystallographic sites in the zeolite lattice.

REFERENCES

1. Taylor, T. J., Bakhmutov, V. I., Gabbai, F. P. 2006. *Angew. Chem. Int. Ed.* 45: 7030.
2. Strange, J. H., Mitchell, J., Webber, J. B. W. 2003. *Magn. Reson. Imag.* 21: 221.
3. Ishikiriyama, K., Todoki, M. 1995. *Coll. Inter. Sci.* 171: 103.
4. Maunu, S. L. 2002. *Prog. Nucl. Magn. Reson. Spectrosc.* 40: 151.
5. Aksnes, D. W., Kimtys, L. 2004. *Solid State Nucl. Magn. Reson.* 25: 146.
6. Xie, X., Satozawa, M., Kunimori, K., Hayashi, S. 2000. *Microp. Mesopor. Mater.* 39: 25.
7. Song, Y. Q. 2009. *Prog. Nucl. Magn. Reson. Spectrosc.* 55: 324.
8. Kuethe, D. O., Montano, R., Pietra, T. 2007. *J. Magn. Reson.* 186: 243.
9. Stupic, K. F., Cleveland, Z. I., Pavlovskaya, G. E., Meersmann, T. 2006. *Solid State Nucl. Magn. Reson.* 29: 79.
10. Flaum, M., Hirasakia, G. J., Flaum, C., Straley, C. 2005. *Magn. Reson. Imag.* 23: 337.
11. Caprihan, A., Clewett, C. F. M., Kuethe, D. O., Fukushima, E., Glass, S. J. 2001. *Magn. Reson. Imag.* 19: 311.

12. Simonutti, R., Comotti, A., Bracco, S., Sozzani, P. 2001. *Chem. Mater.* 13: 771.
13. Davis, M. A., Kaseman, P. J., Parvani, D. C. et al. 2010. *J. Phys. Chem. A* 114: 5503.
14. Luo, Q., Yang, J., Hu, W., Zhang, M., Yue, Y., Ye, C., Deng, F. 2005. *Solid State Nucl. Magn. Reson.* 28: 9.
15. Bonhomme, C., Coelho, C., Baccile, N., Gervais, C., Azais, T., Babonneau, F. 2007. *Acc. Chem. Res.* 40: 738.
16. Brouwer, D. H., Darton, R. J., Morris, R. E., Levitt, M. H. 2005. *J. Am. Chem. Soc.* 127: 10365.
17. Blumel, J. 2008. *Coord. Chem. Rev.* 252: 2410.
18. Shpeizer, B. G., Bakhmutov, V. I., Clearfield, A. 2006. *Micropor. Mesopor. Mater.* 90: 81.
19. Duer, M. J. 2002. *Solid-State NMR Spectroscopy: Principles and Applications*. Oxford, U.K.: Blackwell Science.
20. Ganapathy, S., Gore, K. U., Amoureux, J. P. 2003. *Solid State Nucl. Magn. Reson.* 24: 184.
21. Renz, M., Blasco, T., Corma, A., Fornes, V., Jensen, R., Nemeth, L. 2002. *Chem. Eur. J.* 8: 4708.
22. Shpeizer, B. G., Bakhmoutov, V. I., Zhang, P., Prosvirin, A. V., Dunbar, K. R., Thommes, M., Clearfield, A. 2010. *Colloid. Surface. Physicochem. Eng. Aspect.* 357: 105.
23. Fredoueil, F., Massiot, D., Poojary, D. M., Bujoli-Doeuff, M., Clearfield, A., Bujoli, B. 1998. *Chem. Commun.* 175.
24. Skibsted, J., Jacobsen, C. J. H., Jakobsen, H. J. 1998. *Inorg. Chem.* 37: 3083.
25. Shubin, A. A., Lapina, O. B., Courcot, D. 2000. *Catal. Today* 56: 379.
26. Ashbrook, S. E. 2009. *Phys. Chem. Chem. Phys.* 11: 6892.
27. Wang, Q., Jordan, E., Shantz, D. F. 2009. *J. Phys. Chem. C* 113: 18142.
28. Bohlmann, W., Michel, D., Roland, J. 1999. *Magn. Reson. Chem.* 37: S126.
29. Bakhmutov, V. I. 2011. *Chem. Rev.* 111: 530–562.

7 Paramagnetic Effects in Solid-State NMR

Theoretically, separating the solids into two classes, diamagnetic and paramagnetic, is obvious. However, in practice, it is well known that the synthesis of materials, involving a number of various organic and inorganic reagents, generally leads to systems still containing paramagnetic impurities of a different nature. When concentrations of such impurities are small, they do not significantly affect NMR parameters: isotropic chemical shifts, magnetic shielding anisotropy, and line shapes. The presence of the impurities can be detected only on the level of nuclear relaxation, occurring often via a specific relaxation mechanism, spin diffusion (see the following). We accept such systems as diamagnetic ones. At the same time, when the concentrations of paramagnetic centers become ≥0.5–1 wt%, such solids can show a specific NMR behavior and we accept these systems as paramagnetic.

Magnetic moment of electron is larger by 662 times than that of proton. Therefore, unpaired electrons in paramagnetic centers give rise to strong local magnetic fields, which can mask original NMR spectroscopic properties of objects investigated, such as isotropic and anisotropic chemical shift values, line-shapes of resonances, and their T_1, T_2 relaxation times [1]. Thus, generally, the presence of paramagnetic centers in solids (molecular oxygen, different paramagnetic impurities, paramagnetic metal cations, etc.) is not desirable particularly in accurate structural studies of materials by NMR or in probing their dynamic behavior.

On the other hand, as we will show later, the influence of paramagnetic centers will depend on their concentrations and electron relaxation times. Therefore, their optimal combinations can lead to a number of "positive" effects even in regular NMR experiments. These effects help to significantly shorten relaxation delays in long-term solid-state NMR experiments and increase NMR sensitivity by optimizing ^{1}H T_1 relaxation [2,3]. For example, Wickramasinghe et al. [2] have demonstrated a simple approach to ^{13}C high-resolution solid-state NMR of biomolecules in microcrystals by reducing the ^{1}H T_1 relaxation times in the presence of paramagnetic relaxation reagents. The authors have shown that ^{1}H T_1 times shorten from 0.4–0.8 s to 60–70 ms, for example, in a sample of lysozyme in the presence of 10 mM of paramagnetic complex Cu(II)Na_2EDTA. The latter does not change the spectral resolution in ^{13}C CP MAS NMR. At the same time, it results in faster signal accumulation giving total sensitivity enhancements by factors of 1.4–2.9. This condition reduces the experimental time for a given signal-to-noise ratio by a factor of 2.0–8.4. The sensitivity can also be increased dynamic nuclear polarization [4,5] or by applications of contrast media in the magnetic resonance imaging experiments [6].

Sometimes, however, the concentration of paramagnetic centers is significant and development of new NMR approaches and new strategies is needed for structural characterizations of such solids. For example, the ^{13}C MAS NMR experiments, including very fast magic angle spinning (≥20 kHz) and *recoupling-based polarization transfer*, are recently reported by Wickramasinghe and Ishii [7]. It has been shown that these experiments, performed in paramagnetic complexes, such as Cu(DL-Ala)$_2$, distinguish groups ^{13}CH, $^{13}CH_2$, $^{13}CH_3$, and $^{13}CO_2$ in the one-dimensional (1D) NMR spectra, simplifying the spectral assignments, or even enabling determination of distances between nonbonded C–H pairs by quantitative analysis of cross-peak intensities in the two-dimensional (2D) NMR spectra.

The strong dipolar interactions between target nuclei and the electronic magnetic moments produce the chemical shift anisotropies discussed in the next section, which can be significantly larger than those shown in Chapter 1. The anisotropy can be even larger than the radio frequency amplitudes, giving rise to technical problems in NMR connected with signal registrations.

Kervern et al. have successfully solved this problem by applications of *short high-power adiabatic RF pulses* for population inversion in paramagnetic complexes, containing Fe(II), Yb(III), and Tb(III) [8], where target nuclei generally show very large chemical shift anisotropies. Magnetic shielding anisotropy can be commonly around 500–1000 ppm, as a function of paramagnetic ions. Such anisotropy magnitudes lead to a technical problem for broadband spin excitation and inversion. For example, the authors have emphasized that to obtain a proton band of 1000 ppm (or of 500 kHz at the magnetic field of 11.7 T), a square pulse of length **t** should be equal to **625 ns (90°)!**

Figure 7.1 illustrates the principle in application of the adiabatic pulses that can be used for systems with very large shift anisotropies under very high spinning rates. Here, in the reference frame, rotating synchronously with the instantaneous RF, the RF field is defined as

$$\omega_1(t) = \gamma B_1(t) \tag{7.1}$$

and the resonance offset is written as

$$\Delta\omega(t) = \gamma \Delta B_1(t) \tag{7.2}$$

At the instantaneous frequency offset between the shift and the irradiation frequency they combine into an effective field

$$\omega_{eff}(t) = \gamma B_{eff}(t) \tag{7.3}$$

which is inclined at angle Θ relatively to the X coordinate. In addition, the authors have shown how these pulses can be used in double-echo refocusing and TEDOR experiments for paramagnetic molecular systems of iron, ytterbium, and terbium, where shift anisotropies can be up to 1200 ppm.

The application of very fast spinning, a wideband sensitivity enhancement, and simultaneous adiabatic spin-locking cross polarization have been demonstrated by Peng and coauthors in NMR studies of some Cu(II) paramagnetic complexes [9]. However, these methods, being rather specific, cannot be applied for strongly paramagnetic and amorphous materials. These materials are considered in detail in the next sections.

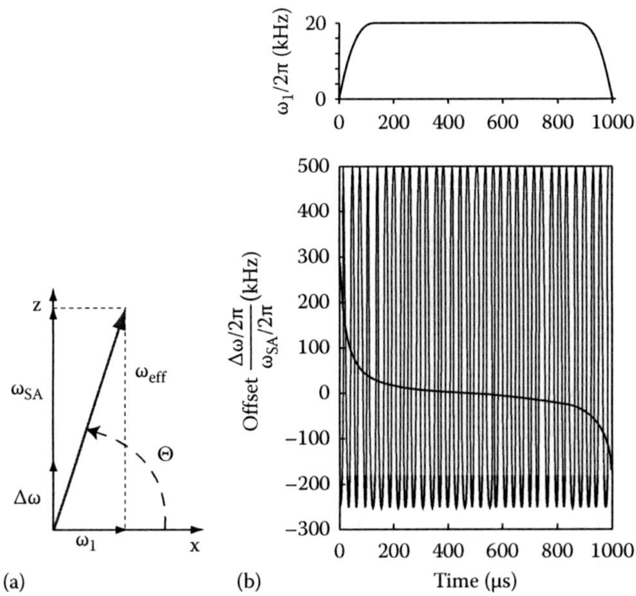

FIGURE 7.1 (a) The effective field vector ω_{eff} and components of ω_{eff} in a frequency modulated frame. (b) Amplitude (top) and frequency (bottom) modulations for a standard radio frequency pulse and short high-power pulses. (Reproduced with permission from Kervern, G. et al., *Chem. Phys. Lett.*, 435, 157, 2007.)

7.1 THEORETICAL ASPECTS

7.1.1 Chemical Shifts and Magnetic Shielding Tensors in the Presence of Unpaired Electrons

A very high magnetic moment of an unpaired electron, located on a paramagnetic center, affects a resonance frequency of target nuclei via effective magnetic field B_{EF}:

$$B_{EF} = \frac{2\pi A \mu_B g B_0 S(S+1)}{3\gamma_N kT} + \frac{\mu_0 \mu_B^2 g^2 B_0 S(S+1)(3\cos^2\theta - 1)}{3kTr^3 + B_{BMS}} + BMS \tag{7.4}$$

where

- A is the isotropic electron–nucleus hyperfine coupling constant [10]
- μ_B is the Bohr magneton, measured as $9.2741 \times 10^{-24}\, J \times T^{-1}$
- r is the electron–nucleus distance
- θ is the angle between the electron–nucleus vector and the external magnetic field B_0
- g is the electron g factor

The first term in Equation 7.4 produces the so-called Fermi-contact (FC) chemical shifts, rationalized by McConnell and Chesnut [11], which is caused by *direct delocalization* of the unpaired electron density on the nuclei. Due to this delocalization, the Fermi-contact shifts are dependent on hyperfine coupling constant "A," independent (directly) of electron–nucleus distance and can reach 3000 ppm or even higher as a function of nuclei and paramagnetic centers. The magnetic shielding tensors in paramagnetic solids are often axially symmetric. They can have very large anisotropies covering ranges up to 3500–4000 ppm. It is obvious that this situation takes place when atoms containing target nuclei chemically bind to paramagnetic ions. Again the aforementioned magnitudes create technical difficulties in detection of such nuclei.

In practice this common information can be illustrated by chemical shifts caused by Fermi contact interaction, recently obtained experimentally for ^{31}P and ^{7}Li nuclei in some paramagnetic phosphates shown in Tables 7.1 and 7.2 [12]. These data clearly demonstrate significant influence of the nature of nuclei and show the experimentally measured line widths giving an idea what widths can be expected in the presence of the FC interaction.

TABLE 7.1
^{31}P Chemical Shifts Measured in Paramagnetic Phosphates and Line Widths ($\Delta\nu$) Determined as Full Width at Half Maximum (Temperature 320 K)

Paramagnetic Compound	^{31}P δ(iso) ppm	^{31}P ($\Delta\nu$) ppm
Heterosite $FePO_4$	5,770	1,217
Rhomb $Li_3Fe_2(PO_4)_3$	14,350	1,839
Strengite	15,800	1,078
Phosphosiderite	16,680	742
$LiFeP_2O_7$	14,398	2,147
	8,230	1,460

Source: Kim, J. et al., *J. Am. Chem. Soc.*, 132, 16825, 2010.

TABLE 7.2
^{7}Li Chemical Shifts (Referred to an LiCl Solution) Measured in Paramagnetic Phosphates and Line Widths ($\Delta\nu$) Determined as Full Width at Half Maximum (Temperature 320 K)

Paramagnetic Compound	^{7}Li δ(iso) ppm	^{7}P ($\Delta\nu$) ppm
Rhomb $Li_3Fe_2(PO_4)_3$	164	1768
$LiFeP_2O_7$	162	2146
Mono $Li_3Fe_2(PO_4)_3$	190	2571
	90	
	40	

Source: Kim, J. et al., *J. Am. Chem. Soc.*, 132, 16825, 2010.

It should be emphasized that there is a fundamental difference between the FC shifts, observed in nonmetallic molecular complexes, and the so-called *Knight shifts*, detected in metallic systems. The Knight shifts, for example, for ^{27}Al or ^{29}Si nuclei in systems URu_2Si_2 and UNiAl, can be found in the works of Bernal et al. [13] and Nowak and Troc [14], respectively. General expression for Knight (or orbital) shifts is shown in the equation in Figure 7.5, where the magnetic field is applied along the z direction. The negative term in this equation is diamagnetic contribution and the positive term is paramagnetic part, corresponding to a contribution to the magnetic susceptibility of ions [15]. According to Abragam [16], in contrast to nonmetallic systems, the conducting electrons in metals are not localized, and thus each nuclear spin "sees" magnetic field from all the electrons.

The field, which produces the so-called pseudocontact chemical shifts (PCS), is not noted in Equation 7.4. In fact, such shifts can be observed in paramagnetic solids, containing *anisotropic* unpaired electrons (e.g., electrons in metal ions Dy^{3+}, Tb^{3+}, or Fe^{3+}). The values of pseudocontact shifts [17] are given by Equation 7.5:

$$\delta(\text{PCS}) = \left(\frac{1}{12\pi}\right) r^{-3} \left\{\Delta\chi_{\text{AX}}(3\cos^2\theta - 1) + \left(\frac{3}{2}\right)\Delta\chi_{\text{RH}} \sin^2\theta \cos 2\phi\right\} \tag{7.5}$$

where

r is the distance between the paramagnetic center and the nucleus

θ and ϕ are the angles describing the location of the nuclear spin relative to the principal axes (where the paramagnetic center is at the origin) of the magnetic susceptibility tensor χ with the axial (AX) and rhombic (RH) components, respectively

As the equation shows, this term is distance dependent.

The second field, created in the presence of an unpaired electron, corresponds to dipolar electron–nucleus interaction. Equation 7.4 shows that this interaction strongly depends on distance r and is angle dependent, as $3\cos^2\theta - 1$, where θ is defined as an angle formed by the r vector and the external magnetic field [16]. In powder solids, due to the absence of fast high-amplitude molecular reorientations, the spectral effect of this field is obvious: a resonance of target nuclei undergoes strong broadening in a static sample similarly to dipole–dipole internuclear interactions. This dipolar electron–nucleus term is very high and can appear even in solutions as residual dipolar couplings when paramagnetic molecules undergo partial alignment in the magnetic field [18].

The last term, B_{BMS}, represents a demagnetization field that appears in a sample due to bulk magnetic susceptibility (BMS). In general, this effect is insignificant for diamagnetic solids but it can play a very important role in paramagnetic systems. Nevertheless, in practice, this effect is often ignored and precisely for this reason it should be discussed with greater attention. Kubo et al. [19] have investigated BMS effects experimentally and theoretically by calculations on a model basis with random orientations and distributions of crystallites in powder.

Kubo has considered a polycrystalline sample containing paramagnetic ions as shown in Figure 7.2, where crystallites take spherical shapes. Since the NMR

$$\hat{\delta}(r) = \hat{\chi}(S_0) - \sum_i \hat{v}(r, S_i)\hat{\chi}(S_i) \tag{7.6}$$

$$\hat{\delta}_{con} = \frac{2}{3}\hat{\chi}(S_0) \tag{7.7}$$

$$\hat{\delta}_{dip} = (1/4\pi)\sum_{k \in Si} \{3r_k^t r_k / r_k^5 - 1/r_k^3\}\,\hat{\chi}_k^{\ ion} \tag{7.8}$$

$$\hat{\delta}_{BMS}^{polycryst}(r) = (1/4\pi)\sum_i p_j[3(r_j - r)({}^t r_j - {}^t r)/|r_j - r|^5 - 1/|r_j - r|^3]\hat{\chi}^j_v V_{cryst} \tag{7.9}$$

$$\hat{\delta}_{BMS}^{container}(r) = \chi_V^{iso,sample}\{1/3 - \hat{v}(r, S_{container})\} \tag{7.10}$$

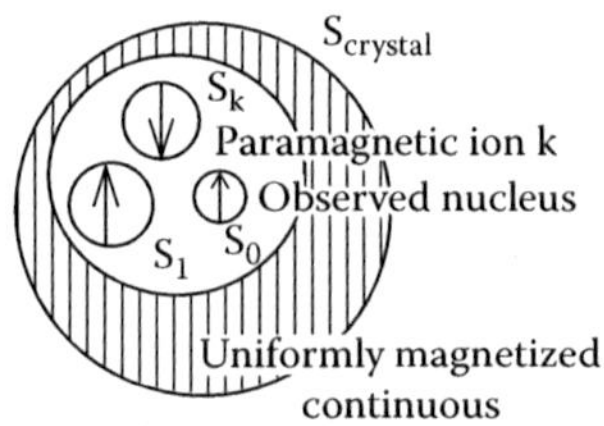

FIGURE 7.2 Kubo's model of bulk magnetic susceptibility in a polycrystalline sample.

resonance frequency at position r inside a sample is proportional to the magnetic induction, a chemical shift tensor can be expressed in Equation 7.6. In turn, the Fermi contact term δ(CON) and paramagnetic dipolar term δ(DIP) are written in the forms of Equations 7.7 and 7.8, respectively. The $\hat{\chi}_k^{ion}$ term in Equation 7.8 is the susceptibility tensor of the k-th paramagnetic ion located in sphere S_1, and r_k is the vector from the observed nuclei to the k-th paramagnetic ion. Since the Fermi-contact and dipolar terms depend on molecular and crystal structures and the rest is a BMS shift tensor, the latter due to the polycrystalline effect can be expressed by Equation 7.9, where r_j corresponds to the position of the center of the j-th crystallite, $\hat{\chi}_V^j$ is the volume susceptibility of the j-th crystallite, and V_{cryst} is the volume of the spherical crystallite, while P_j is a random variable parameter. This parameter takes value of 0 or 1. The effect of a container surface is considered by Kubo as the deviation from a spherical container to give finally the expression (7.10).

A careful analysis of this model has revealed that the BMS shifts can be very large, leading to BMS broadenings in static samples, and can be even dependent on the *shape* of NMR containers. For example, according to the calculations, the demagnetization field is uniform at all the points on the axis of a cylinder container, situated at magic angle with respect to the external magnetic field, but it changes strongly toward walls and ends of the cylinder [19,20]. The theory agrees well with the ^1H and ^{13}C NMR experiments performed for paramagnetic Ln, Pr, and Yb complexes.

Again, since dipolar coupling and BMS are orientation dependent, both of them produce strong broadenings of resonance lines in static paramagnetic samples. In turn, by analogy with dipole–dipole internuclear interactions or chemical shift anisotropy effects, these broadenings transform to intense sideband patterns in the MAS NMR spectra [21], when spinning rates are comparable or larger than line widths in static samples. At the same time, it is difficult to estimate, *a priori*, relative contributions of both effects to these sideband patterns. It is important to emphasize that the presence of paramagnetic centers, itself, can cause significant BMS contributions. As will be shown later, even mechanical mixtures of solids investigated with paramagnetic salts show pronounced BMS effects, when the concentrations of salts are large.

Finally, unpaired electrons can strongly shorten nuclear spin–lattice and spin–spin relaxation times, T_1 and T_2. The latter is particularly remarkable for nuclei undergoing significant Fermi-contact interactions. As a result, the line broadenings can be so strong that the resonances can become even spectrally "invisible."

7.2 NUCLEAR RELAXATION IN THE PRESENCE OF UNPAIRED ELECTRONS

As noted in Chapters 1 and 5, nuclei in diamagnetic solids can relax by different mechanisms. For example, Equations 5.5 and 5.6 show the nuclear relaxation due to dipolar interaction between two identical nuclei. Again we emphasize that this mechanism is particularly powerful at strong dipolar coupling, $\gamma^4\hbar^2r^{-6}I(I + 1)$, short internuclear distances r, and molecular motional frequencies ($1/\tau$) close to the resonance frequency ω_I [22]. In other words, if molecular motions in solids are quite intense, nuclear relaxation can be relatively fast as in liquids.

As emphasized earlier, reorientations of molecules or functional groups in solids are complex and often cannot be characterized by a single molecular correlation time. Generally, various correlation time distributions, Cole–Cole and Davidson–Cole, for example, or a Havriliak–Negami model [23] are used for quantitative interpretations of relaxation data. These and other reasons can lead to the situation when nuclear relaxation becomes multiexponential or nonexponential, particularly in the absence of the spin-diffusion mechanism. In such cases, the relaxation process can formally follow a stretched exponential law: $\exp(-(\tau/T_1)^\beta)$. Tse and Hartmann [24] have shown that β of 0.5–0.6 is typical of powder samples.

Generally the analysis of multiexponential nuclear relaxation, particularly in heterogeneous samples, is not simple due to not only solid-like interactions but also restricted and anomalous molecular diffusion. The detailed theory and algorithms of the treatments of relaxation data can be found in Lamanna [25]. Lamanna has shown that in contrast to liquids, the Laplace transform, where the NMR decays are associated with a continuous distribution of relaxation constants, is not suitable. The main idea consists of the description of multiexponential NMR decays (Figure 7.3) by a sum of the Fredholm integrals, each of which takes into account the correct relaxation or diffusion mechanism in a particular phase. Here, the parameter α may represent different physical properties, such as the pore size or the amount of bound water in a system, or the size of diffusing particles, whereas the shape of W function is determined by the relaxation and diffusion mechanism. Finally, in this expression, the shape of $P(\alpha)$ and $W(\alpha)$ and f depend on features of the physical system that is investigated. As seen, this most correct approach is

$$s(t) = \sum_{k=1}^{K} \int_{\alpha_{min}}^{\alpha_{max}} P_k(\alpha_k)\, f_k(t, W_k(\alpha_k))\, d\alpha_k$$

FIGURE 7.3 The NMR decay can be described by a sum of Fredholm integrals, where each one takes into account the correct relaxation or diffusion mechanism in a particular phase.

difficult to realize. Below we show more practical approaches that can be successful when nuclear relaxation is non-exponential.

7.2.1 Spin Diffusion

In rigid solids, where molecular motions are strongly restricted, nuclear relaxation should be extremely slow. For example, ^{29}Si T_1 relaxation times in silica-based materials and glasses can be more than 100 s or several minutes, when paramagnetic impurities are absent or minimal. In their presence, however, even at small concentrations, the spin–lattice and spin–spin relaxation rates ($1/T_1$ and $1/T_2$) can increase due to spin diffusion, suggested by Bloembergen [26]. Due to mutual flips of spins, shown schematically in Figure 7.4, belonging to dipolar-coupled nuclei and leading to energy-conserving spin transitions, a target spin is "transferred" to a paramagnetic center, where it relaxes very rapidly by electron–nuclear dipolar interaction (see the following). Under these conditions, the observed relaxation T_1^{SD} times are controlled by spin-diffusion coefficients D:

$$\frac{1}{T_1^{SD}} = \left(\frac{1}{3}\right) 8\pi N p C^{1/4} D^{3/4}$$

$$C = \left(\frac{2}{5}\right) \gamma_I^2 \gamma_S^2 \hbar^2 S(S+1) \tag{7.11}$$

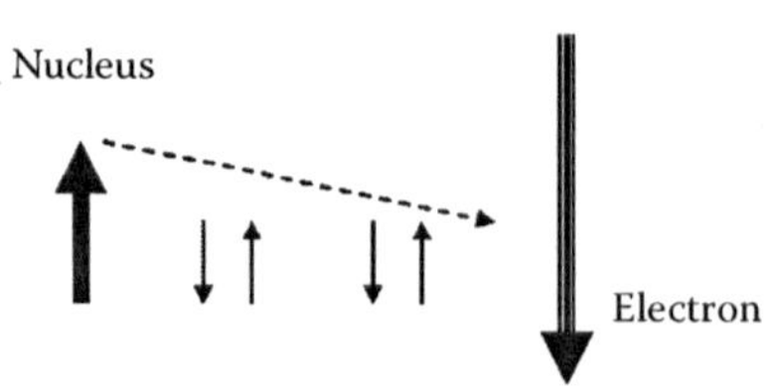

FIGURE 7.4 Schematic representation of spin diffusion, suggested by Bloembergen. Due to mutual flips of spins, or energy-conserving spin transitions, a target spin is "transferred" to a paramagnetic center. Then, the spin undergoes the fast relaxation process via electron–nucleus dipolar interactions. It is important to emphasize that this mechanism phenomenologically takes place even in liquids, however, with a ZERO effectiveness due fast molecular motions.

These coefficients are defined as

$$D = \left(\frac{M^{1/2}}{30}\right)a^2 \tag{7.12}$$

where
Np is the paramagnetic center density
M is a second moment of the dipolar internuclear interaction
a is the internuclear distance

Equation 7.11 shows that effectiveness of the spin-diffusion mechanism is dictated by internuclear dipolar coupling and depends on values γ and the natural abundance of target nuclei. That is why the spin-diffusion mechanism is practically negligible for NMR relaxation of ^{13}C nuclei in polymer systems [23] or for the ^{29}Si NMR relaxation in kaolinite, investigated by Hayashi et al. [27]. In contrast, this spin-diffusion mechanism can dominate in the ^1H or ^{27}Al relaxation. For example, the spin-diffusion coefficients D for protons in talk are measured as 1.4×10^{-13}, 1.7×10^{-13}, 1.0×10^{-13}, and $6.6 \times 10^{-14}\,cm^2/s$ for a sample, spinning at 0, 4, 8, and 12 kHz, respectively [27].

Brown [28] summarizes applications of spin-diffusion NMR experiments and a quantitative analysis of ^1H spin diffusion based on different approaches. It has been emphasized that coefficients D are typically determined around 0.6–0.8 nm^2/ms for protons in different polymers. Chen and Schmidt-Rohr [29] has reported the D value of 0.25 nm^2/ms for polypropylene and Hayashi and Omi [30] have measured the D constants between 9.7×10^{-9} and $1.5 \times 10^{-6}\,m^2/s$ for acidic protons as a function of X in solids $[(NH_4)_{1-x}Rb_x]_3H(SO_4)_2$ $(0 < x < 1)$. The NMR experiments have been performed in the rotating frame.

Finally, as a function of the natures of solids, there are two cases to be distinguished: *rapid spin diffusion* and *diffusion-limited* relaxation. Phenomenologically, nuclear relaxation via fast spin-diffusion (T_1^{SD}) is always *exponential*, whereas a diffusion-limited relaxation is a *nonexponential* process, which can be treated by a stretched exponential function at $0.5 < \beta < 1$, as mentioned earlier.

7.2.2 Nuclear Relaxation via Nucleus–Electron Dipolar and Contact Interactions

Fluctuating magnetic fields created by a large magnetic moment of an unpaired electron, $\mu_S = -g_e\beta_eS$, can affect nuclear relaxation via direct dipolar and/or contact interactions between target nuclei and electrons in paramagnetic centers. The dipolar mechanism, controlled by reorientations of electron–nucleus vectors r, has been formulated by Solomon [31] and can be written for S = 1/2 as follows:

$$\frac{1}{T_1^{DD}} = 0.1\gamma_I^2\gamma_S^2\hbar^2r^{-6}\left\{\frac{\tau_e}{(1+(\omega_I-\omega_S)^2\tau_e^2)}+\frac{\tau_e}{(1+\omega_I^2\tau_e^2)}+\frac{\tau_e}{(1+(\omega_I+\omega_S)^2\tau_e^2)}\right\} \tag{7.13}$$

$$\frac{1}{T_2^{DD}} = \left(\frac{1}{20}\right)\gamma_I^2\gamma_S^2\hbar^2 r^{-6}$$
$$\times\left\{4\tau_e + \frac{\tau_e}{(1+(\omega_I-\omega_S)^2\tau_e^2)} + \frac{3\tau_e}{(1+\omega_I^2\tau_e^2)} + \frac{6\tau_e}{(1+\omega_S^2\tau_e^2)} + \frac{6\tau_e}{(1+(\omega_I+\omega_S)^2\tau_e^2)}\right\} \tag{7.14}$$

In contrast to liquids, where this mechanism also operates, τ_e is an electron relaxation time, if molecular reorientations in solids are slow. Different electron relaxation times can cause different NMR effects. Since the nuclei situated closer to paramagnetic centers should relax much faster (see the r^{-6} factor), the dipolar $T_{1,2}$ times can be rationalized in structural terms.

According to Bloembergen [32], the strong Fermi-contact electron–nucleus coupling can lead to situation when electron-spin flips are directly accompanied by nuclear-spin flips. This mechanism, arising due to delocalization of the unpaired spin density in a nucleus, is given by

$$\frac{1}{T_1^{CON}} = \left(\frac{2}{3}\right)S(S+1)\left(\frac{A}{\hbar}\right)^2\left(\frac{\tau_{E2}}{(1+\omega_I^2\tau_{E2}^2)}\right) \tag{7.15}$$

$$\frac{1}{T_2^{CON}} = \left(\frac{1}{3}\right)S(S+1)\left(\frac{A}{\hbar}\right)^2\left(\tau_{E1}+\left(\frac{\tau_{E2}}{(1+\omega_I^2\tau_{E2}^2)}\right)\right) \tag{7.16}$$

where τ_{E1} and τ_{E2} are the longitudinal and transverse electron spin relaxation times, respectively. Again, the nuclear relaxation times are remarkably dependent on electron relaxation times. In general, the Fermi-contact mechanism can dominate at $\omega_S\tau > 1 > \omega_I\tau$ and lead to T_2 times that are much shorter than T_1 times. In the presence of this mechanism, resonance lines broaden significantly.

7.3 PRACTICAL CONSEQUENCES FROM THE THEORY OF PARAMAGNETIC EFFECTS

The theoretical considerations show that the presence of unpaired electrons in samples causes a large number of spectral effects that strongly depend on the nature of target nuclei, electron relaxation times, the strength and the nature of electron–nucleus coupling, and finally on structural factors, characterizing the systems investigated. All of the paramagnetic effects have been described to account for single contact between a nucleus and an electron, separated by the distance r and coupled via constant A. It is obvious that in the presence of many paramagnetic centers, distanced differently relative to the nucleus, the effects should be summed. This is particularly important in increasing concentrations of paramagnetic centers, when the effect will dramatically increase. In turn, under these conditions, the paramagnetic effects will result in low NMR sensitivity, a quite limited spectral resolution, and difficulties in signal assignments due to large paramagnetic shifts and large chemical shift anisotropies. Moreover, even principles in registrations of NMR signals and their interpretations can be changed. The latter depend strongly on features of investigated materials and on their extraordinary nature. This statement can be well illustrated here by metals that are most unusual systems in the context of NMR applications. The consideration of metals in this section is explained by very pronounced electron–nucleus interactions that strongly affect resonance lines registered in NMR spectra.

Traditionally, NMR, as well as other resonance methods, is widely applied to obtain information about electronic properties of metals [15]. In fact, observation of the resonance lines in metals and also measurement of their relaxation rates leads to determination of the density of states at the Fermi level and the character of the various bands (e.g., s or d). Among various problems connected with studies of metals, obtaining information on the electronic properties of small metallic particles and the metal cores in metal-cluster compounds is of great interest for materials chemists. Potentially the systems containing such particles, particularly small particles of metals, such as Pt, Pd, or Rh, can show important catalytic properties. As noted in the literature, the catalytic behavior of these systems depends on the morphology of the individual particles and the structure of their packing. This situation takes place, for example, in zeolites, where the metal particles can be arranged in the matrix of materials.

Generally, the main task in such studies consists of determining the Knight shift (K_{orb}), for example, in the static ^{103}Rh NMR spectra of supported metal catalysts, where registration of the signal, itself, is problematic. It has been noted that decreasing particle size leads to a large increase in line widths of the ^{103}Rh resonances. Because this effect is only weakly temperature dependent, the NMR spectra can be recorded, in practice, *only by point by point experiments* as illustrated in Figure 7.5. It is seen that the spectra are very sensitive to the nature of rhodium composites: disposition of the rhodium resonance and particularly its line width change strongly in going from aluminum to titanium oxide. Generally speaking, this field of investigations is very specific and requires fundamental knowledge in physics. Therefore the reader is referred to Van der Klink and Brom [15] showing all the elements of the theory and also principles of NMR in metallic systems. At the same time, these studies show that the point-by-point method of registering NMR signals can be successfully used to observe directly an extremely wide line expected for nuclei in atoms located close to paramagnetic metal cations or even in the paramagnetic cations themselves. This situation takes place, for example, in crystalline CuF_2 [33]. In this solid, density of paramagnetic ions is too high to observe ^{19}F signals directly in the absence of line-narrowing mechanisms due to the *enormous* dipole fields. According to the x-ray data, a Cu cation in crystalline CuF_2 has two closest fluorine neighbors with different distances, shown in Figure 7.6. The ^{19}F NMR spectrum can be recorded with a very short 90° pulses in a solid-echo pulse sequence due to very short NMR decays. Nevertheless, the whole spectrum is expected to be too broad to use a single frequency at a single fixed magnetic field. Therefore the spectrum has been collected by the point-by-point method with varying the NMR magnetic field for each data point, keeping the NMR frequency at 45 MHz. As seen from Figure 7.6, the final ^{19}F NMR spectrum shows two well-resolved ^{19}F resonances separated by 5000 ppm in a full accordance with the x-ray structure.

$$K_{orb} = -\frac{\mu_0 e^2}{4\pi m}\sum_{i=1}^{\infty} f(\varepsilon_i)\left\langle \phi_i \left| \frac{x^2+y^2}{2|r|^3} \right| \phi_i \right\rangle + \frac{\mu_0 e^2}{4\pi m}\sum_{i,j=1}^{\infty} \frac{I(\varepsilon_i,\varepsilon_j)}{2m}\left\langle \phi_i \left| \frac{l_z}{|r|^3} \right| \phi_j \right\rangle \langle \phi_j | l_z | \phi_i \rangle$$

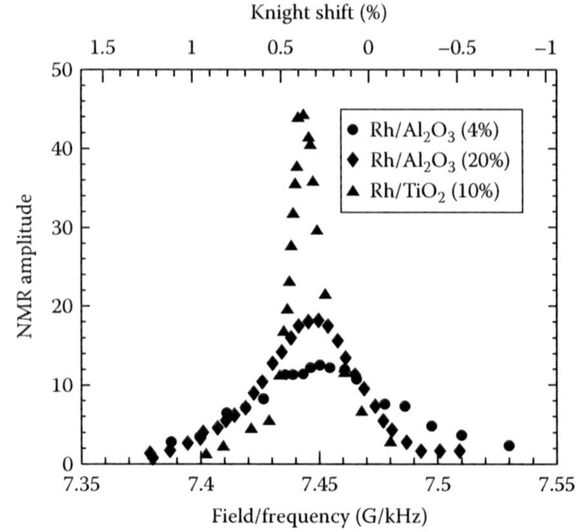

FIGURE 7.5 The static [103]Rh point by point NMR spectra of the oxide supported composites. The Rh–TiO_2 with 10% (Rh) has the largest particles with a diameter of 6 nm and therefore the resonance line is most broadened. (Reproduced with permission from Van der Klink, J.J. and Brom, H.B., *Prog. Nucl. Magn. Reson. Spectrosc.*, 36, 89, 2000.)

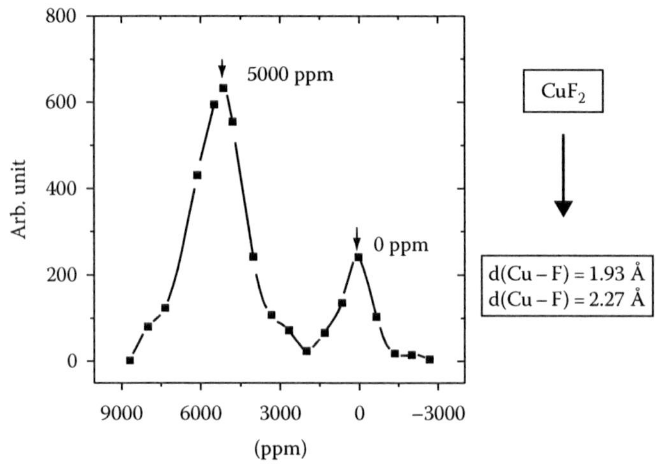

FIGURE 7.6 The ^{19}F NMR spectrum obtained at 45 MHz by the point-by-point method for a crystalline CuF_2 static sample. The compound has two fluorine atoms separated from the Cu atoms. (Reproduced with permission from Lee, C.E. et al., *J. Korean Chem. Soc.*, 38, 628, 1994.)

From theory, spectroscopic effects are strongly dependent on the nature of paramagnetic centers and particularly on electron relaxation times. Therefore, besides the aforementioned unusual registration method, traditional NMR experiments can also be performed to characterize paramagnetic solids. Generally, however, they require modifications of standard pulse sequences in combination with fast sample spinning. In some cases, the experiments can even show better NMR sensitivity. Going back to the work of Wickramasinghe and Ishii [7], sensitivity in the ^{13}C MAS NMR spectra for paramagnetic complexes $Cu(DL\text{-}Ala)_2$ and $Mn(acac)_3$ can be significantly increased by cross-polarization and dipolar INEPT experiments, as illustrated in Figure 7.7. It is seen clearly that a maximum in intensities is reached by the INEPT

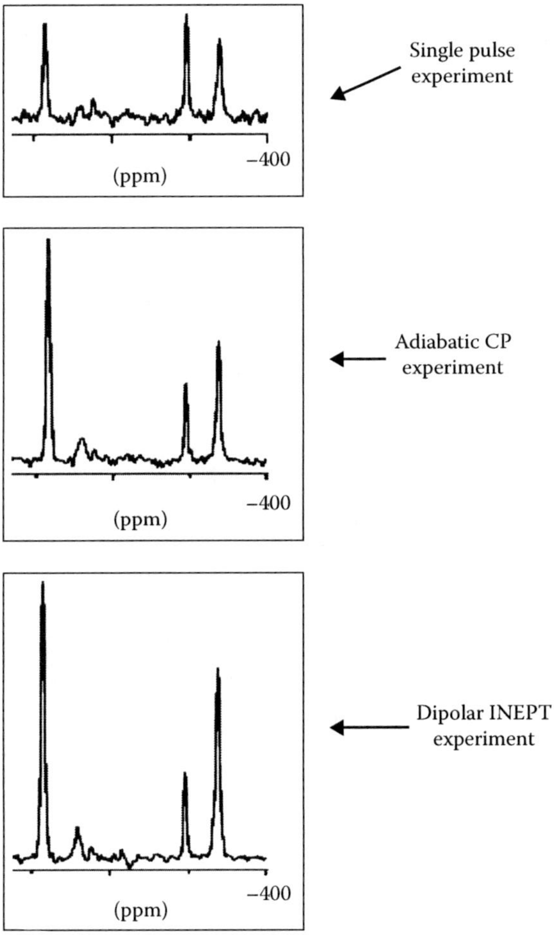

FIGURE 7.7 Comparison of NMR sensitivity in the ^{13}C MAS NMR spectra of paramagnetic complex $Cu(DL\text{-}Ala)_2 \times (H_2O)$ recorded by a one-pulse (90°) sequence, adiabatic cross polarization pulse sequence and by dipolar INEPT experiments. (Reproduced with permission from Wickramasinghe, N.P. and Ishii, Y., *J. Magn. Reson.*, 181, 233, 2006.)

pulse sequence. It has been noted that the dipolar INEPT can be based on 1D and 2D versions, the pulse sequences of which are depicted in Figure 7.8. In the 1D NMR experiment, proton transverse magnetization is transferred by a 90° pulse and then, a *rotor-synchronized* echo pulse is applied to proton and carbon spins during the initial two rotor cycles, to reintroduce proton–carbon dipolar interaction, while ^{1}H anisotropic chemical shifts are canceled. Then, applications of 90° pulses to proton and carbon spins convert *antiphase* transverse magnetization of proton spins to that of carbon spins. When the polarization transfer ends by the second echo sequence, the ^{13}C signal can be detected. As regularly, in the 2D NMR version, an additional evolution time should operate. It has been emphasized that the pulse sequences, shown in Figure 7.8, lead to collection of NMR data in the absence of high power ^{1}H decoupling. It has been noted that the decoupling in paramagnetic systems causes generally "negative" spectroscopic effects. Finally, this strategy, helping in signal assignments and even C–H distance measurements can been recommended for studies of different paramagnetic organometallic solids, often applied in supramolecular chemistry.

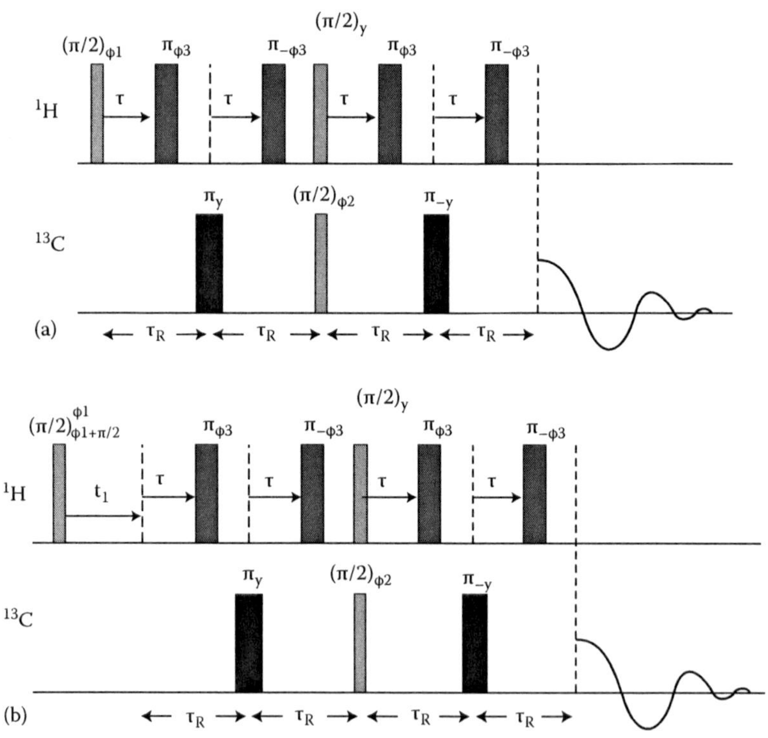

FIGURE 7.8 The pulse sequences that can be applied for paramagnetic solids: (a) 1D version of ^{13}C dipolar INEPT; (b) 2D version of heteronuclear (^{13}C–^{1}H) correlation using the dipolar INEPT polarization transfer. (Reproduced with permission from Wickramasinghe, N.P. and Ishii, Y., *J. Magn. Reson.*, 181, 233, 2006.)

7.4 ^{1}H, ^{31}P, ^{13}C, ^{2}H, ^{29}SI, ^{7}LI, AND ^{6}LI NMR SPECTRA OF PARAMAGNETIC SOLIDS

Fermi-contact coupling, dipole–dipole electron–nucleus interaction and the BMS effects can lead to resonances with singularly large frequency shifts, chemical shift anisotropies, and short spin–spin and spin–lattice relaxation times. Under these conditions, the nuclei in static samples of paramagnetic solids will be detected (when it is principally possible) as strongly broadened signals that will transform to very wide and intense sideband envelopes in MAS NMR spectra. Such envelopes are often comparable with the area of spectral excitation. According to Grey et al. [34], all of these factors dramatically decrease signal-to-noise ratios in the final NMR spectra, expressing in practice with a strong increase in numbers of scans.

The second problem in interpretation of NMR data collected for paramagnetic solids is that the contact or pseudocontact isotropic chemical shifts are *temperature dependent*. The contact and pseudocontact shifts change with temperature proportionally to factor T^{-1} and T^{-2}, respectively. These effects can be well observed in NMR experiments on solids containing paramagnetic lanthanide ions. Such ions are generally characterized by short electron relaxation times, and in their presence, resonances of target nuclei are sharp. For example, Grey et al. [34] have reported the pronounced temperature dependences of ^{89}Y chemical shifts in MAS NMR spectra obtained for different mixed oxides. The practical conclusion is obvious: comparing the spectral parameters obtained for different paramagnetic compounds can be correct only when they have been collected at the same temperature. It should be added that the contact and pseudocontact isotropic chemical shifts depend on the temperature differently and this fact can used to distinguish between them.

Finally target nuclei can become "invisible" in regular solid-state NMR spectra, when the electron–nucleus distances are very short and paramagnetic effects are too strong. In this situation, the NMR spectra can show a remarkable loss in signal sensitivity, observed even for nuclei with high natural abundances such, as ^{1}H or ^{31}P.

Peeters et al. [35] have reported on the significant intensity loss in **the ^{31}P MAS NMR spectra** of aluminophosphates doped with ions of cobalt. Moreover, the authors have found that amounts of undetectable phosphorous can reach up to 30%. It is obvious that such effects can be observed only in samples containing a convenient internal standard. On the level of interpretation, this loss in the intensity has been attributed to locations of the paramagnetic centers in the first and second coordination phosphorous spheres [35,36]. Finally similar effects in the ^{31}P NMR spectra of aluminophosphate materials modified by ions of Ni, Co, Fe, and Mn have been observed by Mali et al. [37].

It is important to emphasize that an intensity loss cannot be prognosticated *a priori*. For example, the ^{31}P resonances in cobalt polyoxometalates have been well observed in ^{31}P MAS NMR spectra by Flambard et al. [38]. It has been found that the phosphorous nuclei can show the large paramagnetic shifts of a Fermi-contact nature and resonances of ^{31}P nuclei situated close to the cobalt centers can exhibit the CSA values up to 3300 ppm versus 500–700 ppm for ^{31}P nuclei, which are remote from the paramagnetic centers.

Such large CSA magnitudes give rise to ineffective excitation of the large Larmor frequency diapasons by the single-pulse NMR experiments. This ineffectiveness additionally results in significant baseline distortions [39]. Generally the spin-echo pulse sequences (180° – τ – 90°) can be more useful under these conditions. Figure 7.9 depicts the ^{31}P spin-echo MAS NMR spectra recorded for paramagnetic complex αβ-$[Co_3Na(H_2O)_2(P_2W_{15}O_{56})_2]^{17-}$ at a spinning rate between 30 and 35 kHz. As can be seen, even at the very high spinning rates, the spectra exhibit the intense sideband patterns that cover the diapason of 1700 ppm.

In addition, the ^{31}P MAS NMR experiments in Figure 7.9 illustrate the temperature dependent Fermi-contact chemical shifts of ^{31}P nuclei. However, here, this temperature effect appears even at "a constant temperature." As can be seen, the isotropic chemical shifts attributed to two different phosphorous sites (see the dotted lines in the spectra) change with the spinning rates. The reason is raising the temperature in a sample on increasing the spinning rates from 30 to 35 kHz.

Similarly, an isotropic **^{13}C chemical shift**, measured by Blumel and coworkers [40] in the MAS NMR spectra of solid paramagnetic complex Cp_2Cr, changes from −258 to −231 ppm, when the spinning rate increases from 3 to 15 kHz. It is obvious that such a change, which is more than 20 ppm, cannot be negligible in interpretation of the spectra. It has been also noted that due to fast proton relaxation in complex Cp_2Cr, the cross-polarization ^1H–^{13}C was not efficient (*generally speaking, ineffectiveness of the cross-polarization can even be used as a good test for the presence of paramagnetic impurities*). The ^{13}C MAS NMR parameters of other paramagnetic metallocenes Cp_2V, Cp_2Co, and Cp_2Ni can be found in Heise et al. [41], where the ^{13}C isotropic chemical shifts change from −398 to +1594 ppm, as a function of the metal center, and chemical shift anisotropy varies between 828 and 2640 ppm. One may

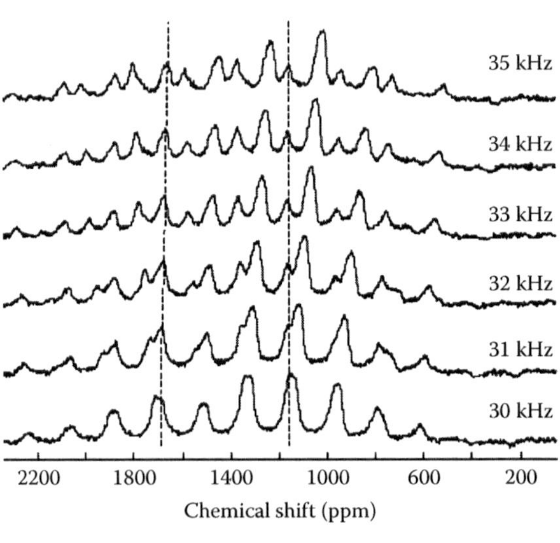

FIGURE 7.9 The ^{31}P spin-echo MAS NMR spectra of paramagnetic complex αβ-$[Co_3Na(H_2O)_2(P_2W_{15}O_{56})_2]^{17-}$ recorded at high spinning rates. The dotted lines show the positions of two different isotropic resonances. (Reproduced with permission from Bakhmutov, V.I., *Chem. Rev.*, 111, 530, 2011.)

think that significant line broadenings observed experimentally in the ^{13}C MAS NMR spectra of the paramagnetic metallocenes play a minor role in the context of signal assignments because the chemical shift scale is very large. At the same time, the strong line broadenings limit applications of quantum chemical calculations for such systems.

In contrast, paramagnetic lanthanide complexes, due to fast electron relaxation, show relatively sharp ^{13}C resonances that are accompanied by very wide sideband patterns [42,43]. Moreover Brough et al. [43] have emphasized that the ^{13}C MAS NMR spectra and ^{13}C chemical shifts in crystalline lanthanide acetates can be quantitatively analyzed or even predicted by considering the interactions between the nuclei and the single nearest paramagnetic ion. Some line broadenings observed in the spectra can be explained in terms of the BMS effects [38].

As shown in Chapter 5, the **^{2}H NMR** powder line shapes are very sensitive to local structures and/or molecular dynamics. For this reason, static ^2H NMR is often applied for studies of diamagnetic solids. Since the ^2H NMR spectra in diamagnetic systems are generally dominated by quadrupole interactions, they are recorded with the ordinary quadrupole-echo pulse sequence [39], focused on only the *dephasing* of the quadrupole interactions. Therefore its applications for paramagnetic solids are not effective. In fact, the ordinary quadrupole-echo pulses lead to distortions in the line shapes, caused by large paramagnetic shifts and their anisotropies. To avoid such spectral distortions, Siminovitch et al. [44] have suggested the *shift-compensated quadrupole echo*. It has been emphasized that this pulse sequence works particularly effectively when electron relaxation times in samples are relatively short.

For longer electron relaxation times, for example, typical of ions Mn^{2+} and taking place in the crystal complex $[Mn(^2H_2O)_6][SiF_6]$ (or ions Gd^{3+}), Mizuno et al. [45] have performed NMR experiments based on the *exorcycled quadrupole-echo sequence* (EQES). The authors have demonstrated that these EQES experiments and the spectral simulations carried out in the frameworks of the developed theory can be a very powerful tool for analysis of dynamics in strong paramagnetic molecular systems. The effectiveness of Mizuno's approach is illustrated in Figure 7.10. First, the spectra have been recorded without distortions in line shapes. Second, under this circumstance the authors have achieved excellent agreement between the experimental and calculated ^2H NMR spectra assuming a motional model: 180° flips of water molecules in a combination with the site jumps of the $[Mn(H_2O)_6][SiF_6]$ complex around C_3-axis. The figure shows the corresponding rate constants obtained at quadrupole coupling of 245 kHz with the asymmetry parameter of 0.01.

The quadrupolar moments of **^{7}Li and ^{6}Li nuclei** are relatively small (e.g., $-4 \times 10^{-30}\,m^2/e$ and $-0.08 \times 10^{-30}\,m^2/e$ for ^7Li and ^6Li, respectively, that are less than $14.66 \times 10^{-30}\,m^2/e$ for ^{27}Al nuclei), their natural abundance is high (92.4% and 7.6%, respectively), and therefore, they are often applied, as targets, for characterizations of different materials, for example, zeolites. These nuclei show sharp lines in the ^7Li, ^6Li MAS NMR spectra and their signal intensities correlate with crystallographic site populations in low-silica zeolite phases [46].

Generally the ^7Li MAS NMR spectra are dominated by quadrupolar interactions and dipolar coupling that lead to broad sideband patterns. The low magnetogyric

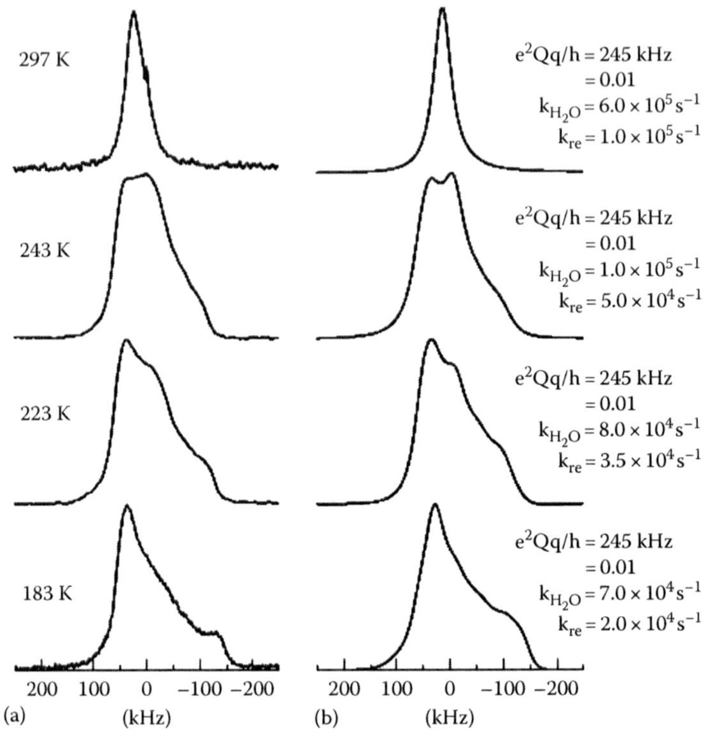

FIGURE 7.10 The temperature dependences of the ^2H NMR spectra of complex $[Mn(H_2O)_6][SiF_6]$ observed experimentally (a) and calculated (b) with the corresponding parameters. (Reproduced with permission from Bakhmutov, V.I., *Chem. Rev.*, 111, 530, 2011.)

ratios of ^6Li nuclei and lower quadrupole moment provide better spectral resolution of different local lithium environments.

The magnetic properties of $^{6,7}Li$ nuclei are also quite convenient to collect NMR data for paramagnetic solids, where lithium resonances are again sensitive to environments. Chazel et al. have recently performed the ^6Li, ^7Li static and MAS NMR experiments on samples of the (LiNiO) family with different component ratios [47]. The studies have resulted in important spectroscopic data. It has been found that Fermi-contact and dipolar interactions with the electron spins on the Ni centers can cause isotropic Li chemical shifts more than +700 ppm. The ^6Li, ^7Li resonances with the *strongly asymmetric shapes* due to anisotropy have been observed in the static NMR experiments.

Since lithium atoms can be chemically added to porous materials, $^{6,7}Li$ nuclei can be used as convenient probes in their investigations. For example, the resonances in ^7Li MAS NMR spectra, belonging to *extra-framework lithium cations* [48,49], have been found to be very sensitive to the presence of paramagnetic gas species in samples. In fact, the NMR signals of the nuclei located close to paramagnetic centers should undergo broadenings and show shorter relaxation times and remarkable paramagnetic shifts.

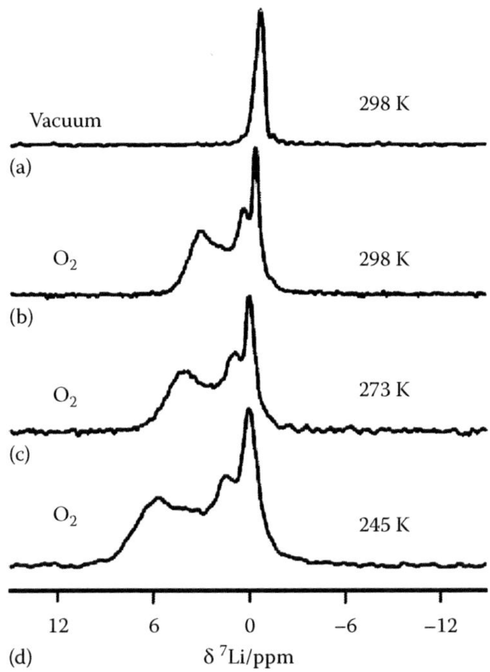

FIGURE 7.11 The ^{7}Li MAS NMR spectra for zeolite Al-ZSM-5: (a) oxygen-free dehydrated sample (300 K), (b) spectra of the oxygen-containing sample at 300 K, (c) at 273 K, and (d) at 245 K. (Reproduced with permission from Bakhmutov, V.I., *Chem. Rev.*, 111, 530, 2011.)

The ^{7}Li MAS NMR spectra of the zeolite Al-ZSM-5 (Figure 7.11) illustrate the influence of oxygen [49]. As we can see, in the oxygen-free system, the ^{7}Li signal appears to be relatively narrow and practically symmetric. However the shape of the resonance changes dramatically in the presence of O_2. It is clear that such effects can be used to characterize accessibility of lithium cations in high-silica zeolites: the Li cations, interacting with oxygen molecules, exhibit anisotropic paramagnetic shifts that are more pronounced at low temperatures.

As we have already noted in previous chapters, the well-established **^{29}Si, ^{27}Al MAS NMR** structural criteria, distinguishing structurally different Si and Al sites in diamagnetic zeolites, are described in Duer's book [39]. The role of these criteria has been considered in Chapter 6. However, many relatives of zeolites have been prepared as solids, exchanged with paramagnetic lanthanide cations to reach specific chemical goals. For example, such systems are often catalytically active and therefore paramagnetic effects for resonances of ^{29}Si, ^{27}Al nuclei are of great interest even in the analytical context. Unfortunately, the zeolites, exchanged with cations Nd^{3+}, Sm^{3+}, Gd^{3+}, and Dy^{3+}, show ^{29}Si, ^{27}Al MAS NMR spectra that are strongly dependent on the nature of cations and generally poorly informative in the structural context [50].

Figure 7.12 illustrates this influence. If the first spectrum, corresponding to the diamagnetic zeolite, is well resolved to show four ^{29}Si resonances, then the ^{29}Si resonances in other spectra broaden in going to Sm and Nd systems and completely

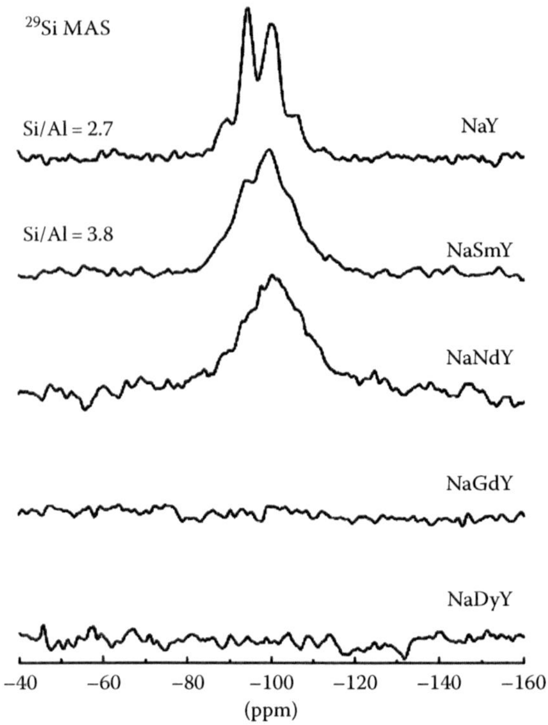

FIGURE 7.12 The ^{29}Si MAS NMR spectra of zeolites, marked as NaY, NaSmY, NaNdY, NaGdY, and NaDyY, prepared by exchange with the corresponding lanthanide ions at similar concentrations. (Reproduced with permission from Bakhmutov, V.I., *Chem. Rev.*, 111, 530, 2011.)

disappear for ions of Gd and Dy. Similar effects are reported for ^{27}Al MAS NMR in these materials.

The ^{29}Si MAS NMR studies of the fully exchanged rare earth zeolites, containing ions Ce^{3+}, Pr^{3+}, Nd^{3+}, Sm^{3+}, and Eu^{3+} were more successful [51]. For example, isotropic *high-field* chemical shifts (up to −146 ppm) of the *Fermi-contact* nature have been measured for the Nd-containing zeolite, whereas a very small and *low-field* shift of +6 ppm has been observed in the Eu-zeolite.

In contrast, recently Stebbins and Kelsey [52] have detected ^{29}Si resonances in aluminum silicate complex ($[Mg,Fe]_3Al_2Si_3O_{12}$) containing 3.5 wt% of FeO, which have been shifted by 25–200 ppm from "normal" silicon shift ranges. According to an analysis, these resonances undergo *pseudocontact* interaction with the unpaired electron spin.

7.5 ^{51}V AND ^{55}MN NMR SPECTRA OF PARAMAGNETIC SOLIDS

Vanadium and manganese atoms are present as structural units in many materials potentially active in catalysis. ^{51}V, ^{55}Mn Nuclei are quadrupolar with the moderate quadrupole moment of -5.2×10^{-30} m^2/e for ^{51}V (i.e., close to that for ^{7}Li and less than that for ^{27}Al) and the high quadrupole moment of $33 \times 10^{-30}\,m^2/e$ for ^{55}Mn nuclei. Thus, ^{51}V, ^{55}Mn NMR spectra of paramagnetic systems can be dominated by quadrupolar interactions and also paramagnetic effects, considered in this section.

Applications of the solid-state ^{51}V NMR techniques for characterizations of heterogeneous catalysts modified with various paramagnetic cations, situated in the matrix surface or incorporated into the bulk of heterogeneous catalysts, have recently been reviewed by Shubin et al. [53]. The NMR data have been represented by different 3d and rare earth vanadates exhibiting strong paramagnetic properties. Generally, a combination of the quadrupolar and paramagnetic effects leads to very intense and wide sideband patterns, illustrated in Figure 7.13. For this reason, the ^{51}V MAS NMR spectra can be analyzed only at high spinning rates up to 30kHz (or higher) by simulations to determine the corresponding spectral parameters. It has been noted that since second-order quadrupolar interactions dominate in the spectra, they should be taken into account for simulation procedures. For example, according to the calculations, compounds $FeVO_4$ and $CeVO_4$ exhibit the vanadium quadrupole coupling constants of 1.7 and 3.5MHz at the isotropic chemical shifts of 17,000 and −559ppm, respectively. As in the case of diamagnetic systems, the chemical shifts refer to external $VOCl_3$.

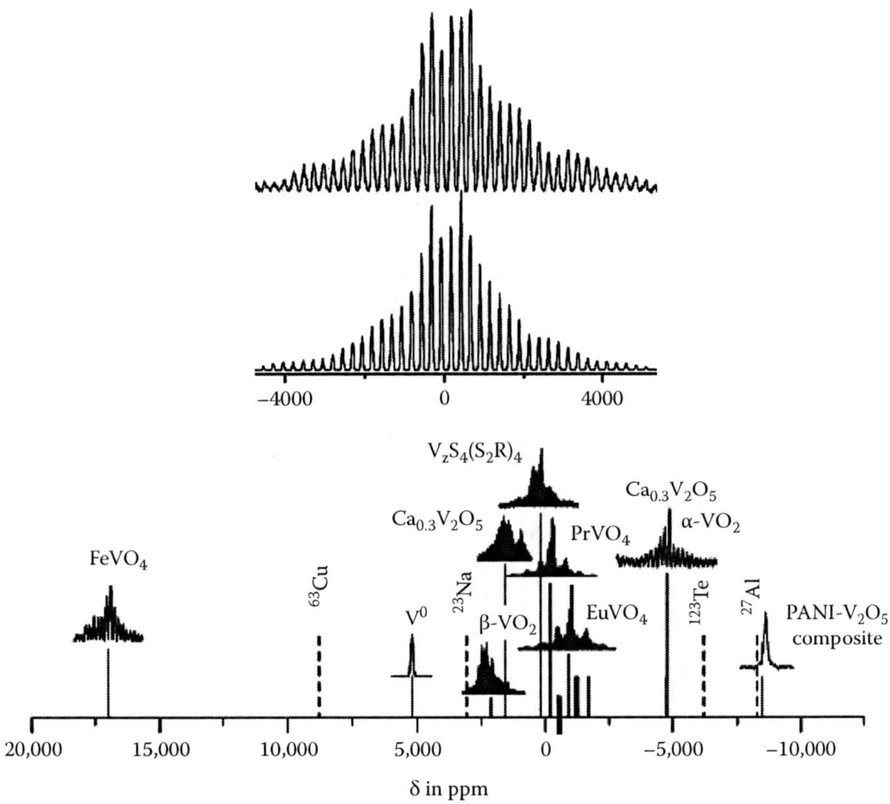

FIGURE 7.13 Top: The experimental and calculated ^{51}V MAS NMR spectra of vanadate $YbVO_4$. Bottom: The chemical shift scale for ^{51}V nuclei in different paramagnetic solids. (Reproduced with permission from Shubin, A.A. et al., *Catal. Today*, 142, 220, 2009.)

The NMR data collected for paramagnetic solids have been summarized as the ^{51}V chemical shift scale in Figure 7.13. The conclusion is obvious: this scale is significantly larger than that for diamagnetic solids, which cover a region of 1200 ppm only. On the one hand, this is a positive fact in the analytical context. On the other hand, the initial search of ^{51}V NMR signals is not simple, requiring carrier frequency changes at accurate tuning, respectively. It has been noted that ^{51}V NMR spectra can be obtained from spin-echo or single-pulse experiments performed at high spinning rates, generally with standard 2.5 mm rotors. Single-pulse experiments can be realized using relatively short relaxation delays of 0.1–3 s and very short radio frequency pulses (e.g., 0.7 μs for a 16° pulse) to avoid the spectral distortions. However, it has been noted that the spin-echo pulse sequence is more preferable.

7.5.1 Analyzing the Nature of Metal Ions by Their Direct Observation in Solid-State NMR Spectra

Many porous materials, interesting from the point of view of catalytic properties, are prepared in the presence of manganese, copper, or nickel cations, which can potentially be paramagnetic centers. Nuclei of these metal ions are pronouncedly quadrupolar and their direct observation is expectedly difficult even when they are in diamagnetic oxidation states (e.g., Cu(I) compounds are always diamagnetic).

^{63}Cu and ^{65}Cu nuclei have very large quadrupole moments of −22.0 and −20.4 × 10^{-30} m^2/e respectively, and probably therefore, the current NMR data for these nuclei are limited. The solid-state ^{63}Cu and ^{65}Cu NMR experiments have been performed by the group of Schurko for a large number of organometallic copper(I) complexes with different copper coordination environments, which have been characterized by central spin transitions [54]. The copper NMR spectra, dominated by strong quadrupolar interactions, show powder patterns with breadths ranging from 760 kHz to 6.7 MHz. The isotropic chemical shifts cover a region between 1050 and −510 ppm with magnetic shield anisotropies of 1000–1500 ppm, while quadrupolar coupling constants are between 22.0 and 71.0 MHz. In order to obtain the high signal-to-noise ratios in the powder NMR spectra, the authors have applied Hahn-echo and/or QCPMG pulse sequences used in a frequency-stepped manner.

A modified approach has been reported by the same group [55], where the pulse sequence wideland uniform rate smooth truncation quadrupole Carr Purcell Maiboom Gill (WURST–QCPMG) has been aimed at rapidly recording the *ultra wide-line* NMR spectra, typical of Cu-containing compounds. This approach has many advantages due to the much wider excitation bandwidth by the WURST pulses. It has been also noted that incorporating a high-power proton decoupling into the WURST pulse sequence can provide an additional signal enhancement. This effect is clearly seen in Figure 7.14 showing the ^{65}Cu NMR spectra recorded for the diamagnetic Cu(I) compound $(PPh_3)_2CuO_2CCH_3$. It is obvious however that the spectrum, stretching between 900 and −1600 kHz, can be analyzed only by calculations. The following parameters have been obtained: ^{65}Cu δ(iso) = 350 ppm, ^{65}Cu QCC = 40 MHz and the asymmetry parameter of 0.35.

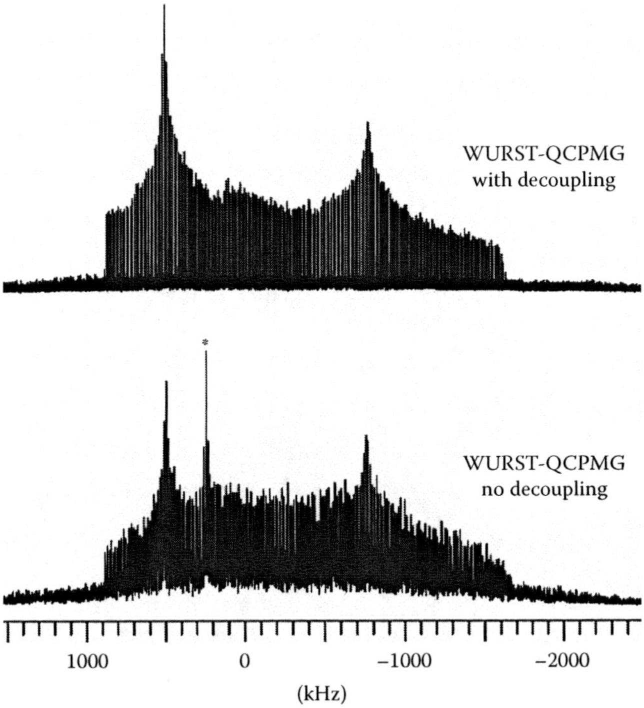

FIGURE 7.14 The ^{65}Cu NMR spectra recorded for compound $(PPh_3)_2CuO_2CCH_3$ by the WURST–QCPMG method with and without proton decoupling. (Reproduced with permission from O'Dell, L.A. et al., *Chem. Phys. Lett.*, 468, 330, 2009.)

It is difficult to expect that NMR studies of such metal ions will be perspective at appearance of an unpaired electron due to a combination of strong quadrupolar interactions and paramagnetic effects. In fact, EPR spectroscopy is preferable in this situation. Nevertheless, there are problems in studies of molecular systems containing manganese ions, which attract much interest from chemists due to their unusual magnetic properties and participation in photosynthesis and the water splitting reactions. For example, approaches to new synthetic photocatalysts require detailed understanding and modeling the oxygen-evolving complex. Generally in spite of EPR or x-ray data, additional spectroscopic properties are needed to accurately describe oxidation states of metal ions and their environments. Potentially, such systems can be characterized by solid-state NMR.

Probing manganese clusters illustrates these characterizations. Among the manganese molecular systems, compound bis(μ-oxo)dimanganese(IV) $[M_2O_2(salpn)_2]$ attracts significant attention, as a model of the O_2 evolving complex recently investigated by Ellis et al. [56]. Generally speaking, the NMR theory assumes that the ^{55}Mn chemical shift tensor (represented by σ) and electric field gradient, described via quadrupole coupling constant QCC and the asymmetry parameter (η_Q) should be very sensitive to oxidation-induced changes for manganese ions. In this connection, the authors [56] have attempted to directly observe ^{55}Mn nuclei in solid-state

NMR experiments, performed on the complex $[Mn_2O_2(salpn)_2]$. The solid-state NMR spectrum of the complex has been successfully obtained only *at low temperature of* 8.5 K and at the external magnetic field of 9.40 T. The CP/QCPMG pulse sequence has been used for its excitation. As Figure 7.15 shows, the spectrum of the Mn(IV)-containing system is very complex and centered between the NMR spectra, specially obtained for samples of species containing manganese centers with Mn(VII) and Mn(0) oxidation states. It has been noted that ions $[MnO_4]^{1-}$ are generally taken as 0 ppm, while compound $Mn_2(CO)_{10}$ exhibits a signal detected between −2325 and −1162.5 ppm. Again, as in the case of copper ions, the shape of the spectrum is very whimsical. However, simulation procedures have revealed the presence

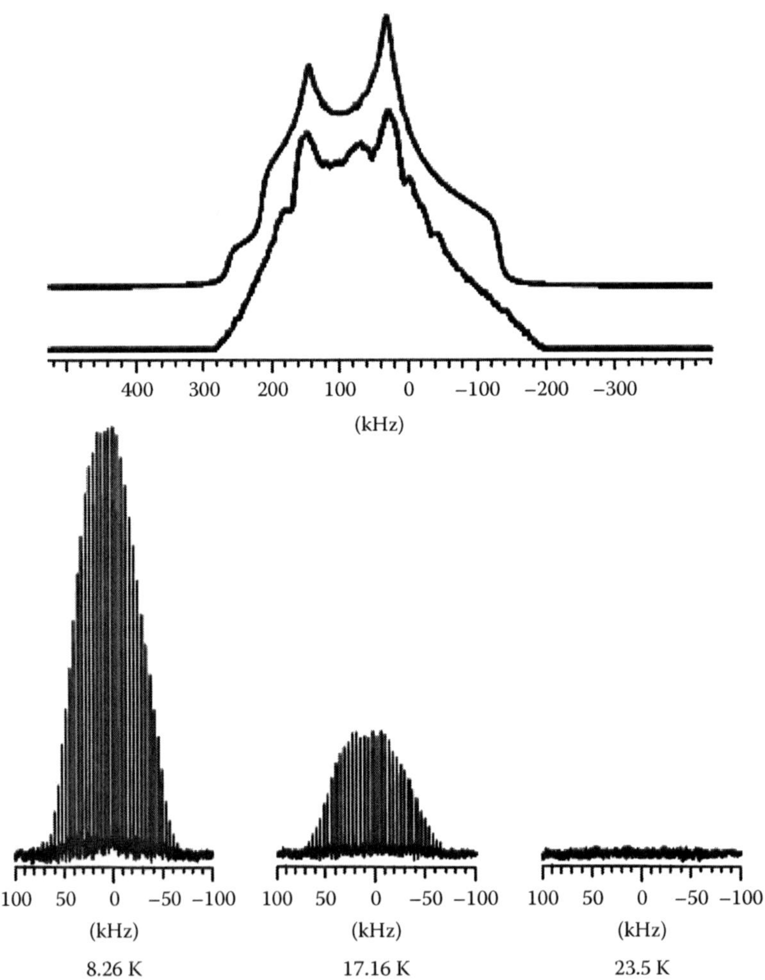

FIGURE 7.15 Top: The experimental and calculated ^{55}Mn NMR spectrum of complex $[Mn_2O_2(salpn)_2]$ obtained at lowest temperature of 8.5 K. Bottom: The temperature dependent ^{55}Mn NMR spectra of complex $[Mn_2O_2(salpn)_2]$. (Reproduced with permission from Ellis, P. et al., *J. Am. Chem. Soc.*, 132, 16727, 2010.)

of a single ^{55}Mn resonance, characterized by quadrupolar coupling constant (QCC) of 24.7 MHz, asymmetry parameter η_Q of 0.43, and very large magnetic shielding anisotropy $\Delta\sigma$ of 3375 ppm.

Finally, dramatic spectral changes have been observed *on increasing the temperature*, when electron–nucleus spin-hyperfine interactions become activated. As can be seen in Figure 4.15, the ^{55}Mn NMR spectrum recorded at 17.2 K corresponds to a *70% loss in signal-to-noise ratio*, while at 23.5 K the manganese signal completely disappears. The latter corresponds to the situation when only 2×10^{-5}% chance of populating a paramagnetic state causes too much broadening and therefore the line becomes undetected.

7.6 SPECIAL NMR TECHNIQUES FOR OBSERVATIONS OF "INVISIBLE" TARGET NUCLEI

One of the most important consequences from theory considered in this chapter, is that target nuclei located in closest environments of unpaired electrons (e.g., at chemical bonding through one or two chemical bonds), are difficult for direct observation by the conventional NMR techniques due to strongest paramagnetic effects. This situation is typical of the CoAPO$_4$-*n* molecular sieves reported by Canesson and Tuel [57]: ^{31}P NMR spectra of static samples show low-intensity broadened lines and demonstrate the strong intensity loss even at relatively low cobalt contents. The ^{31}P MAS NMR spectra of such systems exhibit the wide sideband patterns due to strong dipolar interactions that, however, are not sufficient to draw conclusions about the nature of the frameworks of the materials [36]. Canesson and Tuel [57] have emphasized that the presence of the intense sidebands in the MAS NMR spectra, itself, can be *incorrectly* used as proofs for incorporation of paramagnetic ions into the matrix of materials.

It is obvious that the point-by-point method, considered earlier and applied in such a situation, does not allow accurate descriptions of the line shape. The groups of Tuel [36,57,58] and Mali [37] have suggested and applied a spin-echo mapping technique, involving the Hahn-echo ^{31}P static NMR spectra recorded at different carrier frequencies to finally observe the NMR lines within broad spectral regions. Figure 7.16 shows schematics of the experiment, where a whole echo is acquired. Since the resonance lines are expected to be very broad, echo delays applied in the

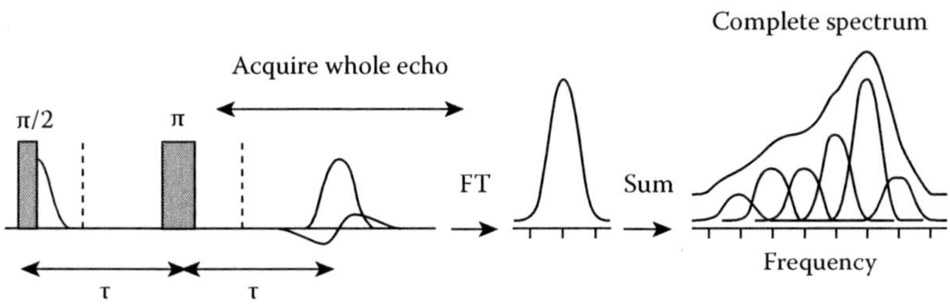

FIGURE 7.16 Schematic representation of the spin-echo mapping experiment used for recording broad ^{31}P NMR spectra. Vertical dashed lines show dead time after radio frequency pulse. (Reproduced with permission from Bakhmutov, V.I., *Chem. Rev.*, 111, 530, 2011.)

pulse sequence should obviously be short to avoid cutting the FID. Technically this condition can be easily realized in the static NMR experiments while any MAS NMR experiments, being synchronized with spinning rates, automatically increase the echo delays. One of the important elements in this mapping experiment is an accurate tuning of the NMR probe for each carrier frequency.

It has been demonstrated that this Hahn-echo technique leads to observation of ^{31}P resonances of nuclei in the closest environments of paramagnetic ions Co^{3+} in the $CoAPO_4$-*n* molecular systems. The resonances can be observed even at relatively high Co contents. In this case the detected lines are exclusively broad and cover a range between about −500 and 10,000 ppm [36]. In addition, the authors have shown that ^{31}P T_2 times measured for the resonances decrease on changing their ^{31}P chemical shifts from a few ms around 0 ppm to 30–40 μs around 8000 ppm. Moreover a *fast* T_2 time decrease has been noted between 0 and 1500 ppm whereas a *slow* decrease was observed for higher shift values. Thus, the presence of ^{31}P resonances with chemical shifts $\delta > 500$ ppm can be considered as direct proof for Co incorporations into the framework of the materials.

Iron phosphates ($FePO_4$) and their lithium derivatives, applied and investigated as important materials in battery technology, also represent a good example of successful applications of spin-mapping approaches to characterizations of paramagnetic systems [12,59]. The static solid-state ^{31}P NMR spectra of the Fe(III) phosphates, $FePO_4$, $Li_3Fe_2(PO_4)_3$, $LiFeP_2O_7$, $FePO_4 \cdot 2H_2O$, and $FePO_4 \cdot 2H_2O$, show extremely large ^{31}P contact shifts in a range of 5,700–20,500 ppm and the lines observed are characterized by widths that can reach magnitudes of more than 4,000 ppm. It should be emphasized that in spite of these line broadenings, the spin-mapping spectra provide identification of different phosphorous sites when they are present in materials. For example, as expected from chemical considerations, compound $LiFeP_2O_7$ shows the ^{31}P NMR spectrum (Figure 7.17) where two resonances are relatively shifted by 6,000 ppm: 8,230 versus 14,398 ppm. As the spectra show, the line shapes of both resonances are not symmetric, particularly for the resonance with δ of 14,398 ppm. In such a situation, experimental determinations of isotropic chemical shifts are impossible. However they can be obtained by spectral simulations performed in the limits of DFT calculations of the Fermi-contact interaction via the equation at the top of the figure.

Interesting phenomenological features in ^7Li static spin-echo NMR experiments performed for the LiNiO materials have been reported by Chazel et al. [47]. The authors have observed "a hole" that appears exactly at the carrier frequency. This phenomenon has been explained in terms of a signal saturation resulting from the *longer second pulse* in the echo pulse sequence. It has been experimentally established that this undesirable effect, leading to phase distortions, can be reduced by decreasing the pulse lengths. In addition, shorter pulses improve the detection of broader signals.

Finally, Wilcke et al. [60] have suggested another NMR technique to detect very wide resonance lines. This technique is based on application of stochastic pulse sequences, which are combined with high magic angle spinning rates. To observe the extremely broad and largely shifted resonances, the authors have used *low-power*, approximately 1° pulses acting stochastically 0° and 180° out of phase

$$\delta_{iso} = \frac{10^6 \Delta\nu}{\nu_0} = \frac{10^6 A_{iso} \chi_M}{N_A \mu_0 g_e \mu_B g_y \mu_N}$$

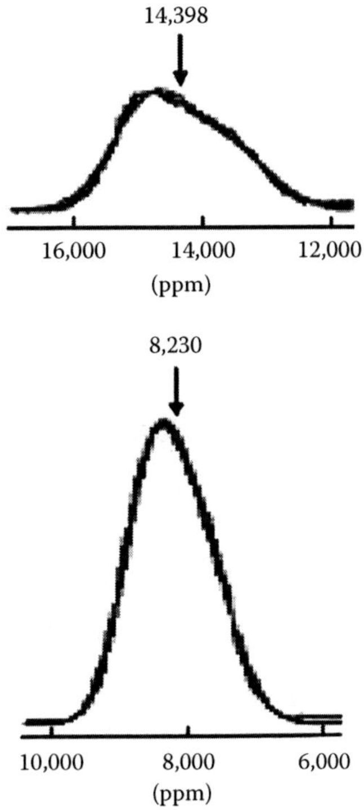

FIGURE 7.17 The spin-echo mapping ^{31}P NMR experiments performed in a static sample of paramagnetic material $LiFeP_2O_7$. The isotropic ^{31}P Fermi-contact chemical shifts are determined by DFT calculations via the equation shown in the top of the figure. (Reproduced with permission from Kim, J. et al., *J. Am. Chem. Soc.*, 132, 16825, 2010.)

relative to one another, and measured the shift of ^{31}P nuclei in paramagnetic solid $LiNiPO_4$ equal to 1726 ppm.

7.7 RELAXATION MEASUREMENTS AND RELAXATION TIMES IN PARAMAGNETIC SOLIDS

Following Bloch, the nuclear spins excited by a radio frequency irradiation reach an initial equilibrium state by two relaxation mechanisms, spin–lattice and spin–spin relaxation [16], where the behavior of macroscopic transverse and longitudinal magnetizations can be measured experimentally as a function of the time (see Chapters 2 and 5). As noted in these chapters, the T_1 and T_2 time constants can be obtained by the inversion-recovery, saturation-recovery or progressive saturation experiments (to measure T_1), where radio frequency pulses can be simple or composite, the Hahn-echo or Carr–Purcell pulse sequences (measuring T_2) and by spin-locking experiments in a rotating coordinate system (to measure $T_{1\rho}$) [61].

If nuclear relaxation in solutions is generally an exponential process, typical of isolated (or weakly coupled) spins at isotropic molecular reorientations with a single correlation time, then even in diamagnetic solids numerous structurally nonequivalent sites can have different motions and will show different nuclear relaxation. Similarly, in paramagnetic materials, where each site has its own relaxation via dipolar interaction with unpaired electrons, the relaxation curves do not follow the exponential law, generally leading to a multiexponential process.

In common cases and independent of the paramagnetic or diamagnetic nature of samples, the relaxation behavior is dependent on the nature of target nuclei, their coupling, and structural features of objects and can be determined, in practice, by fittings of magnetization decays to the corresponding functions. It should be emphasized that even conditions of NMR experiments, implying different pulse sequences, can affect the observed relaxation process. For example, Zhou et al. [62] have observed the ^{13}C spin–lattice NMR relaxation in organic solids as a *triple-exponential* process at ^{1}H radio frequency irradiation of a medium power. At the same time, under strong ^{1}H field irradiation or in its absence, the relaxation process can be reduced to a simple mono-exponential function.

When target nuclei in solids can be characterized by only two distinct environments (including, e.g., two different distances relative to paramagnetic centers) or only two distinct molecular motions, then the relaxation process can become bi-exponential. This most simple case (in the interpretation context) has been well documented in the careful ^{1}H, ^{19}F T_1 relaxation study of solid hexafluoroacetylacetone (at 62.7 K) [63], where the intramolecular ^{19}F–^{1}H dipolar interactions dominate and the bi-exponential T_1 relaxation shows two physically distinguishable CF_3 rotors with different energy activations.

Methodologically it is pertinent to demonstrate how this case can be recognized. Magnetization recovery curves, obtained experimentally by Horsewill and Tomdah [63] for ^{19}F and ^{1}H nuclei, are illustrated in Figure 7.18. As is seen, the inversion-recovery data reveal clearly the presence of two distinguished sections that are observed at short delay times and long τ delays, respectively. Moreover, the corresponding bends in the curves are also evident.

Both total curves can be well fitted to the equations, shown in the figure for ^{19}F (top) and ^{1}H (bottom) that thus describe the relaxation process as the sum of two exponential components. It is important that first, the relaxation rates obtained after treatments differ significantly (more than by 10 times), confirming their reliability, second, magnitudes “a” are calculated to be very similar for ^{19}F and ^{1}H experiments, third, the observation of the bends in the experimental curves seems to be very significant for further quantitative interpretations. Finally, Figure 7.18 shows clearly that the number of the experimental points in the curves plays a major role in identification of relaxation behavior.

It should be noted that the bi-exponential relaxation appears rather rare in studies of different complex systems, and first of all, amorphous materials. Moreover, nonexponential relaxation curves can be often fitted to two exponentials, parameters of which differ slightly. At the same time, the presence of both crystalline and amorphous regions within the bulk of samples, for example, in silica gel systems [64] can be probably responsible for the bi-exponential NMR decays collected by the ^{29}Si T_1 NMR experiments.

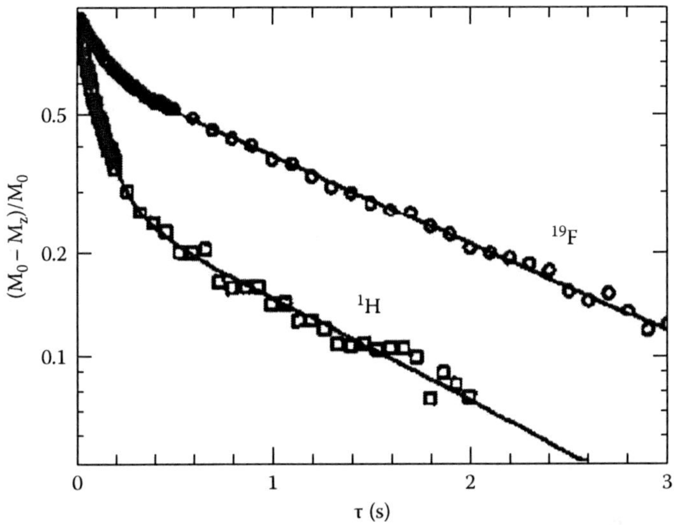

$$\left(\frac{I_0-\langle I_z\rangle}{I_0}\right)_{\substack{\langle I_z\rangle_{\tau=0}=0\\ \langle S_z\rangle_{\tau=0}=S_0}} = a_1^{(I)}\exp(-R_1\tau) - a_2^{(I)}\exp(-R_2\tau)$$

$$\left(\frac{S_0-\langle S_z\rangle}{S_0}\right)_{\substack{\langle I_z\rangle_{\tau=0}=0\\ \langle S_z\rangle_{\tau=0}=S_0}} = c^{(s)}\exp(-R_1\tau) - c^{(s)}\exp(-R_2\tau) \quad (7.17)$$

FIGURE 7.18 Magnetization inversion-recovery, obtained for ^{19}F (○) and ^{1}H (□) nuclei in solid hexafluoroacetylacetone. The solid lines are obtained by fittings to the corresponding equations, shown in the figure. For ^{19}F nuclei: $a_1^{(I)} = 0.67$, $a_2^{(I)} = 0.30$, $R_1^{-1} = 1.74$ s, and $R_2^{-1} = 0.129$ s; for ^{1}H nuclei: $a_1^{(S)} = 0.29$, $a_2^{(S)} = 0.66$, $R_1^{-1} = 1.48$ s, and $R_2^{-1} = 0.107$ s. (Reproduced with permission from Horsewill, A.J. and Tomdah, I.B.I., *Solid State Nucl. Magn. Reson.*, 2, 61, 1993.)

When nuclear relaxation is dominated by spin-diffusion to a paramagnetic center, it becomes mono-exponential even in rigid solids. In this case, a single T_1 value characterizes all of the nuclei in a solid, even chemically nonequivalent, which are situated within the barrier radius effective for spin-diffusion to paramagnetic centers [65,66]. Generally, this is typical of nuclei undergoing strong dipolar coupling, such as protons, phosphorous nuclei or ^{19}F. Therefore, Alaimo and Roberts [65] have emphasized that generally nuclear relaxation in materials follows a stretched exponential, $\exp(-(\tau/T_1)^\beta)$, where the β parameter, measured experimentally, can vary between 0 and 1 from a highly nonexponential relaxation to a mono-exponential relaxation behavior. In these terms, a stretched exponential with the β value of 0.5 corresponds to nuclear spin–lattice relaxation occurring via paramagnetic centers without spin-diffusion [22], while intermediate values between 0.5 and 1.0 can be attributed to a "diffusion-limited" mechanism of relaxation via random paramagnetic impurities [67]. On the other hand, it seems to be reasonable that a stretched exponential function can approximate a superposition of mono-exponential decays in the presence of *smooth relaxation-time distributions*. For example, a Gaussian T_1 distribution will correspond to the β parameter of 0.67.

Since this situation is not single-valued, reliable identification of a relaxation mechanism is needed for an investigated system. Stebbins and Kesley [52] have shown that this problem in identification of relaxation behavior can appear in NMR probing of silicates, when they contain remarkable amounts of paramagnetic centers, such as MnO, FeO, and CoO oxides. The authors have studied samples of the synthetic and natural pyrope garnets ($[Mg,Fe]_3Al_2Si_3O_{12}$), named here as SP and DW, respectively, which contain 0.01 and 1.4 wt% of paramagnetic oxide FeO. The ^{29}Si and ^{27}Al MAS NMR spectra have been recorded at spinning rates of 20–30 kHz to show a main resonance with δ(iso) of −73 and ~0 ppm, for silicon and aluminum nuclei, respectively. Both the saturation-recovery and inversion-recovery methods have been applied to determine spin–lattice relaxation rates of ^{29}Si and ^{27}Al nuclei. The experimental data have clearly shown that the ^{29}Si relaxation is not exponential. It has been found that such ^{29}Si T_1 data can be fitted to the stretched exponential in linear space, that is, in coordinates "intensity versus delay time τ," because these fittings provide *correct weighting of uncertainties for each point* of the curves. This is a very important point for a researcher. Nevertheless, as noted by the authors, results of the fitting procedures and an analysis of the relaxation curves can be better methodically realized using *log–log plots* shown in Figure 7.19. Here, the SP sample shows only slight deviations, observed at the very shortest τ values, where uncertainties on a log scale are greatest.

The next important recommendation is as follows: the comparison of relaxation curves with different β values can be remarkably simplified by using of a parameter, T(50), corresponding to the situation when the magnetization reaches 50% of the fully recovered value.

In spite of the very large difference in the ^{29}Si T_1 data found in the SP and DW samples, both of them show the β parameter of 0.5. Thus, the spin-diffusion mechanism for ^{29}Si nuclei is negligible at the different concentrations of the FeO oxide. In silicate materials with low paramagnetic contents, spin–lattice relaxation for ^{27}Al nuclei is generally much faster than for ^{29}Si nuclei. The SP and DW samples in Figure 7.19 confirm well this statement for two reasons. First, ^{27}Al nuclei in these samples undergo quadrupolar interactions (e.g., the simulations of the ^{27}Al MAS NMR spectra of the SP sample lead to the aluminum quadrupolar coupling constant of 1.05 MHz at a symmetry parameter of 0) and they can additionally relax via the quadrupolar mechanism. Second, in contrast to ^{29}Si nuclei, the natural abundance of ^{27}Al nuclei is 100%. Under this circumstance, the probability for the spin-diffusion mechanism significantly increases. In fact, the relaxation curves in Figure 7.19 are better treated as mono-exponential, that is, at $\beta = 1$.

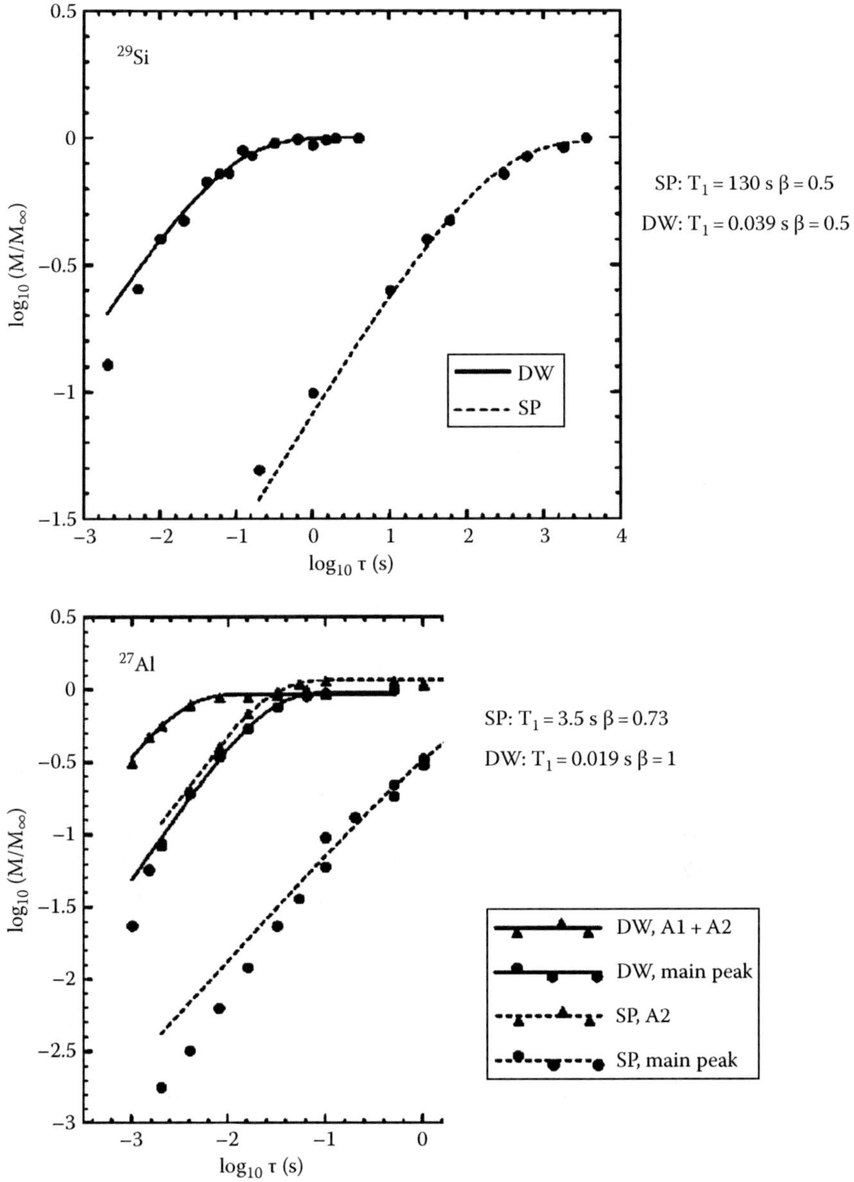

FIGURE 7.19 The T_1 relaxation measurements performed in the synthetic and natural pyrope garnets ($[Mg,Fe]_3Al_2Si_3O_{12}$) (SP and DW, respectively) for ^{29}Si (top) and ^{27}Al (bottom) nuclei (main and additional peaks) represented in log–log coordinates. Solid lines are obtained by fittings to the stretched exponential function in linear space: "intensity versus delay time τ." (Reproduced with permission from Stebbins, J.F. and Kelsey, K.E., *Phys. Chem. Chem. Phys.*, 11, 6906, 2009.)

Thus, relaxation measurements in paramagnetic solids and the interpretations of collected data are not simple and the following points should be taken into account.

1. The β parameter should be determined reliably on the basis of a large number of delay times in the relaxation experiments, performed with the carefully adjusted radio frequency pulses [61]. It has been shown in practice, that incorrect pulse adjustments expressed by their elongations, for example, in experiments with silica-based materials doped with the paramagnetic metal ions, can lead to changes in the ^{29}Si T_1 times and the β parameters by more than 30%, while the quality of fitting procedures does not change [68].
2. Even at the well-determined β parameters, discrimination between the different relaxation mechanisms is still problematic. For example, an exponential relaxation that could be evidence for dominating the spin-diffusion mechanism is often observed for ^{29}Si nuclei (as is shown in the following) that are coupled weakly and therefore probability of their mutual spin flops is very low. In contrast, a stretched-exponential relaxation with $\beta > 0.5$ can be caused by the dipolar mechanism or the diffusion-limited mechanism.
3. If T_1 or T_2 values are reliably determined and the spin-diffusion mechanism completely dominates, these relaxation times characterize the NMR properties of solids but not their chemical structure.
4. When the dipolar mechanism dominates, Equations 7.13 and 7.14 describe the relaxation rates due to one of nucleus–electron dipolar contacts. Commonly, however, in the absence of independent data, the number of these contacts remains unknown.
5. Finally, spin diffusion and dipole–dipole relaxation differently depend on concentrations of paramagnetic centers. Therefore, their relative contributions to $T_{1,2}$ times measured experimentally can change at variation in concentrations of paramagnetic ions. In other words, researchers need criteria distinguishing the relaxation mechanisms.

In the classic work [69], Kessemeier and Norberg have carefully investigated a number of static and spinning samples containing paramagnetic centers. It has been shown that the spin-diffusion coefficient, $D(\omega_R)$, decreases in spinning samples and thus the T_1^{SD} times in Equation 7.18 depend on MAS rates ω_R:

$$\frac{1}{T_1^{SD}} = \left(\frac{1}{3}\right) 8\pi N p C^{1/4} D^{3/4}(\omega_R) \tag{7.18}$$

In turn, the T_2^{SD} times increase in spinning samples as follows:

$$T_2^{SD} = \left(\frac{4}{3}\right) \frac{(\omega_R)^2 \tau^2}{\Delta\nu^2 \tau} \tag{7.19}$$

or

$$T_2^{SD} = \left(\frac{4}{3}\right)\frac{(\omega_R)\tau}{\Delta\nu^2} \quad (7.20)$$

It should be noted that these expressions are valid at $\nu_R/\Delta\nu \geq 1$, where spinning rates ν_R ($\omega_R = 2\pi\nu_R$) and line widths $\Delta\nu$ are measured in Hz. Because the dipolar spin–lattice relaxation times T_1^{DD} are independent of magic angle spinning, the variable spinning rate experiments performed in solids can potentially recognize the relaxation mechanisms.

Gil and Alberti [70] have reported on remarkable spinning effects on ^{1}H T_1 values measured in solid organic compounds, for example, in adamantane and glycine. The data have been interpreted in terms of decreasing the spin-diffusion efficiency at faster rotations. Similarly, reducing the ^{1}H spin diffusion by 65% has been found for the high-density polyethylene when the MAS rate has been increased from 2 to 12 kHz [71]. In addition, the authors have predicted decrease in the effective spin diffusivity by >90% at spinning rates of 30 kHz and more.

Hayashi [72] has measured the room-temperature ^{1}H and ^{29}Si T_1 times in a static and spinning sample of talc, where paramagnetic impurities work as relaxation centers. Since spin-diffusion plays an important role in relaxation of protons due to their strong dipolar coupling, the ^{1}H T_1 times remarkably increase in spinning samples. This effect, exceeding strongly errors in the relaxation time measurements is shown in Figure 7.20. In contrast, the ^{29}Si T_1 times in talc did not change practically in going from static to spinning samples. They are measured as 46 and 47 s, respectively, again demonstrating that spin-diffusion is negligible and relaxation occurs via direct ^{29}Si–^{29}Si dipolar interactions and/or interactions electron-^{29}Si.

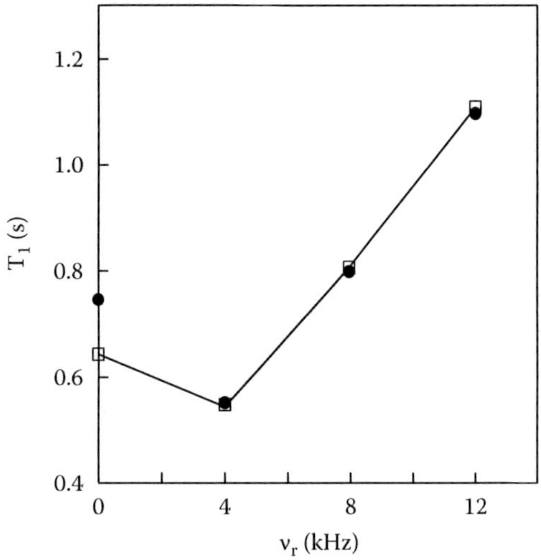

FIGURE 7.20 The spinning rate dependence of the ^{1}H T_1 time in talk, measured at room temperature. (Reproduced with permission from Bakhmutov, V.I., *Chem. Rev.*, 111, 530, 2011.)

Interpretation of spinning effects on T_2 measurements in solids is more complex especially for systems with relatively high molecular mobility [70]. In fact, even in the absence of spin diffusion the dipolar spin–spin relaxation can depend on spinning rates ω_R:

$$\frac{1}{T_2^{DD}} = \left(\frac{1}{5}\right)\gamma^4\hbar^2 r^{-6} I(I+1)\left\{\frac{2\tau}{(1+\omega_R^2\tau^2)} + \frac{\tau}{(1+4\omega_R^2\tau^2)}\right\} \tag{7.21}$$

According to the theoretical considerations and the experiments performed by Haeberlen and Waugh [73], the dipolar T_2^{DD} time becomes spinning dependent, if $\omega_R^2\tau^2 > 1$. This condition is valid at the relatively long correlation time, τ, calculated as $>2 \times 10^{-5}$ s at a spinning rate of 10 kHz. Then, as in the case of the spin-diffusion T_2^{SD} times in Equation 7.19, the dipolar T_2^{DD} time will be proportional to ω_R^2.

7.7.1 Solid-State T_1 Measurements

In practice, the spin–lattice relaxation time measurements by the inversion-recovery, saturation-recovery or progressive saturation methods are experimentally simple even for paramagnetic amorphous materials, where T_1 times are relatively short. The T_1 experiments require a minimum of experience directed to accurate pulse calibrations particularly in the case of a nonexponential relaxation behavior. Since generally nonlinear regression fittings to stretched exponentials, dependent on the β parameter, are more complicated than those to an exponential function, the relaxation data would be collected at high enough signal-to-noise ratios and the y should be treated statistically. Alaimo and Roberts have investigated silica gels with the surface covered by copper ions and demonstrated that the β values obtained in triplicate measurements with >20% errors are statistically unreliable. Such results should be ruled out. Only the β values obtained with smaller errors, for example, as shown in Figure 7.21, can

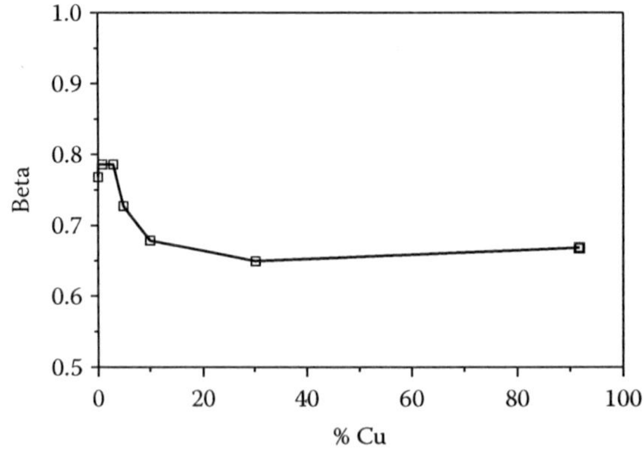

FIGURE 7.21 The β parameters, obtained by triplicate measurements with <20% errors in the ^{29}Si T_1 relaxation studies of silica gels with the surface covered by copper ions at their different concentrations. (Reprinted with permission from Bakhmutov, V.I., *Chem. Rev.*, 111, 530, 2011.)

be truly analyzed in terms of relaxation mechanism and distributions of paramagnetic centers [65]. It is obvious that the T_1 times, resulting from treatments of relaxation data with stretched exponentials, should also be statistically reliable.

7.7.2 Spin–Spin Relaxation Times and Anomalies in Solid-State T_2 Measurements

In general, the spin–spin relaxation times T_2 of nuclei (independently of their nature) in solids are strongly shorter than the T_1 times and therefore their accurate determination becomes more difficult. Nevertheless, in spite of this problem, the T_2 times are often applied for structural studies or probing dynamics of diamagnetic and paramagnetic systems. For example, Oldfield et al. [74] have demonstrated how ^{27}Al T_2 times can be applied for characterizations of the complex commercial zeolite catalysts. The authors have reported on the ^{27}Al T_2 times of 1.1–4.6 ms measured in diamagnetic zeolites and amorphous silica-aluminas extracted from practically exponential NMR decays. It has been found that these aluminum relaxation times depend on hydration of the samples. ^1H T_2 NMR decays have been also exponential in silicate fiberglass supports and resulted in the ^1H T_2 times of 0.2–6.8 ms that characterize the surface silanol groups [75].

More complex ^1H NMR decays, exhibiting several relaxation components in the Hahn-echo and CPMG experiments, have been observed in different elastic polymers by Hayashi and Komori [76]. The authors have improved the formula expressing a proton transverse magnetization decay together with distributed residual dipolar couplings. In addition, they have shown that T_2 times extracted from spin–lattice relaxation experiments in the rotating frame can be successfully used for characterization of the cross-link density in these systems.

A bi-exponential ^{31}P T_2 relaxation has been observed for the NMR-detectable phosphorous nuclei in the materials of the AlPO-5 family [35]. This phosphorous relaxation has been very sensitive to incorporation of paramagnetic ions into the matrix of the materials. For example, the material CoAlPO has shown ^{31}P spin–spin relaxation that consists of *slow* and *fast* T_2 components. These components have been measured as 269 and 0.4 ms, respectively, and used as characteristics of close and remote cobalt environments. Finally, the effects of anisotropic deuterium spin–spin relaxation on line shapes in ^2H NMR spectra of paramagnetic complex $[Mn(H_2O)_6]$ $[SiF_6]$ have been analyzed by Mizuno et al. [45].

Electron relaxation times change in very large limits as a function of the nature of paramagnetic centers. Some magnitudes of relaxation times, τ_e in Equation 7.14, can obviously produce spin–spin relaxation times that are too short and their accurate determination can be difficult in practice. Potentially such a situation can be expected for different materials doped with ions Mn^{2+}, Ni^{2+}, or Cu^{2+}. Therefore, recently the Hahn-echo ^{29}Si T_2 experiments in supermicroporous materials SiO_2–MnO and SiO_2–Al_2O_3–MnO have been performed with τ echo delays, varying from very short magnitudes of 10 μs [77]. These methodologically important experiments have been carried out in static samples because the Hahn-echo MAS NMR requires synchronization with spinning rates, limiting, thus, applications of short delays.

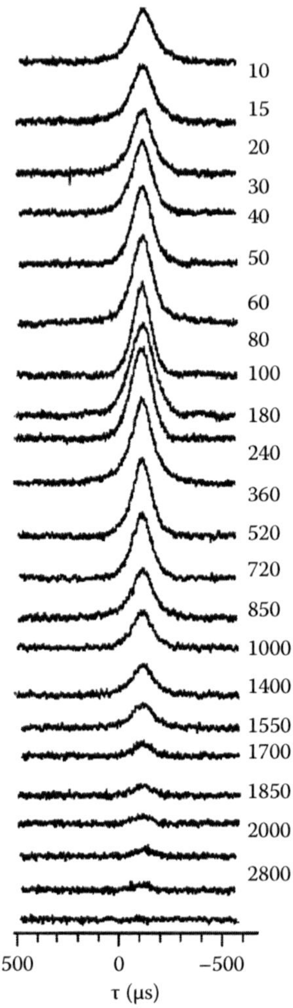

FIGURE 7.22 The ^{29}Si Hahn-echo NMR spectra (chemical shift is measured in ppm) recorded for a static material SiO_2–MnO (15 wt% of Mn) at echo delays from 10 (top) to 2800 (bottom) μs. (Reprinted with permission from Bakhmutov, V.I., *Chem. Rev.*, 111, 530, 2011.)

Figure 7.22 shows the set of the relaxation data where already at a delay τ of 2 ms the ^{29}Si line intensity is going to zero.

According to the principles of registration of signals in solid-state NMR [39], short echo delays in a combination with an instrumental dead time can distort the collected FIDs. Since the effect is difficult to predict theoretically and real instrumental dead time is generally unknown and changes with "age of a spectrometer," the distorted FIDs, in practice, will complicate T_2 determinations. The patterns "intensity-τ" for materials SiO_2–MnO and SiO_2–Al_2O_3–MnO, doped with different Mn concentrations, are shown in Figure 7.23, where the echo-intensity initially increases with increasing τ values and then it "normally" reduces due to spin–spin relaxation. These "normal" τ zones can be well fitted to an exponential function,

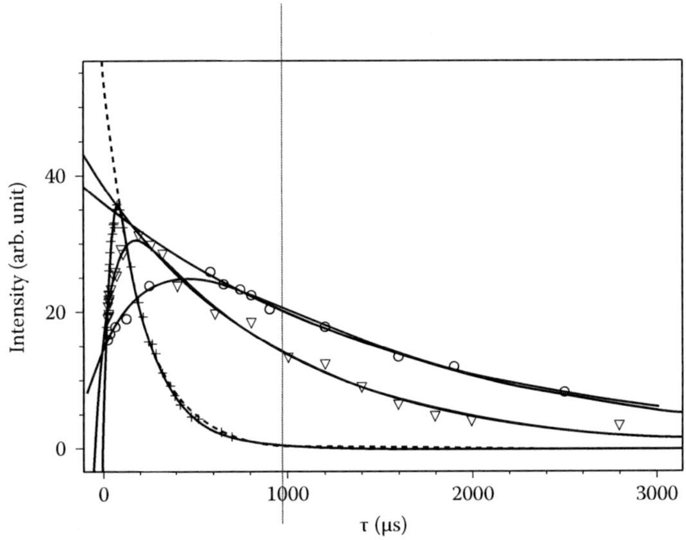

FIGURE 7.23 The ^{29}Si Hahn-echo curves in coordinates "intensity (arbitrary unit) – echo delay (μs)" obtained in materials SiO_2–MnO with 0 (○), 2 (▿), and 12 (+) wt% of manganese. (Reprinted with permission from Bakhmutov, V.I., *Chem. Rev.*, 111, 530, 2011.)

$I(\tau) = I_0 exp(-\tau/T_2)$ (dashed lines in Figure 7.23) to give the ^{29}Si T_2 times between 1.7 and 0.2 ms for the diamagnetic system SiO_2–Al_2O_3 and the paramagnetic material SiO_2–MnO with 13 wt% of Mn, for example.

Menetrier et al. [47] have shown that echo-intensities in paramagnetic solids can be decreased due to saturation at a carrier frequency, which can be minimized by reducing pulse lengths. However, the character of the curves (e.g., in Figure 7.23) has been practically independent of pulse lengths and thus the abnormal points can be definitely connected with the instrumental dead time comparable with the shortest τ magnitudes. Thus, the "abnormal" zones can reflect "recovery of the receiver."

The author [77] has formally described the behavior of the echo intensity assuming that the current intensity can be represented as *magnitude, relaxing exponentially, but reduced by intensity loss that, in turn, decreases also exponentially with time constant* T^R:

$$I(\tau) = I_0\left(a\exp\left(\frac{-\tau}{T_2}\right) - (1-a)\exp\left(\frac{-\tau}{T^R}\right)\right) \tag{7.22}$$

The solid lines in Figure 7.23 correspond to this *formal* description, where at $\tau = 0$ the intensity loss is very large. It has been emphasized that the dashed and solid lines are very similar in the normal τ zones and thus if the first abnormal points are ruled out, the calculated ^{29}Si T_2 times change insignificantly. The latter is very important since the MAS experiments synchronized with spinning rates will be performed precisely in these τ zones. Finally, the T^R constants, obtained as fitting parameters, correlated well with the T_2^* values measured experimentally from the line widths at half-height as $T_2^* = 1/\Delta\nu$ [77]. Anomalies in the spin echo behavior (similar to the mentioned above) at T_2 determinations are well known mainly in solutions due to competition between radiation damping effect and nuclear relaxation [80–82], leading to appearance of an unreal echo. Radiation damping effect originates from feed-back between coil and sample after the action of first pulse. As noted, such feed-back is very probable when samples contain such nuclei as ^1H, showing strongest dipolar coupling. This echo behavior can be also connected with a distribution of resonance frequencies in inhomogeneous systems when the radio-frequency pulses can became imperfect. In such cases, magnetization dephasing during the time between the first and second pulses could be converted into refocusing magnetization leading to the echo formation. However, when the second pulse is not perfect, part of the dephasing magnetization can keep dephasing after the action of the second pulse.

Some solids, where ^{29}Si nuclei are targets at their natural abundance, also show an unusual behavior in ^{29}Si T_2 experiments, which can be found in the works of Gil and Alberti [70] and Levstein and Franzoni [78]. However, here, the anomalies of other nature are connected with appearance of *long-lived spin-echoes* detected as long magnetization tails. Levstein and Franzoni [78] have explained the nature of this phenomenon in the limits of two conditions: inhomogeneity of the applied radio frequency field and the absence of spin diffusion, when flip-flop interactions are not effective.

In summary, it should be noted again that high concentrations of paramagnetic centers generally complicate NMR studies of solids and prevent the application of various advanced techniques developed for diamagnetic systems. Even the cross-polarization NMR experiments providing great advantage in observations of nuclei at their small natural abundance are not effective due to "negative" paramagnetic effects. Finally, the NMR technique based on dynamic nuclear polarization leading to sensitivity enhancement cannot be applied for paramagnetic materials, according to the DNP principles that can be found in Wind et al. [4].

REFERENCES

1. Sharp, R., Lohr, L., Miller, L. 2001. *Prog. Nucl. Magn. Reson. Spectrosc*. 38: 115.
2. Wickramasinghe, N. P., Kotecha, M., Samoson, A., Past, J., Ishii, Y. 2007. *J. Magn. Reson*. 184: 350.
3. Wang, W., Hu, J. Z., Alderman, D. W., Pugmire, R. J., Grant, D. M. 1995. *Solid State Nucl. Magn. Reson*. 5: 257.
4. Wind, R. A., Duijvestun, M. J., Van Der Lugt, C., Manenschijn, A., Vriend, J. 1985. *Prog. Nucl. Magn. Reson. Spectrosc*. 17: 33.
5. Reynhard, E. C. 2003. *Concept. Magn. Reson. A* 19A: 44.
6. Woods, M., Woessner, D. E., Sherry, A. D. 2006. *Chem. Soc. Rev*. 35: 500.
7. Wickramasinghe, N. P., Ishii, Y. 2006. *J. Magn. Reson*. 181: 233.
8. Kervern, G., Pintacuda, G., Emsley, L. 2007. *Chem. Phys. Lett*. 435: 157.
9. Peng, W. K., Samoson, A., Kitagawa, M. 2008. *Chem. Phys. Lett*. 460: 531.
10. Poole, C. P. 1983. *Electron Spin Resonance*. New York: Wiley.
11. McConnel, H. M., Chesnut, D. B. 1958. *J. Chem. Phys*. 28: 107.
12. Kim, J., Middlemiss, D. S., Chernova, N. A., Zhu, B. Y. X., Masquelier, C., Grey, C. P. 2010. *J. Am. Chem. Soc*. 132: 16825.

13. Bernal, O. O., Rodrigues, C., Martinez, A. 2001. *Phys. Rev. Lett.* 87: 196402.
14. Nowak, B., Troc, R. 2000. *Solid State Nucl. Magn. Reson.* 18: 53.
15. Van der Klink, J. J., Brom, H. B. 2000. *Prog. Nucl. Magn. Reson. Spectrosc.* 36: 89.
16. Abragam, A. 1961. *Principles of Nuclear Magnetism.* Oxford, U.K.: Oxford University Press.
17. Bertini, I., Luchinat, C., Parigi, G. 2002. *Prog. Nucl. Magn. Reson. Spectrosc.* 40: 249.
18. Clore, G. M., Iwahara, J. 2009. *Chem. Rev.* 109: 4108.
19. Kubo, A., Spaniol, T. P., Terao, T. 1998. *J. Magn. Reson.* 133: 330.
20. Barbara, T. M. 1994. *J. Magn. Reson. A* 109: 265.
21. Aime, S., Bertini, I., Luchinat, C. 1996. *Coord. Chem.* Rev. 150: 221.
22. Bloembergen, N., Purcell, E. M., Pound, R. V. 1948. *Phys. Rev.* 73: 679.
23. Forte, C., Geppi, M., Malvaldi, M., Mattoli, V. 2004. *J. Phys. Chem. B* 108: 10832.
24. Tse, D., Hartmann, S. R. 1968. *Phys. Rev. Lett.* 21: 511.
25. Lamanna, R. 2005. *Concept. Magn. Reson. A* 26A: 78.
26. Bloembergen, N. 1949. *Physica* 15: 386.
27. Hayashi, S., Hayamizu, K., Akiba, E. 1992. *J. Phys. Chem.* 96: 10928.
28. Brown, S. P. 2007. *Prog. Nucl. Magn. Reson. Spectrosc.* 50: 199.
29. Chen, Q., Schmidt-Rohr, K. 2006. *Solid State Nucl. Magn. Reson.* 29: 142.
30. Hayashi, S., Omi, H. 2010. *Solid State Nucl. Magn. Reson.* 37: 69.
31. Solomon, I. 1955. *Phys. Rev.* 99: 559.
32. Bloembergen, N. 1957. *J. Chem. Phys.* 27: 572.
33. Lee, C. E., Lee, C. H., Kim, J. H., Kim, K. S., Kim, K. 1994. *J. Korean Chem. Soc.* 38: 628.
34. Grey, C. P., Smith, M. E., Cheethem, A. K., Dobson, C., Duptreet, R. 1990. *J. Am. Chem. Soc.* 112: 4670.
35. Peeters, M. P. J., Van De Ven, L. J. M., Haan J. W., Van Hooff, J. H. C. 1993. *Colloid. Surface. Physicochem. Eng. Aspect.* 72: 87.
36. Canesson, L., Boudeville, Y., Tuel, A. 1997. *J. Am. Chem. Soc.* 119: 10754.
37. Mali, G., Ristić, A., Kaucic, V. 2005. *J. Phys. Chem. B* 109: 10711.
38. Flambard, A., Ruhlmann, L., Canny, L., Thouvenot, R. 2008. *C. R. Chimie* 11: 415.
39. Duer, M. J. 2002. *Solid-State NMR Spectroscopy: Principles and Applications.* Oxford, U.K.: Blackwell Science.
40. Blumel, J., Herker, M., Hiller, W., Koehler, F. H. 1996. *Organometallics* 15: 3474.
41. Heise, H., Kohler, F. H., Xie, X. 2001. *J. Magn. Reson.* 150: 198.
42. Clayton, A. N., Dobson, C. M., Grey, C. P. 1990. *Chem. Commun.* 72.
43. Brough, A. R., Grey, C. P., Dobson, C. M. 1993. *J. Am. Chem. Soc.* 115: 7318.
44. Siminovitch, D. J., Rance, M., Jeffrey, K. R., Brown, M. F. 1984. *J. Magn. Reson.* 58: 62.
45. Mizuno, M., Itakura, N., Endo, K. 2005. *Chem. Phys. Lett.* 416: 358.
46. Feuerstein, M., Lobo, R. F. 1998. *Chem. Mater.* 10: 2197.
47. Chazel, C., Menetrier, M., Croguenne, L., Delmas, C. 2005. *Magn. Reson. Chem.* 43: 849.
48. Accardi, R. J., Lobo, R. F. 2000. *Micropor. Mesopor. Mater.* 40: 25.
49. Terskikh, V. V., Ratciffe, C. I., Ripmeester, J. A., Reinhold, C. J., Anderson, P. A., Edwards, P. P. 2004. *J. Am. Chem. Soc.* 126: 11350.
50. Nery, J. G., Giotto, M. V., Mascarenhas, Y. P., Cardoso, D., Zotin, F. M. Z., Sousa-Aguiar, E. F. 2000. *Micropor. Mesopor. Mater.* 41: 281.
51. Plevert, J., Okubo, T., Wada, Y., O'Keefe, M., Tatsumi, T. 2001. *Chem. Commun.* 2112.
52. Stebbins, J. F., Kelsey, K. E. 2009. *Phys. Chem. Chem. Phys.* 11: 6906.
53. Shubin, A. A., Khabibulin, D. F., Lapina, O. B. 2009. *Catal. Today* 142: 220.
54. Tang, J. A., Ellis, B. D., Warren, T. H., Hanna, J. V., Macdonald, C. L. B., Schurko, R. W. 2007. *J. Am. Chem. Soc.* 129: 13049.
55. O'Dell, L. A., Rossini, A. J., Schurko, R. W. 2009. *Chem. Phys. Lett.* 468: 330.

56. Ellis, P., Sears, J. A., Yang, P. et al. 2010. *J. Am. Chem. Soc.* 132: 16727.
57. Canesson, L., Tuel, A. 1997. *Chem. Commun.* 241.
58. Tuel, A., Canesson, L., Volta, J. C. 1999. *Colloid. Surface. Physicochem. Eng. Aspect.* 158: 97.
59. Wilcke, S. L., Lee, Y.-J., Cairns, E. J., Reimer, J. A. 2007. *Appl. Magn. Reson.* 32: 547.
60. Wilcke, S. L., Cairns, E. J., Reimer, J. A. 2006. *Solid State Nucl. Magn. Reson.* 29: 199.
61. Bakhmutov, V. I. 2005. *Practical NMR Relaxation for Chemists*. Chichester, U.K.: Wiley.
62. Zhou, J., Fu, R., Hu, J., Li, L., Ye, C. 1997. *Solid State Nucl. Magn. Reson.* 7: 291.
63. Horsewill, A. J., Tomdah, I. B. I. 1993. *Solid State Nucl. Magn. Reson.* 2: 61.
64. Pfleiderer, B., Albert, K., Bayer, E., van den Ven, L., de Haan, J., Cramers, C. 1990. *J. Phys. Chem.* 94: 4189.
65. Alaimo, M. H., Roberts, J. E. 1997. *Solid State Nucl. Magn. Reson.* 8: 241.
66. Rorschach, H. E. 1964. *Physica* 30: 38.
67. Narayanan, A., Hartman, J. S., Bain, A. D. 1995. *J. Magn. Reson. Ser. A* 112: 58.
68. Bakhmutov, V. I. 2008. *Solid State Nucl. Magn. Reson.* 34: 197.
69. Kessemeier, H., Norberg, R. E. 1967. *Phys. Rev.* 155: 321.
70. Gil, A. M., Alberti, E. 1998. *Solid State Nucl. Magn. Reson.* 11: 203.
71. Jia, Z., Zhang, L., Chen, Q., Hansen, E. W. 2008. *J. Phys. Chem. A* 112: 1228.
72. Hayashi, S. 1994. *Solid State Nucl. Magn. Reson.* 3: 323.
73. Haeberlen, U., Waugh, J. S. 1969. *Phys. Rev.* 185: 420.
74. Oldfield, E., Haase, J., Schmitt, K. D., Schramm, S. F. 1994. *Zeolites* 14: 101.
75. Romanenko, K. V., Lapina, O. B., Sominova, L. G., Fraissard, J. 2002. *Phys. Chem. Chem. Phys.* 5: 2686.
76. Hayashi, S., Komori, Y. 2009. *Solid State Nucl. Magn. Reson.* 36: 167.
77. Bakhmutov, V. I. 2009. *Solid State Nucl. Magn. Reson.* 36: 164.
78. Franzoni, M. B., Levstein, P. R. 2005. *Phys. Rev. B* 72: 235410.
79. Bakhmutov, V. I. 2011. *Chem. Rev.* 111: 530–562.
80. Mao, X. A., Guo, J. X., Ye, C. H. 1994. *Chem. Phys. Lett.* 221: 65.
81. Mao, X. A., Guo, J. X., Ye, C. H. 1994. *Chem. Phys. Lett.* 218: 249.
82. Augustine, M. P. Hahn, C. L. 2001. *Conc. Magan. Reson.* 13: 1.

8 Strategy in NMR Studies of Amorphous Porous Paramagnetic Materials

The previous chapter has shown how the presence of unpaired electrons affects isotropic and anisotropic chemical shifts of target nuclei, their line shapes in static and spinning regimes, and spin–lattice and spin–spin relaxation times. Some practical aspects in NMR studies of paramagnetic solids have also been considered. This chapter formulates the common strategy in NMR investigations of amorphous porous paramagnetic systems and reviews the available literature, which is focused on distributions of paramagnetic ions through the volume of a material and on incorporation of the ions into the matrix of the material. In the context of solid-state NMR, paramagnetic amorphous porous materials seem to be most difficult for studies and therefore the strategy applied for such complex objects can be successfully used for other systems. This chapter focuses on the following:

1. Interpretation of sideband patterns observed in magic angle spinning (MAS) NMR of paramagnetic solids
2. Direct spectral NMR observation of atoms binding to paramagnetic ions
3. Relaxation times dependent on concentrations of paramagnetic ions
4. Relaxation models providing reliable structural conclusions

8.1 INTENSE SIDEBAND PATTERNS IN MAS NMR SPECTRA OF PARAMAGNETIC AMORPHOUS MATERIALS AND THEIR ANALYSIS

When the supermicroporous silica-based materials or the aluminophosphates, described in works [1,2], respectively, are doped with paramagnetic metal ions (Mn^{2+} or Co^{2+}), and then their single-pulse ^{27}Al, ^{29}Si, or ^{31}P MAS NMR spectra exhibit broadened resonances, accompanied by numerous intense sidebands similar to those shown in Figure 7.9. This is in full contrast to the well-resolved ^{27}Al or ^{29}Si MAS NMR spectra of diamagnetic materials that do not exhibit intense sidebands under the same conditions. It has been noted in the literature that such intense sidebands, for example, observed in the ^{31}P MAS NMR spectra of cobalt-aluminophosphates, have initially been considered as a proof for incorporations of cobalt atoms into the framework of the materials. However later Tuel et al. [2,3] have observed similar spectral behavior in the Co-*impregnated* aluminophosphates and concluded that the phenomenon cannot generally make the discrimination between the framework and nonframework cobalt atoms. Moreover, the intense sidebands can appear in ^{29}Si MAS NMR spectra of silica-based materials even on their mechanical mixing with paramagnetic Mn or Ni salts due to large BMS effects (as shown as follows). Intuitively it is clear that in the absence of quantitative analysis of the observed side-band effects, reliable structural conclusions are impossible.

NMR sideband patterns can be quantitatively analyzed, for example, in solid complexes containing paramagnetic lanthanide ions, which do not lead to remarkable line broadenings due to relatively short electron relaxation times. Recently it has been demonstrated by Polenova et al. [4] that local geometry of the lanthanide ions in a series of oxo-tungstates can be accurately solved by analysis of anisotropic electron–nuclear dipolar interactions that completely dominate in the observed ^{31}P MAS NMR spectra. Clearly, the validity of this approach can be proofed only for crystal systems providing independent structural data. Then the approach can be applied for amorphous solids.

The anisotropy of electron–nuclear dipolar interactions for ^{31}P nuclei and also isotropic chemical shifts are determined directly from NMR spectra, which can be recorded at varying spinning rates. From these MAS NMR experiments, shown in Figure 8.1, for example, in compound $K_{14}(H_3O)_3[Er(\alpha_1\text{-}P_2W_{17}O_{61})_2]$ 4KCl $64H_2O$, containing europium paramagnetic ions, there are two different phosphorous atoms with the remarkable chemical shift difference. The NMR parameters can also be calculated theoretically on the basis of a point–dipole approximation [4].

In this model, the Hamiltonian for nuclei with spins of 1/2 (^{31}P) in a paramagnetic system takes the form

$$H = H_{Zeeman} + H_{RF} + H_{DIP}^{Paramgn} + H_{CSA} + H_{FC} + H_{DIP} \tag{8.1}$$

representing the Zeeman part and contributions from the radio frequency field, the electron–nuclear dipolar interaction, the chemical shift anisotropy, the Fermi contact shift, and the nuclear dipolar interaction. The electron–nuclear dipolar interaction is expressed as $H_{DIP}^{Paramgn}$ in the form shown in this figure. In turn, the electron–nuclear dipolar coupling tensor is expressed in a matrix representation, where r is the electron–nuclear distance, δαβ is the Kronecker delta function, and eα and eβ are the (x, y, z) components of the electron–nuclear dipolar vector. Finally, the total electron–nuclear dipolar Hamiltonian can be obtained by summing up the dipolar matrices that correspond to the individual electron–nuclear spin pairs.

When one of the terms in Equation 8.1 dominates, $H_{DIP}^{Paramgn}$, the calculations of the anisotropy can be carried out to account for the paramagnetic centers located within 100 Å relative to phosphorous atoms, to provide good agreement with experimentally determined magnitudes. The authors [4] have emphasized that for objective reasons, the phosphorous–paramagnetic ion distances extracted from NMR data differ slightly from those determined by the x-ray data (Table 8.1). Generally the NMR distances are more elongated.

It has been noted also that this approach, applicable to nuclei other than ^{31}P and for other noncrystalline inorganic solids, is principally limited by features of lanthanide ions: due to very large magnetic moments of Gd, Dy, and Tb, they complicate observations of ^{31}P resonances causing the much shortened ^{31}P T_2 times [4].

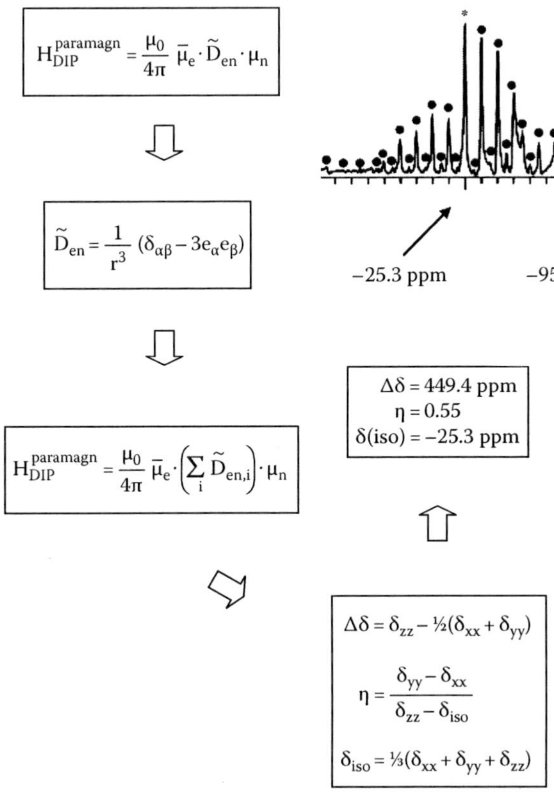

FIGURE 8.1 The ^{31}P MAS NMR spectrum of compound $K_{14}(H_3O)_3[Er(\alpha_1\text{-}P_2W_{17}O_{61})_2]$ 4KCl $64H_2O$, recorded at 15 kHz (there are two sets of NMR signals corresponding to the P1 site and P2 site with different intensities). A theoretical model used for calculations of electron–nuclear dipolar coupling tensors.

TABLE 8.1
Distances between Phosphorous Atoms, Noted as P1 and P2, and Lanthanide Centers in Some Paramagnetic Solids Determined by X-Ray and NMR Methods

Compound	X-Ray Distance (Å)	NMR Distance (Å)
$K_{14}(H_3O)_3[Nd(\alpha_1\text{-}P_2W_{17}O_{61})_2]$ 4KCl $64H_2O$	P1: 4.67	P1: 4.96
	P2: 6.23	P2: 6.30
$K_{14}(H_3O)_3[Er(\alpha_1\text{-}P_2W_{17}O_{61})_2]$ 4KCl $64H_2O$	P1: 4.56	P1
	P2: 6.03	P2: 6.56
$K_{14}(H_3O)[Nd(H_2O)_3\ (\alpha_2\text{-}P_2W_{17}O_{61})_2]$ 4KCl nH_2O	P1: 4.44	P1: 4.77
	P2: 8.09	P2: 6.51

Source: Huang, W. et al., *J. Am. Chem. Soc.*, 130, 481, 2008.

Similarly, Goldfarb [5] has analyzed the intense sideband patterns observed in the ^{27}Al and ^{31}P MAS NMR spectra of the paramagnetic materials MnAlPO$_5$ prepared at different concentrations of manganese ions. These ions have been identified as Mn^{2+} by the EPR spectroscopy. Again, the anisotropic dipolar electron–nucleus interactions have been calculated and fitted to the experiments. On the basis of distances Mn…P, it has been demonstrated that the Mn^{+2} cations occupy framework sites as well as extra framework positions in these materials.

One of the main points in applications of the earlier calculations is *the complete dominance of dipolar electron–nucleus interaction* in the observed sideband patterns. However, in general, the validity of this domination is not always obvious. Moreover, the domination requires independent experimental and/or theoretical proofs. Interesting and complex behavior of the sidebands has recently been illustrated by the single-pulse ^{27}Al and ^{29}Si MAS NMR spectra of supermicroporous silica-based Si/Al/Mn materials obtained at variation in Si/Mn ratios [1]. Figure 8.2 (bottom) shows the single-pulse ^{29}Si MAS NMR spectrum of the calcined Si/Al/Mn porous system, prepared at a Si/Mn ratio of 8, that is, at high manganese concentration. The spectrum shows a distribution of the sideband intensities that differs slightly from symmetrical distribution. Remember that generally an axial symmetry is expected for Fermi-contact interaction at the Mn–O–Si sites, or a strong asymmetry, which has been observed for dipolar interaction in paramagnetic lanthanide materials (see Figure 8.1) [4].

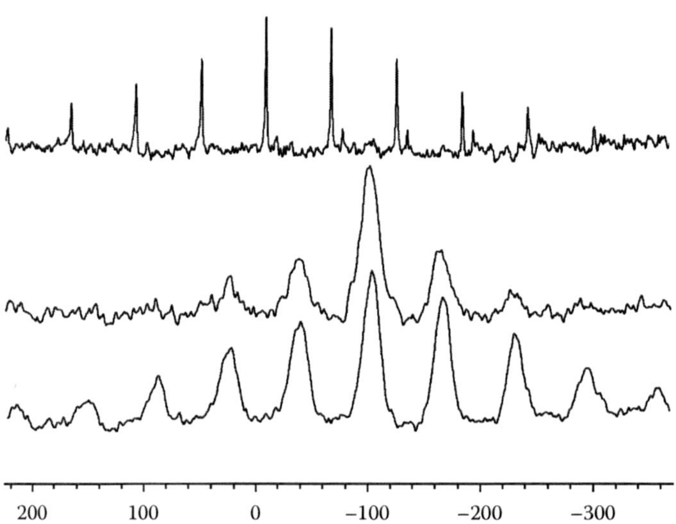

FIGURE 8.2 The single-pulse ^{29}Si MAS NMR spectra, recorded at spinning rates of 5 kHz: (top) tetrakis(trimethylsilyl)silane, mixed with 50 wt% of $Mn(CH_3COO)_2 \times 4H_2O$, where lines at −10.3 and −135.9 ppm correspond to isotropic chemical shifts; (middle) a diamagnetic Si/Al sample mixed mechanically with 30 wt% of $Mn(CH_3COO)_2 \times 4H_2O$; (bottom) a calcined Si/Al/Mn sample prepared at Si/Mn = 8. (Reproduced with permission from Bakhmutov, V.I., *Chem. Rev.*, 111, 530, 2011.)

The isotropic chemical shifts of ^{27}Al and ^{29}Si nuclei, determined in this SiO_2–Al_2O_3–MnO system (and its numerous relatives) are very similar to those observed usually for calcined diamagnetic Si/Al materials [6–8]. The chemical shifts correspond to the silica framework with four-coordinated Al atoms incorporated into the silica matrix in accord with the well-known criteria. Thus, the target ^{27}Al and ^{29}Si nuclei of the silica matrix are remote from the paramagnetic centers and intensities of the spinning sidebands could be increased by Kubo's BMS effects [9]. Abidi et al. [10] have also discussed the BMS influence on the solid-state NMR spectra of silica xerogels doped with low Mn^{2+} concentrations. A direct proof follows from the medium ^{29}Si MAS NMR spectrum in Figure 8.2: the intense sidebands are observed in a mechanical mixture of the diamagnetic silica-based Si/Al material with $Mn(CH_3COO)_2 \times 4H_2O$, where close Fermi- and dipolar contacts ^{29}Si–Mn are excluded, and thus, the BMS is completely responsible for the appearance of sidebands. Very similar spectroscopic effects have recently been reported by Tušar et al. [11], where mesoporous silicates and their manganese-containing relatives have been carefully investigated by solid-state NMR. The ^{29}Si MAS NMR spectra recorded at a spinning rate of 8 kHz for three samples: as-synthesized manganese silicate, template-free manganese silicate, and a mixture of the silicate and ground crystals of manganese acetate, have shown the same ratio between the center-band intensity and the intensities of sidebands. It is interesting that distribution of the sidebands was identical to that in Figure 8.2 and no hyperfine interactions were detected.

According to Kubo's model [9], considered in the previous chapter, the BMS effect in powder samples is governed by distances between paramagnetic ions and "centers of crystallites." In full agreement with this model, the ^{29}Si and ^{27}Al MAS NMR spectra of the mixture containing a diamagnetic Si/Al sample and 30 wt% of $Mn(CH_3COO)_2 \times 4H_2O$ (middle ^{29}Si spectrum in Figure 8.2), recorded at the same spinning rates and plotted *on frequency scales*, have shown identical distributions of sidebands [1], in spite of remarkable difference in the γ values of ^{29}Si and ^{27}Al nuclei. Finally, Kubo's model agrees with the ^{29}Si MAS NMR spectrum in Figure 8.2 (top) observed for a 1:1 solid mixture (by weight) of tetrakis(trimethylsilyl)silane and $Mn(CH_3COO)_2 \times 4H_2O$, where the ^{29}Si sideband patterns are very similar for two ^{29}Si nuclei, one of which is peripheral. Again, the sideband intensities are distributed identically for ^{29}Si and ^{13}C nuclei of the mixed tetrakis(trimethylsilyl)silane, when the spectra are plotted on frequency scales. It is interesting that the BMS does not significantly change the ^{29}Si T_1 times that remain in the aforementioned mixture as long as in the individual tetrakis(trimethylsilyl)silane. Thus one can assume that recording the MAS NMR spectra for different target nuclei at the same spinning rates and their plotting *on frequency scales* can be used as a test for the presence of the BMS.

In addition, as noted in the previous chapter, the demagnetizing BMS field and its effects on NMR frequencies of target nuclei depend on even shapes of NMR containers and their orientations in the external magnetic field [9,12]. Barbara [13] has demonstrated theoretically that, for example, in a water-containing cylinder placed in a magnetic field of a 500 MHz NMR spectrometer, the ^{1}H resonance can be broadened by 1.8 kHz. However, despite this, in practice, the sideband BMS effects are often ignored in structural NMR studies of materials.

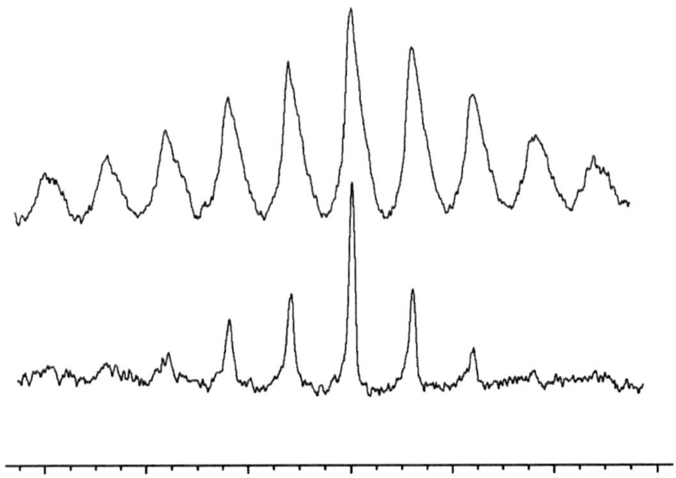

FIGURE 8.3 The single-pulse ^{27}Al (top) and ^{29}Si (bottom) MAS NMR spectra, recorded for a calcined Si/Al/Mn material (wt% of Mn ≈ 4) at a spinning rate of 12 kHz. The spectra are depicted on the frequency scale, where the lines, corresponding to isotropic ^{27}Al and ^{29}Si chemical shifts, are taken as 0 Hz.

Unfortunately, separations of dipolar and BMS contributions to sideband patterns and their calculations are impossible in the absence of the accurately determined coordinates of the corresponding paramagnetic centers. The sideband analysis in paramagnetic systems will therefore usually be difficult and unreliable because of superposition of the both contributions. For example, $SiO_2/Al_2O_3/MnO$ materials doped with Mn^{2+} ions incorporated into the silica matrix [1,14] show the ^{29}Si and ^{27}Al sideband patterns in the MAS NMR spectra plotted *on frequency scales*, which depend on the nature of the nuclei. The widths of the sideband patterns and the sideband intensities in Figure 8.3 reduce significantly with decreasing the γ_N magnitudes from ^{27}Al to ^{29}Si nuclei proofing, thus, the presence of the electron–nucleus dipolar contribution to the spinning sidebands.

8.2 DIRECT DETECTION OF THE NUCLEI CLOSEST TO PARAMAGNETIC IONS IN POROUS MATERIALS BY THE HAHN-ECHO MAPPING NMR EXPERIMENTS

As noted in Chapter 2, proving the incorporation of metal paramagnetic ions into the matrix of porous materials is one of the important and difficult steps in their structural studies. The NMR spectroscopic proof can be provided by direct spectral observation of nuclei that are closest neighbors of paramagnetic centers. It follows from theory that target nuclei with a low natural abundance, for example, ^{15}N, ^{13}C, or ^{29}Si, and paramagnetic ions Mn^{2+} or Ni^{2+} having relatively long electron relaxation times in contrast to paramagnetic lanthanide ions [13], represent a most difficult case especially for materials with high contents of paramagnetic centers.

The supermicroporous silica-based materials synthesized and studied by Shpeizer et al. [14] have shown the single-pulse ^{29}Si MAS NMR spectra similar to those in Figure 8.3, where intense ^{29}Si sideband patterns are caused by combinations of the BMS and electron–nucleus dipolar effects. The spin-echo mapping NMR technique, discussed in Chapter 7 and generally applied for static samples, has resulted in the Hahn-echo NMR spectra exhibiting the very broadened and practically symmetric ^{29}Si resonances with line widths reaching up to 40kHz [15]. At the same time, the static resonances have transformed to combinations of broad lines and isotropic resonances with δ around −110ppm, accompanied by intense sideband patterns in the ^{29}Si Hahn-echo MAS NMR spectra recorded at various carrier frequencies. One of the experiments is illustrated for the system SiO_2–Al_2O_3–MnO (Si/Mn = 3.5, Si/Al = 21) in Figure 8.4. The spectral assignment is obvious: the sideband pattern typical of paramagnetic systems [5,12] can be attributed to ^{29}Si nuclei situated in the silica-matrix and remote from the manganese, while the wide line belongs to the nuclei located in closest environments of paramagnetic centers [16–18].

A number of interesting and useful factors for performances of such experiments have been emphasized [18]. *First*, there is an insignificant temperature dependence of the spectra collected, where sideband intensities slightly increase from +50°C to −70°C on changing the electron relaxation rate. *Second*, the broad resonance is observed much better at a carrier frequency, shifted by 600–800ppm from TMS in Figure 8.5. From the spectrum, the wide component looks asymmetric and stretches from +1400 (taken as δ_{11}) to −400ppm (δ_{33}) and shows a maximum ~ at +100ppm (δ_{22}), roughly corresponding to an isotropic chemical shift of +300ppm. It should be noted that this value is not large relative to magnitudes observed for nuclei ^{31}P and ^{13}C in paramagnetic systems, containing Mn, Co, Fe, and Ni [2,19,20], but it

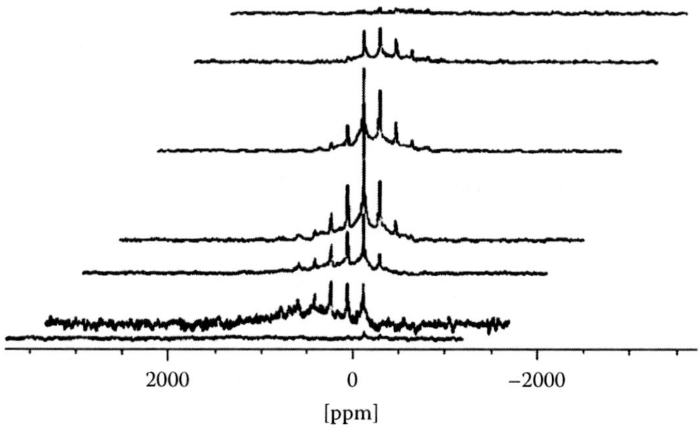

FIGURE 8.4 The ^{29}Si Hahn-echo MAS NMR spectra of silica-based material SiO_2–Al_2O_3–MnO (Si/Mn = 3.5, Si/Al = 21) spinning at 14kHz, recorded with the echo-delay corresponding to single rotational cycle at a carrier frequency centered at (from top to bottom): −1200, −800, −400, 0, +400, +800 (scaled by a factor of 4), and +1200ppm from TMS. (Reproduced with permission from Bakhmutov, V.I., *Chem. Rev.*, 111, 530, 2011.)

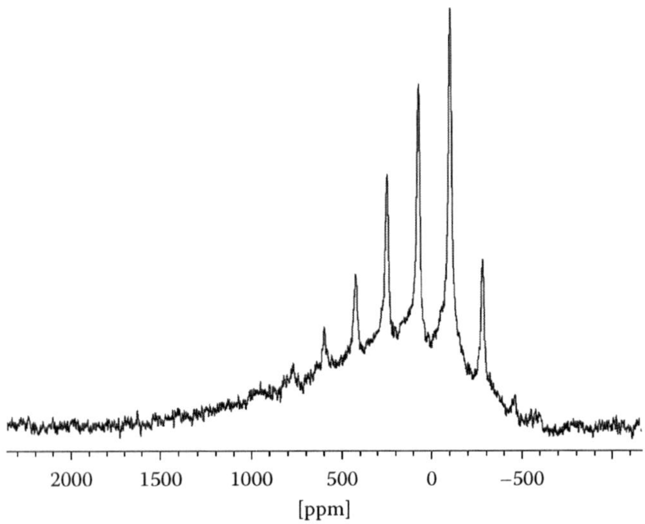

FIGURE 8.5 The ^{29}Si Hahn-echo MAS NMR spectrum recorded for sample SiO_2–Al_2O_3–MnO (Si/Mn = 2.6) with carrier frequency at 600 ppm, a spinning rate of 14 kHz, number of accumulations at 500,000, and a total relaxation delay of 0.01 s. (Reproduced with permission from Bakhmutov, V.I., *Chem. Rev.*, 111, 530, 2011.)

is significantly larger than paramagnetic shifts of ^{29}Si nuclei measured in zeolites containing paramagnetic lanthanide ions [21]. *Third*, the broad resonance can be completely "filtered" by increasing the echo delays in the ^{29}Si Hahn-echo MAS NMR spectra (Figure 8.6). Since the Hahn-echo MAS experiments, synchronized with spinning rates, are phase sensitive, the aforementioned factors confirm the

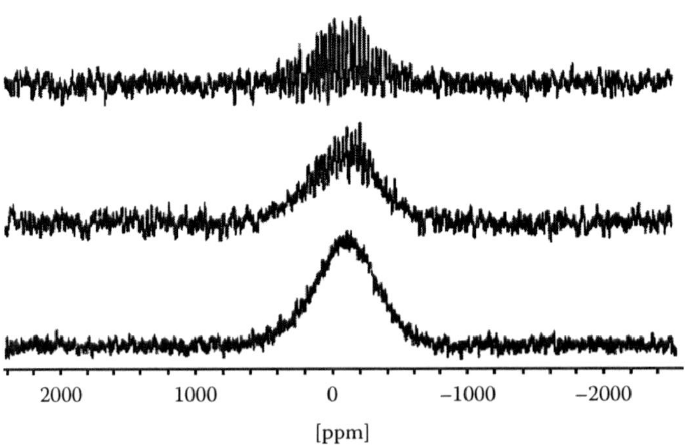

FIGURE 8.6 The ^{29}Si Hahn-echo MAS NMR spectra of silica-based material SiO_2–Al_2O_3–MnO (Si/Mn = 3.5, Si/Al = 21) recorded at a spinning rate of 2.5 kHz with the echo-delay corresponding to three rotational cycles (top), at a spinning rate of 2.5 kHz with the echo-delay corresponding to one cycle (middle), and in static sample (bottom). (Reproduced with permission from Bakhmutov, V.I., *Chem. Rev.*, 111, 530, 2011.)

nonartifact nature of the wide resonances. This point is very important because the phase and baseline correction procedures in the software package of an NMR spectrometer help to make corrections in a separated spectral range. In other words, the wide line observed, for example, in Figure 8.4 (bottom) can be "successfully corrected" and eliminated.

Similarly to the spin-echo ^{27}Al MAS NMR and static ^{27}Al NMR experiments performed by Schmitt et al. [22] in probing the framework/extra-framework ^{27}Al sites in zeolite catalysts, the aforementioned filtration of wide lines by the ^{29}Si Hahn-echo MAS experiments was very effective even at small spinning rates as shown in Figure 8.6 for a spinning rate of 2.5 kHz. One can assume that only after that the sideband patterns are ready to be analyzed quantitatively in terms considered in Section 8.1.

The single-pulse ^{27}Al MAS NMR spectra of the numerous porous silica-based Al-containing materials doped with high concentrations of ions Mn^{2+}, Ni^{2+}, or Co^{2+} [17,19], show the framework Al atoms observed as signals accompanied by intense sidebands with isotropic chemical shifts of $\delta \approx 50$–60 ppm, while the extra framework ^{27}Al signals are unobservable. Generally these spectra reveal a remarkable intensity loss with respect to diamagnetic Si/Al materials. Again, the Hahn-echo ^{27}Al MAS NMR spectra obtained at different carrier frequencies operating, for example, between 2400 and −2400 ppm have provided direct observation of the extra framework Al species as extremely wide lines [19]. The conclusion is obvious: the paramagnetic ions are located mainly in pores of materials and they are chemically binding to Al–O units of the extra framework Al atoms.

Finally, it is pertinent to formulate the obvious limitations in application of the Hahn-echo MAS NMR mapping for characterizations of paramagnetic materials.

Limitation 1: Echo signals corresponding to very broad lines with short T_1 and T_2 times should be collected properly at short echo delays. Since synchronization with spinning rates, required in the MAS experiments, automatically increases the τ values, there are evident difficulties in quantitative determinations of relative integral intensities corresponding to the broad line and sideband pattern, shown, for example, in Figure 8.5.

Limitation 2: Detection of wide signals in the mapping Hahn-echo MAS NMR experiments is directly related to absolute amounts of the nuclei closely situated relative to paramagnetic ions: *incorporation of paramagnetic ions into the matrix of the materials is difficult to observe, when the ions can be incorporated only at small concentrations (2–5 wt%) and target nuclei have a low natural abundance (^{13}C or ^{29}Si).* Therefore the absence of wide resonances, for example, in the mapping Hahn-echo ^{29}Si MAS NMR spectra of paramagnetic microporous materials SiO_2–Al_2O_3–NiO or SiO_2–Al_2O_3–CuO [23] is not valuable in a chemical sense and investigations of such paramagnetic systems require independent data that can be obtained by FTIR [24] and EPR [25] techniques or by NMR relaxation measurements.

For example, this ^{29}Si spectroscopic feature is observed in mesoporous silicates and their manganese-containing relatives [11]. It has been demonstrated that only a combination of pulsed EPR (with electron spin-echo envelope modulation) and XANES/EXAFS experiments suggests that 80% of manganese excite in the Mn^{3+}form (located in extra-framework) while 20% belong to Mn^{2+} ions.

8.3 NMR RELAXATION APPROACHES TO STRUCTURE OF POROUS AMORPHOUS PARAMAGNETIC SILICA-BASED MATERIALS: FROM EXPERIMENTS TO MODELS AND INTERPRETATIONS

The chemistry of porous materials, generally prepared by the sol–gel method in the presence of transition or main-group metal ions, shows that probably due to high-temperature calcinations of such systems, the amount of ions incorporated into the matrix cannot often reach more than 2–5 wt% [26,27]. Due to such small concentrations a structural analysis by solid-state NMR will obviously be difficult. In addition, organic templates applied in the sol–gel synthesis can reduce metal ions under high-temperature calcinations [28].

The mapping Hahn-echo MAS ^{29}Si NMR experiments performed for one of these systems, SiO_2–Al_2O_3–MnO (2.5 wt% of Mn^{2+}), are exampled in Figure 8.7, where the sideband patterns belong to ^{29}Si nuclei definitely remote from paramagnetic ions and no evidence indicating the presence of fragments Si–O–Mn is observed in spite of the actual incorporation of these ions into the silica matrix [14]. In fact, the ^{29}Si isotropic chemical shift does not change in the presence of the manganese and no wide line is observed. Because this structural situation is not rare in the chemistry of porous systems, it is necessary to show how the incorporation can be determined by relaxation studies performed for the nuclei that are *detected directly but remote from paramagnetic ions*.

Here, it would be emphasized that while NMR spectra of paramagnetic materials have been constantly in focus of researchers, accurate NMR relaxation measurements and their quantitative interpretations have been relatively limited. Nevertheless,

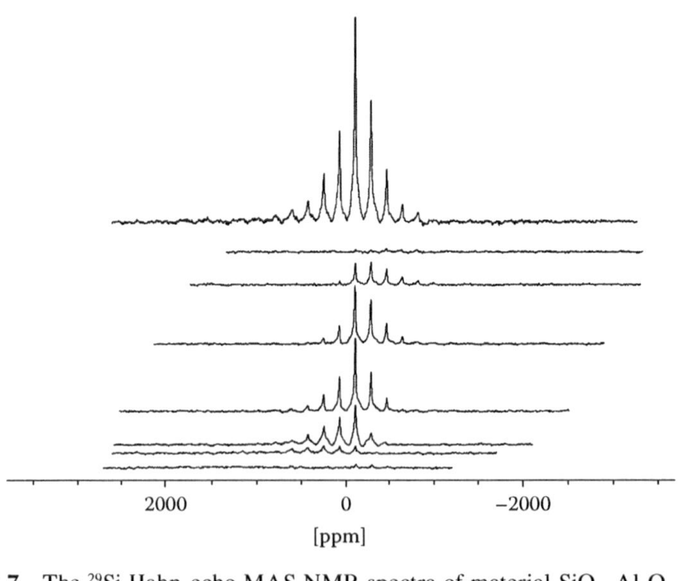

FIGURE 8.7 The ^{29}Si Hahn-echo MAS NMR spectra of material SiO_2–Al_2O_3–MnO containing 2.5 wt% of Mn^{2+}, recorded at different carrier frequencies similar to those in Figure 8.4. (Reproduced with permission from Bakhmutov, V.I., *Chem. Rev.*, 111, 530, 2011.)

recently a series of works aimed at probing the long-range electron–nucleus interactions in amorphous solids has resulted in formulation of true NMR relaxation criteria that are based on T_1, T_2 measurements for isotropic resonances, and their sidebands that better describe systems under investigation.

8.3.1 Factors Affecting Relaxation Times in Porous Solids

Theoretically, nuclear relaxation in solids seems to be extremely sensitive to the presence of paramagnetic centers: T_1, T_2 relaxation times depend on concentrations and the nature of paramagnetic ions, and on their distributions through the volume of samples and electron relaxation times. *A priori*, significance of these factors is unknown, thus complicating an analysis of relaxation data. Moreover, in practice, some additional relaxation effects can be expected due to the structural inhomogeneous nature of amorphous solids and influence of their chemical prehistory on relaxation times.

Scholz and Thomas [29] have demonstrated that the ^{29}Si spin–lattice times in silicate glasses are exclusively long and can reach several minutes. However, they reduce dramatically in the presence of oxide MnO. Probing the glasses with various concentrations of MnO, the authors have experimentally established that the ^{29}Si relaxation rates, $1/T_1$, have been proportional to the MnO contents by assuming homogeneous distributions of paramagnetic ions. Sen and Stebbins [30] have revealed influence of phase separations on the ^{29}Si spin–lattice relaxation times measured in silicate glasses and induced by direct dipolar interactions with paramagnetic ions. Relaxation time effects related to migrations of paramagnetic centers at thermal treatments of materials have been demonstrated by McHenry et al. [31].

The unexpectedly strong influence of chemical prehistory on the ^{29}Si T_1 relaxation times has been well illustrated by measurements in the supermicroporous materials $SiO_2Al_2O_3NiO$ [23]. For example, it has been found that ^{29}Si T_1 times in the *freshly prepared* materials with 4.8 wt% of the nickel but calcined at 450°C and 540°C are different and measured as 1.0 and 0.64 s, respectively. However, after 1 month, the ^{29}Si T_1 times reduce to the same values (0.24 s) for both systems. Then, after 2 and even 3 months, the ^{29}Si T_1 values do not change in the 10% limit. Generally the 10% errors in relaxation time measurements are accepted as a good accuracy in relaxation time measurements. The effects have been reasonably interpreted by very slow migrations of the nickel centers along pores, resulting in their more homogeneous redistributions through the *volumes* of the samples and providing, thus, a larger number of close contacts Ni…Si than those in materials immediately after preparation.

Similarly, the porous systems $SiO_2Al_2O_3MnO$ also demonstrate remarkable influence of the sample prehistory: the materials calcined at 130°C and 650°C have shown reducing the ^{29}Si T_1 times by a factor of 4 [17].

The aforementioned data are important in the methodological context: comparing the relaxation data, own or available in the literature, can be correct only after accounting for the aforementioned effects. Finally, it should be emphasized that the prehistory of the materials *does not affect the ^{29}Si MAS NMR spectra* and only T_1 times can show the processes possible in sample.

8.3.2 Spin–Lattice Relaxation Times in Porous Silica-Based Materials and Concentrations of Paramagnetic Ions

The theory shows that relaxation rates by the spin-diffusion mechanism and via dipolar electron–nucleus interactions depend on a number of paramagnetic centers. Consequent variations in their concentrations could be one of the simplest tools in studies of paramagnetic materials. In fact, the formalism of the equations [32] in Chapter 7, corresponding to spin diffusion and dipolar relaxation, respectively, describes increasing the relaxation rates ($1/T_1$) proportionally to N_p or N_p^2, where N_p is the number of paramagnetic centers at their homogeneous distribution through the sample volume

$$\frac{1}{T_1} = \left(\frac{16}{9}\right)\pi^3 N_p^2 \left(\frac{2}{5}\right)\gamma_I^2\gamma_S^2\hbar^2 S(S+1)\, r^{-6}\, \frac{\tau_e}{(1+\omega_I^2\tau_e^2)} \tag{8.2}$$

In practice, however, the situation is often more complex. For example, Scholz and Thomas [29] have observed a linear increase in the ^{29}Si $1/T_1$ rates of silicate glasses proportionally to manganese concentration [MnO]. Nevertheless, the data have been interpreted in terms of homogeneous distributions of ions Mn^{2+}.

The supermicroporous silica-based materials $SiO_2Al_2O_3NiO$ [23], where ^{29}Si spin–lattice relaxation could be dominated by dipolar electron–nucleus interactions, show a relaxation rate that increases practically linearly for the first four points in Figure 8.8. Then, it reaches a maximum and decreases again at higher nickel contents. It has been noted that a sample of diamagnetic $SiO_2Al_2O_3$ carefully mixed with $Ni(CH_3COO)_2$ corresponded to the point that is out of the curve. Even in the absence

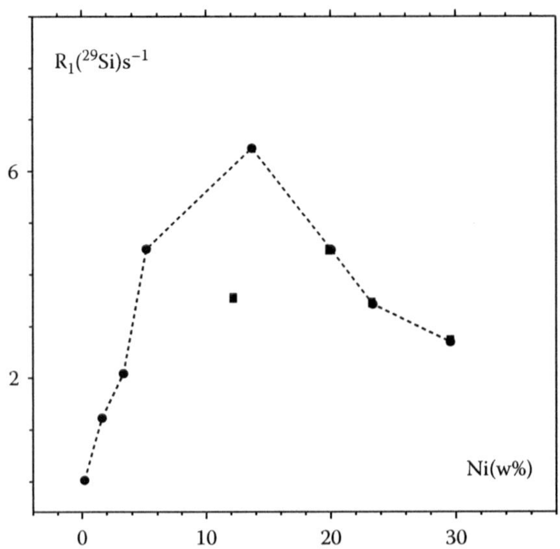

FIGURE 8.8 The ^{29}Si spin–lattice relaxation rates as a function of the nickel loading in calcined supermicroporous materials $SiO_2Al_2O_3NiO$. (Reproduced with permission from Bakhmutov, V.I., *Chem. Rev.*, 111, 530, 2011.)

of accurate knowledge on relaxation mechanisms, such dependences can be a good criterion for changing the nature of paramagnetic ions in materials prepared at higher nickel loadings. A partial reduction of Ni^{2+} by template organic molecules has been suggested to give a nonparamagnetic state, for example, Ni^0. This transformation has been independently confirmed by variable-temperature magnetic susceptibility measurements: the Ni species were identified as NiO and Ni^0 observed at high nickel loadings and aggregated into nanoparticles. The Ni^0 nanoparticles were responsible for the room-temperature ferromagnetic behavior of the materials prepared with high nickel loadings, whereas the low-temperature ferromagnetism observed in the materials with lower nickel loadings was caused by NiO nanoparticles.

At the same time, it is obvious that these simple ^{29}Si T_1 measurements did not provide valuable conclusions on the structure of the matrix and locations of paramagnetic species. However, some conclusions can be drawn by additional relaxation experiments involving other nuclei.

8.3.3 ^{29}Si Spin–Lattice and Spin–Spin Relaxation in Static and Spinning Porous Amorphous Paramagnetic Silica-Based Materials

Among the numerous factors affecting relaxation behavior of nuclei that interact with unpaired electrons in amorphous porous solids, a single fact is obvious: paramagnetic metal ions significantly shorten the T_1 and T_2 times of target nuclei [33]. As we have shown in previous chapters, this relaxation time shortening can even be used to reduce time in NMR experiments.

However, then, quantitative interpretations of NMR relaxation data, T_1 and T_2, in structural terms, where incorporation of paramagnetic metal ions into the matrix of materials is most important, are not simple. Canesson et al. [34] have emphasized that "a decrease in both T_1 and T_2 relaxation times cannot be regarded as characteristic of the presence of framework paramagnetic species." This statement is particularly valuable when relaxation mechanism is not recognized. This section will show ^{29}Si NMR experiments performed in static and spinning paramagnetic systems and directed to discrimination between relaxation mechanisms.

When target nuclei are not quadrupolar, for example, ^{29}Si or ^{13}C, dipole–dipole coupling is relatively weak (e.g., in contrast to ^1H or ^{19}F nuclei) and molecular motions are limited, *a priori* their T_1, T_2 NMR relaxation is a sum of contributions due to direct electron–nucleus dipolar interaction (DD) and spin diffusion (SD) to paramagnetic impurities:

$$\frac{1}{T_{1,2}} = \frac{1}{T_{1,2}^{DD}} + \frac{1}{T_{1,2}^{SD}} \tag{8.3}$$

Mortuza et al. [32] have noted that due to the relatively low natural abundance of ^{29}Si nuclei (4.7%), their homonuclear dipolar interaction is attenuated and ^{29}Si spin-diffusion coefficient D is very small and determined as $\sim 10^{-19}\,m^2\,s^{-1}$ (compare with the D constants of ^1H nuclei in Chapter 7). In other words, the ^{29}Si spin-diffusion mechanism can be effective in solids, where ^{29}Si relaxation times are very long (as long as 20 s or more) and amounts of paramagnetic centers are very small. Since the probability of mutual ^{29}Si spin flips remains low and increase in concentrations of paramagnetic centers shortens the ^{29}Si T_1 times, the ^{29}Si spin-diffusion mechanism cannot dominate. Maiti and McGarvey [35] have experimentally confirmed this general statement by NMR experiments on ^{13}C nuclei. They have shown that spin-diffusion contributions to the T_1 NMR relaxation of ^{13}C nuclei in some organic solids are practically negligible *even at their enrichment up to 30% or higher.*

At the same time, Alaimo and Roberts [36] have performed accurate ^{29}Si T_1 measurements for silica gels with the surface functionalized by aromatic sulfonic acid and ions Cu^{2+}, depicted in Figure 8.9. It should be noted that the ^{29}Si NMR spectra were well resolved even in the presence of larger concentrations of Cu^{2+} ions. The collected data have been interpreted in terms of the *diffusion-limited*

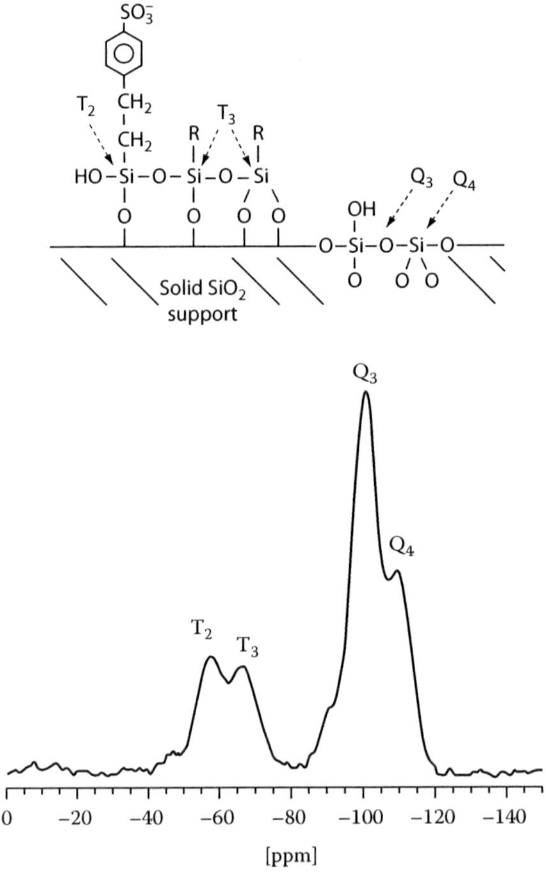

FIGURE 8.9 The ^{29}Si CP MAS NMR spectrum (bottom) of the aryl-sulfonate silica gel (top), which is obtained with a 90% Cu^{2+} loading. The signal assignments correspond to labels at the top of the figure. These signals have been characterized by T_1 experiments. (Reproduced with permission from Alaimo, M.H. and Roberts, J.E., *Solid State Nucl. Magn. Reson.*, 8, 241, 1997.)

mechanism because ^{29}Si NMR decays treated with a stretched exponential have resulted in relatively high β values (see Chapter 7). This conclusion, however, was not verified by experiments at different spinning rates, which are definitely needed in this context [37,38].

Recently such experiments have been carefully carried out for the supermicroporous silica-based material $SiO_2Al_2O_3MnO$ [15]. The system has been synthesized by Shpeizer et al. at Si/Mn and Si/Al ratios of 3.5 and 21, respectively. The inversion-recovery and saturation-recovery experiments performed in a static sample and a sample spinning at 2.5–14 kHz have resulted in ^{29}Si T_1 stretched exponential data with the β parameters between 0.6 and 0.7 and absolutely *independent* of spinning rates (Figure 8.10). Now there is no doubt that ^{29}Si spin-diffusion is ineffective, the ^{29}Si spin–lattice relaxation is completely governed by direct dipolar coupling between electrons and nuclei and thus the relaxation data can be treated quantitatively in structural terms.

The importance of NMR relaxation experiments performed at different spinning rates for interpretations of relaxation data has been additionally illustrated by NMR behavior of the diamagnetic manganese-free material $SiO_2Al_2O_3$ [15]. It was obvious, however, that this porous solid was not free from oxygen. The material has shown the ^{29}Si T_1 stretched-exponential relaxation that has been well treated with the β parameter of 0.80. This magnitude is very close to that corresponding to an exponential function and hence a long ^{29}Si T_1 time of 25 s measured for this material could be attributed to the spin-diffusion mechanism. However, again, the ^{29}Si T_1 time has been *independent* of spinning rates and thus ^{29}Si spin-diffusion has been

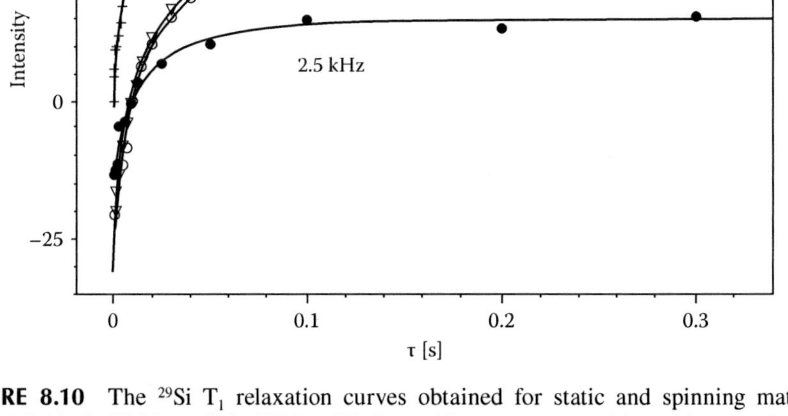

FIGURE 8.10 The ^{29}Si T_1 relaxation curves obtained for static and spinning material $SiO_2Al_2O_3MnO$ (Si/Mn = 3.5, Si/Al = 21). Intensities are measured in arbitrary units and scaled to avoid the curve overlaps. The measurements at MAS are performed for isotropic resonances. (Reproduced with permission from Bakhmutov, V.I., *Chem. Rev.*, 111, 530, 2011.)

negligible. In other words, discussion of the T_1 and β parameters in the absence of the variable-spinning measurements will be unreliable.

Since spin–spin relaxation times are often applied for characterizations of different materials, T_2 NMR experiments in static and spinning samples of porous silica-based materials are also of great interest. The ^{29}Si T_2 times in the diamagnetic material $SiO_2Al_2O_3$, have been short and measured as 0.61, 0.68, 0.8, and 0.79 ms in a static sample and at 6, 10, and 14 kHz spinning rates, respectively. In other words, they are practically *independent of spinning rates* [15].

Assuming the ^{29}Si–^{29}Si dipolar relaxation mechanism or relaxation by direct dipolar interaction between ^{29}Si nuclei and uncontrolled paramagnetic impurities (e.g., O_2), the ^{29}Si T_1/T_2 measurements have resulted in estimation of correlation times τ of 2–4 × 10^{-7} s. According to Equation 7.21, such magnitudes are actually *too small* to provide a ^{29}Si T_2 dependence on spinning rates in full agreement with the aforementioned experiments.

In contrast, the paramagnetic porous materials $SiO_2Al_2O_3MnO$ have shown the ^{29}Si T_2 curves that strongly depend on spinning rate in Figure 8.11. In going from the static experiment to a spinning rate of 14 kHz, it has been found the ^{29}Si T_2 time values increase linearly with increasing the spinning rates. Moreover, it has been emphasized that the dependences observed have been surprisingly similar to those predicted theoretically in terms of spin-diffusion in Equations 7.18 through 7.20 under conditions $\nu_R/\Delta\nu \gg 1$ and observed experimentally for ^{31}P and ^{27}Al nuclei (actually capable of spin-diffusion) in rigid solids by Kessemeier and Norberg [37] and Schmitt et al. [22].

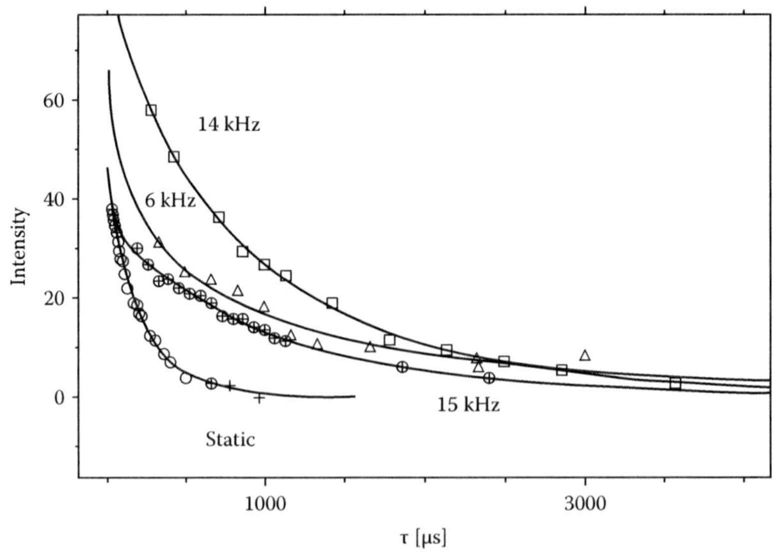

FIGURE 8.11 The ^{29}Si T_2 relaxation curves obtained for static and spinning material $SiO_2Al_2O_3MnO$ (Si/Mn = 3.5, Si/Al = 21). Intensities are measured in arbitrary units and scaled to avoid the curve overlaps. The measurements at MAS are performed for isotropic resonances. (Reproduced with permission from Bakhmutov, V.I., *Chem. Rev.*, 111, 530, 2011.)

In other words, there is a contradiction: on the one hand, the ^{29}Si T_1 times refute spin-diffusion and on the other hand, the ^{29}Si times suggest its presence. Careful analysis of these "spinning" ^{29}Si T_2 effects observed for various materials of the SiO_2–Al_2O_3–MnO and SiO_2–MnO families containing different manganese concentrations has revealed their nature: in the absence of spin-diffusion, spinning rates can affect the ^{29}Si T_2 times due to large BMS contributions, which can reach 55%–63% in static systems [39]. Because these BMS contributions will complicate quantitative analysis of nuclear relaxation, the authors have suggested their minimization by high spinning.

Finally, it should be noted that the aforementioned ^{29}Si T_2 elongating effect in paramagnetic solids differs from the linear increase in ^{29}Si T_2 time with increasing spinning rates in the NMR experiments performed by Pruski and coworkers [40] on samples of polycrystalline silicon and the AL-MSM mesoporous zeolite. The effects observed were explained by the presence of residual homo-dipolar coupling silicon–silicon, changing with spinning rates.

8.3.4 T_1 Criteria for Locations of Paramagnetic Ions: Relaxation of Isotropic Resonances

In general, nuclear relaxation in paramagnetic materials can be a tool for structural studies, if relaxation times are completely dominated by direct dipolar interactions between target nuclei and paramagnetic centers, and the spin-diffusion mechanism is negligible. In practice this situation is often realized for rare nuclei relaxing nonexponentially. Table 8.2 illustrates the T_1 relaxation times measured for isotropic resonances in some spinning solids studied at consequent variations in concentrations of paramagnetic centers. As can be seen, dramatic reduction in the T_1 times is observed on going from diamagnetic solids to paramagnetic systems. For example, the ^{29}Si T_1 time decreases by a factor of 45 or 64 in the kaolinites or the porous systems SiO_2–Al_2O_3–NiO, respectively (it is significant that the longest ^{29}Si T_1 times were measured for the resonances at −101.1 to −109.9 ppm as the magnitudes of 7.2 to 10 h in isotopically enriched SiO_2 glasses [41]). Then, however, the T_1 dependences look nontrivial. In spite of this, if the nature of paramagnetic particles and electron relaxation times are known, for example, from variable-temperature magnetic susceptibility measurements, showing the spin-states of the magnetic centers [44], or from the EPR spectroscopy [44], there is a unique opportunity to determine contents of paramagnetic centers expressed as Np in Equation 8.2 assuming their homogenous distributions though the volume of materials. This relatively simple approach is especially valuable for probing the natural molecular systems [43].

Another strategy should be used for characterization of synthetic porous paramagnetic materials, where concentrations of paramagnetic metal ions are already determined independently by convenient methods, whereas their locations relative to the matrix of materials remain unknown. This task has been formulated for a series of the supermicroporous silica-based materials SiO_2–Al_2O_3–NiO prepared at nickel loadings between 0 and 30 wt% [23]. These systems have shown TEM micrographs

TABLE 8.2
T_1 Times Measured for Isotropic Resonances in Paramagnetic Solids, Where Nuclear Relaxation Is Governed by Dipole–Dipole Nucleus–Electron Interactions

Material/Ion	T_1 (s)/Nucleus	Content of Paramagnetic Ions (wt%)	Reference
Pr_3N-$CoAPO_5/Co^{+2}$	9/^{31}P	0	[42]
	0.5/^{31}P	0.05[a]	
	0.23/^{31}P	0.10[a]	
	0.13/^{31}P	0.13[a]	
	0.04/^{31}P	0.22[a]	
Co-SAPO-34/Co^{2+}	1.8/^{31}P	0.38[b]	[32]
	1.8/^{31}P	0.59[b]	
	0.043/^{31}P	1.6[b]	
SiO_2–Al_2O_3–NiO/Ni^{2+}/Ni^0	55/^{29}Si	0	[23]
	0.86/^{29}Si	1.35	
	0.50/^{29}Si	2.95	
	0.23/^{29}Si	4.80	
	0.16/^{29}Si	13.4	
	0.40/^{29}Si	29.3	
SiO_2–Al_2O_3–MnO/Mn^{2+}			
SiO_2–Al_2O_3–MnO	23.5/^{29}Si	0	[39]
SiO_2–Al_2O_3–MnO	0.066/^{29}Si	1.2	
SiO_2–Al_2O_3–MnO	0.027/^{29}Si	2.5	
	0.019/^{29}Si	19.8	
Aromatic sulfonic acid silica gel/Cu^{2+}	23.1/^{29}Si	0	[36]
	22.3/^{29}Si	1[c]	
	3.2/^{29}Si	3[c]	
	3.17/^{29}Si	5[c]	
	2.49/^{29}Si	10[c]	
Kaolinites/Fe^{3+}	2300/^{29}Si	<0.01	[43]
	17.5/^{27}Al		
	2.5/^{27}Al		
	51/^{29}Si	0.46	
	5.7/^{27}Al[d]		
	0.8/^{27}Al[d]		
	3.3/^{29}Si	1.15	
	1.9/^{27}Al[d]		
	0.25/^{27}Al[d]		
	5.3/^{29}Si	1.25	
	2.0/^{27}Al[d]		
	0.30/^{27}Al[d]		

[a] Co/P ratio.
[b] Cobalt concentrations as N_p (nm^{-3}).
[c] Surface coverage by Cu^{2+} in %.
[d] Bi-exponential behavior.

similar to those in Chapter 2 revealing the nickel particles distributed homogenously through the volumes of materials. It has been demonstrated that the localization of the nickel ions can be successful by comparing the relaxation behavior of two target nuclei, one of which belongs to sites situated in pores. The T_1 MAS NMR relaxation measurements have been performed for ^1H and ^{29}Si resonances characterizing water molecules located in pores, and the silica matrix, respectively. It has been found that spin–lattice NMR relaxation of ^1H and ^{29}Si nuclei is nonexponential and completely governed by direct dipolar interactions with paramagnetic centers. The validity of the statement on the dipolar relaxation mechanism, operating for both nuclei, has been additionally supported by a linear correlation between the ^1H and ^{29}Si T_1 relaxation rates in Equation 8.4.:

$$R_1(^1H)(s^{-1}) = 1170R_1(^{29}Si)(s^{-1}) - 720 \tag{8.4}$$

Then the authors [23] have used $R_1(^1H)/R_1(^{29}Si)$ ratios determined in the materials SiO_2–Al_2O_3–NiO containing different nickel concentrations. These ratios were significantly larger than those expected theoretically at equal internuclear distances r(Si…Ni) and r(H…Ni) revealing, thus, accumulation of the nickel centers within pores. It is obvious that the same approach (or similar one) can be applied for other nuclei relaxing by the dipolar mechanism.

In special cases, the ^{29}Si T_1 relaxation of isotropic resonances in paramagnetic porous silica-based systems can be bi-exponential. Figure 8.12 shows the partially relaxed ^{29}Si MAS NMR recorded for a SiO_2–Al_2O_3–MnO material containing 3.8 wt% of manganese at different τ values and scaled to the sideband intensities. It is seen that *all the sidebands relax identically* but significantly faster than the central isotropic line.

A simple model suggesting the dipolar electron–nucleus relaxation mechanism could explain this result. In the frameworks of the model, the ^{29}Si MAS NMR spectrum of this material represents a superposition of resonances from the ^{29}Si nuclei located in the first, second, and third coordination spheres of Mn^{2+} (see the middle part in Figure 8.12). The ^{29}Si nuclei from the first coordination sphere, Si(1), belonging to the Si–O–Mn units are obviously invisible. The nuclei from the second sphere, Si(2), show a resonance accompanied by intense sidebands. Finally, a resonance belonging to the more remote nuclei located in the third coordination sphere, Si(3), exhibits the less intense sidebands. In such a situation, a main contribution to relaxation rates, experimentally measured for the sidebands in the total spectrum, is from the ^{29}Si nuclei located in the second coordination sphere. Then, the central resonance, relaxing as a sum of ^{29}Si magnetizations, Si(2) and Si(3), could show two components with a short and long relaxation time. In these terms, the inversion-recovery curves of the central lines are treated to give a long T_1 time of 0.76 s while the short component is determined as 0.0097 s. It has been demonstrated that this simple model can be adequate, when the thickness of the silica walls is small and created by 3–5 layers of condensed silica units, and ions of manganese are situated on the silica surface.

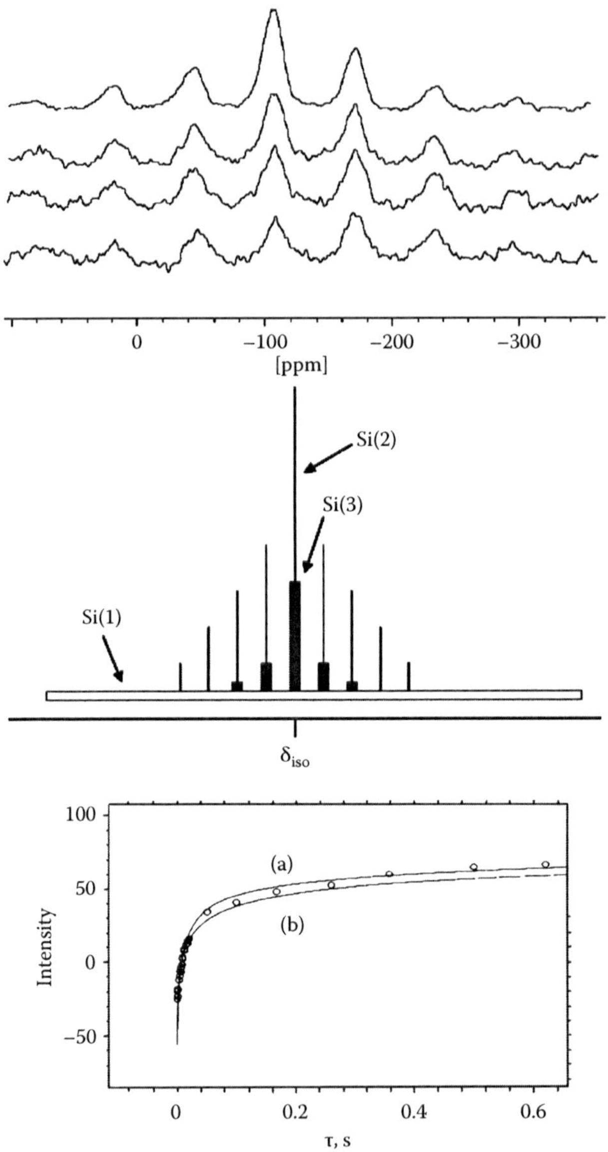

FIGURE 8.12 Top: The completely and partially relaxed ^{29}Si MAS NMR spectra of material SiO_2–Al_2O_3–MnO material, containing 3.8 wt% of manganese, obtained at τ value 0.02, 0.015, and 0.012 s (from top to bottom). Middle: A schematic presentation of the ^{29}Si MAS NMR spectrum as a superposition of resonances from the ^{29}Si nuclei located in the first (Si(1)), second (Si(2)), and third (Si(3)) coordination spheres of Mn^{2+}. Bottom: The ^{29}Si inversion-recovery curve obtained for the central resonance of the material (intensity in arbitrary units versus delay time τ (in seconds) fitted to a stretched exponential function, solid line (b), and to a bi-exponential function, (a). (Reproduced with permission from Bakhmutov, V.I. et al., *Magn. Reson. Chem.*, 44, 861, 2006.)

8.3.5 T_1 Criteria Based on Relaxation of Sideband Patterns in MAS Spectra of Porous Paramagnetic Materials

Theoretically, incorporation of paramagnetic metal ions into the matrix of porous materials or their extra-framework nature can be determined by quantitative analysis of T_1 relaxation times measured experimentally (see, e.g., the short ^{29}Si T_1 component in a bi-exponential relaxation) in terms of the distances "target nucleus–unpaired electron." Scheme 8.1 represents a silica-based material doped with ions Mn^{2+}, where Si(1), Si(2), and Si(3) symbolize the silica matrix, whereas Mn^{2+}(1) and Mn^{2+}(2) correspond to ions incorporated into the matrix and accumulated in pores, respectively. It is obvious that ^{29}Si(1) nuclei are spectroscopically "invisible" because of the strong paramagnetic effects. The latter is particularly valid at low concentrations of the manganese. Nuclei ^{29}Si(2) and ^{29}Si(3) (or more remote nuclei) will show a very wide line in a static ^{29}Si NMR spectrum (as shown schematically in Figure 8.13) due to various Si…Mn distances and the angle dependent term ($3\cos^2\theta - 1$) important for dipole–dipole electron–nucleus interaction in Equation 8.1. On spinning, this wide line transforms to a broad sideband pattern, where the central line in Figure 8.13 is an isotropic resonance.

Accurate T_1 NMR measurements performed in different spinning solids have revealed a very interesting experimental fact: T_1 times of isotropic resonances were remarkably longer than those measured for sidebands. This effect detected for ^{1}H

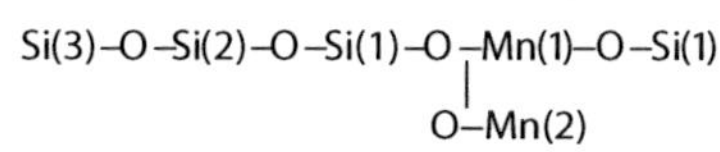

SCHEME 8.1 A structural motive of a silica-based material doped with ions Mn^{2+} Si(1), Si(2) and Si(3) are located in the silica matrix, whereas Mn^{2+}(1) and Mn^{2+}(2) correspond to ions incorporated into the matrix and accumulated in pores, respectively.

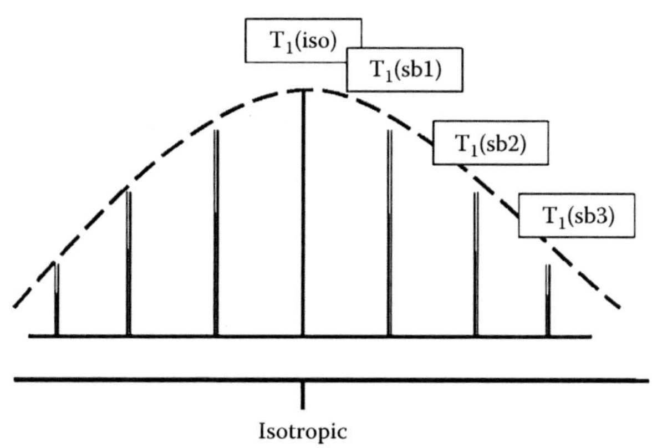

FIGURE 8.13 A static ^{29}Si NMR spectrum (dashed line) of a silica-based material doped with ions Mn^{2+} (schematically), transforming to the pattern at spinning with different T_1 times measured for the isotropic resonance and its sidebands. (Reproduced with permission from Bakhmutov, V.I., *Chem. Rev.*, 111, 530, 2011.)

nuclei has been explained by Hayashi [38] in terms of the presence of T_1 distributions. Suttler and coworkers have observed similar T_1 differences for isotropic resonances and sidebands of ^{31}P nuclei [45].

Physically such a T_1 behavior is reasonable. In fact, according to the authors [1], ^{29}Si nuclei in paramagnetic silica-based materials are differently distanced from paramagnetic ions and the lines remote from the isotropic signal in Figure 8.13 will belong to nuclei undergoing stronger dipolar nuclear–electron interaction. As a result, they will relax significantly faster and the latter can be used for a quantitative T_1 time analysis.

Following Solomon's theory of nucleus–electron dipolar interaction (Equations 7.13 and 7.14) [43], the spin–spin and spin–lattice relaxation times of target nuclei in paramagnetic solids can be quantitatively analyzed in terms of distances r, *if the τ_e parameter is accurately determined.* It has been noted that the relatively long electron relaxation times τ_e can be measured by advanced EPR techniques, developed and applied in works of Atsarkin et al. [46] and Tanaka et al. [47]. In general, however, these EPR experiments are difficult and they are strongly limited by τ_e magnitudes.

Another approach can be based on accurate T_1 and T_2 NMR determinations and exclusion of unknown distances r from Equations 7.13 and 7.14. This method, described by Carper [48], has recently been applied for the supermicroporous materials SiO_2–Al_2O_3–MnO and SiO_2–MnO prepared at large variations in contents of Mn^{2+} ions. The nature of paramagnetic ions has been independently studied by variable-temperature magnetic susceptibility measurements [39]. The relaxation times of electrons in Mn^{2+} ions have been determined in the aforementioned systems between 0.6 and 0.8×10^{-8} s by the ^{29}Si R_1, R_2 NMR relaxation rates via Equation 8.5 [15]:

$$\frac{R_2^D}{R_1^D} = 2\tau_{1e}^2\omega_I^2 + 1.5 \tag{8.5}$$

where ω_I is the Larmor frequency of ^{29}Si nuclei.

Equation 8.6, where T_1 and r(Si…Mn) are measured in seconds and Å, respectively, can be easy deduced from Solomon's Equation 8.5 by addition of the well-known factor $(\mu_0/4\pi)^2$, an Mn^{+2} electron spin of 5/2, $\tau_{1e} = 1 \times 10^{-8}$ s, and the ^{29}Si resonance frequency of 79.46 MHz:

$$\frac{1}{T_I} = \frac{43.5}{(0.1\, r(\text{Si}\ldots\text{Mn}))^6} \tag{8.6}$$

Finally the expression can be applied for calculations of distances Mn…Si.

On the one hand, the available x-ray data can be applied to predict ^{29}Si T_1 times for the different $Mn^{2+}(1)$ coordination spheres in Scheme 8.1. On the other hand, ^{29}Si T_1 values can be simulated by variations in numbers of dipolar contact Mn…Si. Influence of the Mn(2) accumulated in pores and distanced from the silica lattice can also be estimated. Similar expressions can also be deduced for other target nuclei.

The criterion in the quantitative analysis of relaxation data is simple: ions Mn^{2+} will be incorporated into the silica matrix if ^{29}Si T_1 times measured for isotropic resonances in ^{29}Si MAS NMR spectra correspond to distances Mn…Si of 7–8 Å typical of the second manganese coordination sphere (see Si(2) in Scheme 8.1).

Obviously this criterion is particularly reliable if the collected NMR relaxation data are bi-exponential [1]. In fact, these data provide accurate determinations of short and long ^{29}Si T_1 time components where the shortest times can be involved in quantitative treatments. However, the bi-exponential relaxation is not typical of silica-based materials. In fact, ^{29}Si nuclei in the silica matrix in Scheme 8.1, which are differently distanced from Mn^{2+} paramagnetic centers, will relax differently even at relatively long distances Si…Mn. For example, a distance Si(1)…Mn(1) of 20 Å predicts ^{29}Si T_1 values of 1.5, 3.5, and 7.3 s for Si(1), Si(2), and Si(3) nuclei, respectively [39]. In combination with very similar chemical shifts of such remote nuclei in ^{29}Si MAS NMR spectra, the ^{29}Si relaxation process becomes theoretically multi-exponential [49].

In practice, however, the relaxation curves are well treated with a stretched-exponential function. Here, the resulting T_1 time is a representative value of whole spins, and T_1 relaxation times measured in MAS NMR experiments for isotropic resonances and their sidebands can differ due to the presence of T_1 distributions [1,38].

In the case of porous silica-based materials, it has been assumed that ^{29}Si T_1 distributions can be wider when paramagnetic centers are incorporated into the silica matrix. The pattern in Figure 8.14 obtained at different spinning rates for the super-microporous material SiO_2–Al_2O_3–MnO (2.5 wt% of Mn^{2+}) [39] illustrates well this statement. As can be seen, the ^{29}Si T_1 time reduces from the isotropic resonance, taken as 0 kHz, to sidebands, shifted to a low- and high-field. It has been noted that

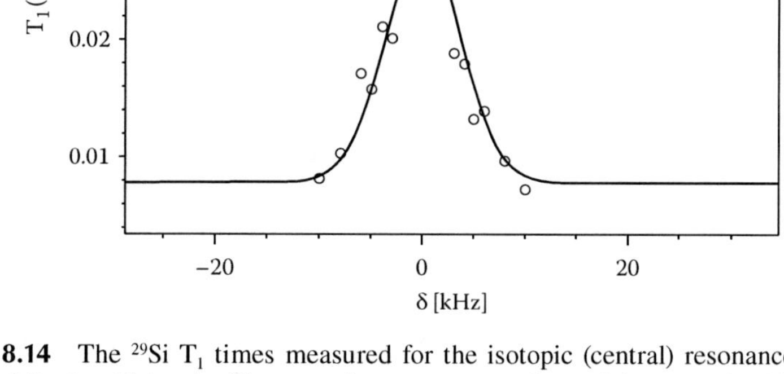

FIGURE 8.14 The ^{29}Si T_1 times measured for the isotopic (central) resonance, taken as 0 kHz, and for its sidebands. The experiments were performed for material SiO_2–Al_2O_3–MnO (2.5 wt% of Mn and 2 wt% of Al) spinning at different rates. Solid line corresponds to a Gaussian distribution. (Reproduced with permission from Bakhmutov, V.I., *Chem. Rev.*, 111, 530, 2011.)

the data correspond to a Gaussian distribution (solid line) and thus the pattern in Figure 8.14 is a ^{29}Si T_1 image of the ^{29}Si static resonance in static regime (compare with Figure 8.13).

Since the ^{29}Si T_1 data exhibit shorter T_1 times and electron relaxation time τ_e is determined as 1×10^{-8} s, the shortest T_1 components can be analyzed quantitatively, for example, in terms of Equation 8.2, where N_p (expressed as $(4/3)\pi r_{AV}^3 N_p = 1$) is the number of paramagnetic centers distributed homogeneously through the materials, and r_{AV} is the radius of the sphere around the paramagnetic center [29]. This approach applied for the systems SiO_2–Al_2O_3–MnO with 2.5 and 4.8 wt% of the manganese gave the r_{AV} parameter of 8.1 and 7.6 Å, respectively, that is, very close to a structural situation where Si atoms are located in the second coordination sphere of Mn^{2+}.

The treatments of short ^{29}Si T_1 components seem to be quite accurate but they cannot be performed in the absence of independent data on the nature of paramagnetic centers. In fact, electron relaxation times depend on oxidation states of metal ions and their environments. Good examples are the Al-free mixed oxides SiO_2–MnO, which at preparation by the sol/gel method and calcination are contaminated by oxide M_3O_4 [39]. It is obvious that in the presence of ions Mn^{2+} and Mn^{3+} the aforementioned T_1/T_2 method for determinations of electron relaxation times and the T_1 analysis will be incorrect. Nevertheless, it has been shown that at minimal contaminations by of M_3O_4, (e.g., 0.21 wt% of M_3O_4 versus 2 wt% of Mn^{2+}) the ^{29}Si T_1 time analysis is valid leading to a value in the range of 13–14 Å for the closest Mn… Si distances. In other words, only extra-framework manganese has been identified.

Finally, to avoid misunderstandings, it should be emphasized that the presence of T_1 distributions, in itself, is not a criterion for incorporation of paramagnetic ions into the matrix of materials, whereas their quantitative analysis, based on the well-determined electron relaxation times, can provide such a criterion.

8.3.6 BMS Effects and NMR Relaxation of Sideband Patterns in MAS Spectra of Porous Paramagnetic Materials

Numerous NMR relaxation experiments performed for the supermicroporous paramagnetic materials SiO_2–Al_2O_3, SiO_2–Al_2O_3–MnO, SiO_2–MnO, and SiO_2–Al_2O_3–NiO [1,14,17,39] have demonstrated that ^{29}Si T_1 time distributions, and different T_1 values measured for isotropic ^{29}Si resonances and their sidebands in MAS NMR experiments are a common phenomenon in NMR of paramagnetic solids. It has been found that this phenomenon is observed especially well for systems with relatively low concentrations of paramagnetic metal ions. In contrast the porous material SiO_2–Al_2O_3–MnO containing up to 20 wt% of Mn^{2+} shows ^{29}Si T_1 data where relaxation times of the isotropic resonance (0.019 s) and the sidebands (14 kHz, 0.012–0.013 s) are practically equal [15,39].

The occurrence of these virtually identical ^{29}Si T_1 times, important in the methodological context, requires an explanation. Recently the explanation has been found due to the careful measurements of ^{29}Si T_1 and T_2 times in spinning (at different rates) and static samples: *a ^{29}Si T_1 distribution can be masked by the strong BMS effects.* The masking mechanism has been simulated quantitatively [39]. It has been shown that in systems where T_1 distributions exist, a 63% BMS contribution to sideband intensities obtained in the inversion-recovery experiments, leads to practically equal T_1 times measured for all the lines in the sideband pattern. It is obvious that these strongly paramagnetic systems restrict application of T_1 relaxation investigations. In such cases, however, direct detections of nuclei located in the first coordination spheres of metal ions by the Hahn-echo MAS NMR experiments at different frequencies are more preferable.

8.4 PROTOCOL FOR QUANTIFICATION OF "INVISIBLE NUCLEI" IN MAS NMR

As demonstrated in Chapter 7, strong paramagnetic effects can result in a significant intensity loss for observed nuclei that are situated in close environments of paramagnetic ions. In some sense, quantification of this effect can characterize investigated paramagnetic solids by localization of the paramagnetic metal ions in their NMR spectra. The protocol of such quantifications has been suggested by Ziarelli et al. [50].

Basically, any quantitative determinations of signal intensities in solid-state NMR imply direct excitation of nuclei, for example, by single pulse experiments (or other pulse sequences) with complete nuclear relaxation in each cycle. The determination of integral intensities for target nuclei requires obviously the presence of a convenient reference signal.

In principle, an internal reference could simplify quantitative measurements in solid-state NMR, first of all, by avoiding the experimental problems connected with variable sample volumes and inhomogeneity of the B_1 field. However, because of the well-known complexity in preparing perfectly homogeneous mixtures of the samples, the external reference method is preferable. Unfortunately, this method has a significant disadvantage since its precision strongly depends on the overall instrumental stability especially in electronics. The effect coming from such instabilities can be reduced by using the so-called electronic reference to access in vivo concentration (ERETIC) approach, where a low-power radio frequency shaped pulse mimicking the form of an NMR signal is used during the acquisition time. Then, after Fourier transformation, a reference peak appears at a frequency defined by the researcher. The relative intensity of this peak can be adjusted by changing the corresponding pulse power level.

The second source of errors in the external reference method is connected with the rotor packing problem. Such errors can be minimized by the rotor packing design aimed at the optimal filling condition. The packing design can be realized as shown in Figure 8.15. It is seen that for the best filling condition, they use a 4 mm zirconia rotor that is open at both ends. Then two Kel-F inserts equipped with the corresponding rings can be applied for closing. It is important to add that the spacers together with the inserts can placed in the quantitative volume region of any NMR probe. Finally excitation profiles and data processing can be found in details in Ziarelli et al. [50].

Cardarelli et al. [51] have recently illustrated application of this technique for location of cobalt ions in Co-zeolites and cobalt-containing amorphous silicoalumina materials on the basis of ^{27}Al and ^{29}Si MAS NMR. The authors have applied

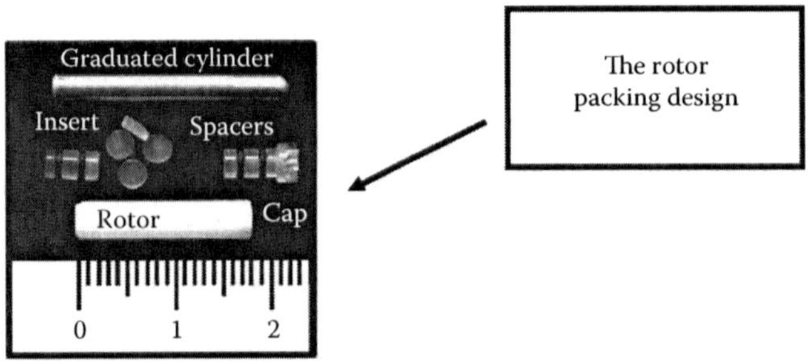

FIGURE 8.15 Tools used for the packing design allowing quantitative solid-state NMR measurements. (Reproduced with permission from Bakhmutov, V.I., *Chem. Rev.*, 111, 530, 2011.)

TABLE 8.3
Quantitative Analysis of Co-Containing Zeolites and Silicoalumina Materials (ASA) by Fractions (in %) of Nuclei Affected by Paramagnetic Interaction

System	Nucleus	Fermi-Contact Interaction	Pseudocontact Interaction	Unaffected
Co-H-MFI	^{27}Al	12	57	31
Co-H-MOR	^{27}Al	58	30	12
Co-H-MFI	^{29}Si	4	29	67
Co-H-MOR	^{29}Si	9	71	20
Co-ASA	^{29}Si	19	48	33

Source: Caldarelli, S. et al., *J. Phys. Chem. C*, 115, 10569, 2011.

this procedure to estimate the number of tetrahedral sites in the framework of the materials affected by large paramagnetic shifts due to the covalent bonding to paramagnetic cobalt through oxygen. When the fractions of invisible nuclei have been obtained, the remaining intensity in the MAS spectra has been renormalized. The latter leads to an overall description of the paramagnetic interactions. The data are represented in Table 8.3. It has been emphasized that quantitative measurements in MAS NMR, according to the optimized protocol, produce errors smaller than 1%.

8.5 CONCLUDING REMARKS

The present chapter shows that chemistry of porous materials modified by paramagnetic metal ions is an area of increasing activity and requires development of new reliable approaches to a structural analysis of synthesized systems. The available literature illustrates how a protocol of solid-state NMR studies changes in going from diamagnetic to paramagnetic solids, when distribution of metal ions is the focus of a researcher. Using porous silica-based materials doped with paramagnetic ions, as practical examples, we demonstrated that incorporation of ions into the matrix or their accumulation within the cavities of materials can be distinguished by the following experiments:

1. Observation of "invisible" target nuclei by the Hahn-echo NMR experiments (static and MAS), performed at different carrier frequencies
2. Quantification of "invisible nuclei" in MAS NMR
3. Editing the Hahn-echo MAS NMR experiments
4. T_1 and T_2 NMR relaxation measurements in static and spinning paramagnetic materials

On the basis of relaxation studies performed by different authors, it has been shown that NMR relaxation data often show T_1 time distributions, which are observed as different T_1 values obtained in MAS NMR experiments for isotropic resonances and their sidebands. Such T_1 distributions, being a common phenomenon in paramagnetic solids, can be masked by the bulk magnetic susceptibility (BMS) effects increasing with concentrations of paramagnetic centers. The presence of T_1 distributions, in itself, is not a criterion for incorporation of paramagnetic ions into the matrix or its surface. However, a quantitative analysis of the experimentally observed short T_1 components, based on the well-determined electron relaxation times, can provide such a criterion.

REFERENCES

1. Bakhmutov, V. I., Shpeizer, B. G., Clearfield, A. 2006. *Magn. Reson. Chem.* 44: 861.
2. Canesson, L., Tuel, A. 1997. *Chem. Commun.* 241.
3. Tuel, A., Canesson, L., Volta, J. C. 1999. *Colloid. Surface. Physicochem. Eng. Aspect.* 158: 97.
4. Huang, W., Schopfer, M., Zhang, C., Howell, R. C., Todaro, L., Gee, B. A., Francesconi, L. C., Polenova, T. 2008. *J. Am. Chem. Soc.* 130: 481.
5. Goldfarb, D. 1989. *Zeolites* 9: 509.
6. Munoz-Aguado, M. J., Gregokiewitz, M., Bermejo, F. J. 1995. *J. Non-Cryst. Solids* 189: 90.
7. Mokaya, R. 2001. *Micropor. Mesopor. Mater.* 44–45: 119.
8. Simonutti, R., Comotti, A., Bracco, S., Sozzani, P. 2001. *Chem. Mater.* 13: 771.
9. Kubo, A., Spaniol, T. P., Terao, T. 1998. *J. Magn. Reson.* 133: 330.
10. Abidi, N., Deroide, B., Zanchetta, J. V., Menorval, L. C., Espinose, J. B. 1998. *J. Non-Cryst. Solids* 231: 49.
11. Tušar, N. N., Ristic, A., Mali, G. et al. 2010. *Chem. Eur. J.* 16: 5783.
12. Tucker, M. C., Dieff, M. M., Richardson, T. J., Finones, R., Reimer, J. A., Cairns, E. J. 2002. *Electrochem. Solid-State Lett.* 5: A95.
13. Barbara, T. M. 1994. *J. Magn. Reson. A* 109: 265.
14. Shpeizer, B. G., Bakhmoutov, V. I., Zhang, P., Prosvirin, A. V., Dunbar, K. R., Thommes, M., Clearfield, A. 2010. *Colloid. Surface. Physicochem. Eng. Aspect.* 357: 105.
15. Bakhmutov, V. I. 2008. *Solid State Nucl. Magn. Reson.* 34: 197.
16. Grey, C. P., Dobson, C. M., Cheethman, A. K., Jakeman, R. J. B. 1989. *J. Am. Chem. Soc.* 111: 505.
17. Bakhmutov, V. I., Shpeizer, B. G., Clearfield, A. 2006. *Magn. Reson. Chem.* 44: 985.
18. Mali, G., Ristić, A., Kaucic, V. 2005. *J. Phys. Chem. B* 109: 10711.
19. Bakhmutov, V. I., Shpeizer, B. G., Clearfield, A. 2007. *Magn. Reson. Chem.* 45: 118.

20. Clayton, A. N., Dobson, C. M., Grey, C. P. 1990. *Chem. Commun.* 72.
21. Plevert, J., Okubo, T., Wada, Y., O'Keefe, M., Tatsumi, T. 2001. *Chem. Commun.* 2112.
22. Schmitt, K. D., Haase, J., Oldfield, E. 1994. *Zeolites* 14: 89.
23. Bakhmutov, V. I., Shpeizer, B. G., Prosvirin, A. V., Dunbar, K. M., Clearfield, A. 2009. *Micropor. Mesopor. Mater.* 118: 78.
24. Berlier, G., Prestipino, C., Rivallan, M., Bordiga, S., Lamberti, C., Zecchina, A. 2005. *J. Phys. Chem. B* 109: 22377.
25. Cordoba, G., Arroyo, R., Fierro, J. L. G., Viniegra, M. 1996. *J. Solid State Chem.* 123: 93.
26. Sanchez, C., Soller-Illia, G. J. A., Ribot, F., Lalot, T., Mayer, C. R., Cabuil, V. 2001. *Chem. Mater.* 13: 3061.
27. Sayari, A. (ed.) 2000. *Studies in Surface Science and Catalysis: Nanoporous Materials II, Proceedings of the 2nd Conference on Access in Nanoporous Materials*, Vol. 129, pp. 1–889. Amsterdam, the Netherlands: Elsevier Science.
28. Shunn, R. A., Baker, R., Cole, R. J., Gilbert, J. D., Madden, D. P. 1974. *Inorg. Synth.* 15: 5.
29. Scholz, K., Thomas, B. 1995. *Solid State Nucl. Magn. Reson.* 4: 309.
30. Sen, S., Stebbins, J. F. 1994. *Phys. Rev. B* 50: 822.
31. McHenry, M. R., Silbernagel, B. G., Wernick, J. H. 1972. *Phys. Rev. B* 5: 2958.
32. Mortuza, M., Dupree, G. R., Holland, D. 2001. *J. Non-Cryst. Solids* 281: 108.
33. Abragam, A. 1961. *Principles of Nuclear Magnetism.* Oxford, U.K.: Oxford University Press.
34. Canesson, L., Boudeville, Y., Tuel, A. 1997. *J. Am. Chem. Soc.* 119: 10754.
35. Maiti, B., McGarvey, B. R. 1984. *J. Magn. Reson.* 58: 37.
36. Alaimo, M. H., Roberts, J. E. 1997. *Solid State Nucl. Magn. Reson.* 8: 241.
37. Kessemeier, H., Norberg, R. E. 1967. *Phys. Rev.* 155: 321.
38. Hayashi, S. 1994. *Solid State Nucl. Magn. Reson.* 3: 323.
39. Bakhmutov, V. I., Shpeizer, B. G., Prosvirin, A. V., Dunbar, K. R., Clearfield, A. 2009. *Solid State Nucl. Magn. Reson.* 36: 129.
40. Wiench, J. W., Lin, V. S. Y., Pruski, M. 2008. *J. Magn. Reson.* 193: 233.
41. Malfait, W. J., Halter, W. E., Verel, R. 2008. *Chem. Geol.* 256: 69.
42. Chen, S. H., Sheu, S. P., Chao, K. 1992. *J. Chem. Commun.* 1504.
43. Hayashi, S., Hayamizu, K., Akiba, E. 1992. *J. Phys. Chem.* 96: 10928.
44. Bourdreaux, E. A., Mulay, L N. 1976. *Theory and Application of Molecular Magnetism.* New York: Wiley.
45. Suttler, B., Taylor, R. E., Hossner, L. R., Ming, D. W. 2002. *Soil Sci. Soc. Am. J.* 66: 455.
46. Atsarkin, V. A., Demidov, V. V., Simon, F., Gaal, R., Moritomo, Y., Conder, K., Janossy, A., Forro, L. 2003. *J. Magn. Magn. Mater.* 258–259: 256.
47. Tanaka, S., Mizushima, K., Iida, S. 1980. *J. Phys. Soc. Jpn.* 49: 569.
48. Carper, W. R. 1999. *Concepts Magn. Reson.* 11: 51.
49. Chudek, J. A., Hunter, G., McQuire, G. W., Rochester, C. H., Smith, T. F. S. 1996. *J. Chem. Soc. Faraday Trans.* 92: 453.
50. Ziarelli, F., Viel, S., Sanchez, S., Cross, D., Caldarelli, S. 2007. *J. Magn. Reson.* 188: 260.
51. Caldarelli, S., Pizzala, H., Arrighi, L., Ziarelli, F., Busca, G. 2011. *J. Phys. Chem. C* 115: 10569.
52. Bakhmutov, V. I. 2011. *Chem. Rev.* 111: 530–562.

Index

A

Absorption, 49, 51
  energy, 9
  line shape, 13
Abundance, nuclear, 5–7, 12
Accumulation, 77, 191
Acquisition time, 11, 85
Activation energy, 15, 140
Albite glass, 120
^{27}Al NMR, 239
^{27}Al–^{17}O cross polarization, 81
^{27}Al relaxation
  diamagnetic materials, 153
  paramagnetic materials, 199
^{27}Al triple-quantum NMR, 105; *see also* Multiple quantum, experiments
Aluminophosphates, 205, 231
Aluminosilicate glasses, 44, 178; *see also* Glasses
Aluminum
  extra framework, 46
  framework, 53
Amplifiers, 27
Amplitude, 10, 16, 35, 58, 77, 120
Andrew, magic angle, 3, 13, 28, 69
Angle, acting, 13, 17, 28, 76; *see also* Pulse
Angle-flipping experiments, 58
Angular moment, 5, 22
Anisotropy
  chemical shift, 84–87, 123, 130 (*see also* Chemical shift)
  motions, 16, 131
Asymmetry
  correlation time distribution, 151
  parameter, 25, 70–72, 86, 112, 117
Autocorrelation function, 16; *see also* Function, spectral density

B

BABA sequence, 85
Baseline, 13, 31, 75, 141
Bio-degradation, 43; *see also* Wood, morphology
Biological solids, amorphous, 37
Bloch equation, 12, 142; *see also* Relaxation
Blombergen-Purcell-Pound (BPP), function, 12, 17
Bohr magneton, 193
Boltzmann constant, 9
Boron nitride, caps, 30
Bulk diffusion coefficient, 160

C

Calibration, pulses, 148, 172, 224
Carbon nanotubes, 41, 51, 106
Carrier frequency, 13, 26, 76, 212–216, 227
Carr-Purcell-Meiboom-Gill, experiment, 106, 109, 117, 212
Cellulose, 42
Charge, density, 21
  distribution, 24
Chemical exchange, 19, 156; *see also* Dynamics; Exchange
Chemical shift, 3, 5, 17–22
  anisotropy, 17, 84–85
  asymmetry, 61
  diamagnetic term, 21, 64
  Fermi-contact term, 193–196, 256
  isotropic, 21–22, 111–113, 117
  paramagnetic term, 21, 64
  tensor, 58–61, 120, 125
Closed system, 172
^{13}C NMR, 76–78, 131
  cellulose, 127–129
  relaxation, 155
  wood, 43, 127
Cobalt materials, 205, 215, 225, 255
CODEX, 157
Coils, 29–30, 68, 74
Cole-Davidson, distribution, 151; *see also* Correlation time distributions
Composite materials, 158
Composite pulses, 146
Connectivity, 112
Contact time, 77–79, 128–130
Coordination numbers, 51, 53
Correlation spectroscopy, 40, 102, 109, 114, 125
Correlation time
  distributions, 151–152, 197, 224
  molecular motions, 2, 15–18, 25
Coupling
  dipolar, 14, 24, 33, 77, 100–102
  quadrupolar, 3, 17, 25, 128
  spin-orbital, 23
  spin-spin, 5, 22, 104
CP rate constant, 78; *see also* Cross polarization
CRAMPS, 101

Cross-polarization, 35, 38, 76–79
  adiabatic, 109
Cryogenic probes, 34
Crystals, bulk, 116
Cupper fluoride, 201
Curie, law, 53
Current-switched-field-cycling, 162

D

DAS experiments, 121
Dead time
  base-line distortion, 31
  spectrometer, 31, 75
Deconvolution, 38, 52, 58, 128, 176–178
Decoupling, 23, 29, 67, 77, 81, 101
Degradation, wood, 43, 127
Dephasing,76, 128–130, 207
Deuteration, 102
Deuterium NMR, 84, 129–130, 161, 169
DFT calculations, 64, 116, 179, 216
Dielectric relaxation, 41
Diffusion coefficient, 136, 161, 171, 222
Diffusion-limited relaxation, 220, 244
Digitizer, 28
Dihedral angle, 89
Dipolar relaxation, 149–150; *see also* Relaxation
Dipole-dipole interactions, 29, 85, 93
Dirac function, 23
2D NMR, 32, 85, 106, 117, 192, 203
DOR experiments, 105
DOSY experiments, 136
Double quantum
  experiments, 84, 101–103
  transitions, 8
Double resonance, 49, 88
DRAMA sequence, 85
DUMBO sequence, 101
Dwell time, 31
Dynamic
  angle spinning, 121
  NMR, 141
  nuclear polarization, 34, 191
Dynamics
  amorphous systems, 45, 66, 94, 152
  gas molecules, 175
  liquids in pores, 160–161
  polymers, 41, 145, 155

E

Echo, 31, 75, 109–110, 117, 170; *see also* Hahn echo
Electric field gradient, 24, 26, 116, 213
  axially-symmetrical, 25
  electron term, 25
  nuclear term, 25
Electron
  imbalance, 21
  magnetic moment, 12, 14
  nucleus coupling, 193, 195, 215
  relaxation time, 200, 205, 207, 247, 252
Elementary charge, 25
ENDOR, 49, 53
Energy levels, 5, 8–9, 23, 70
Enhancement, sensitivity, 35, 78, 117
Ensemble, nuclear, 9, 83
EPR, 53, 213, 234, 239
EQES, 207
ERETIC, 255
Ernst, rule, 10, 17, 76
EXAFS, 49, 51, 53
Exchange, 19, 53, 140–144; *see also* Chemical exchange
  2D NMR, 157
  rate, 143
Exponential relaxation, 146–149, 245; *see also* Relaxation
External standard, 21, 111, 117

F

Fast motion, 140, 161; *see also* Motion correlation time
Fermi-contact shift, 193–196, 206
  temperature-dependent, 206, 214
Ferromagnetism, 243
^{19}F-^{19}F distance, 109
^{19}F-^{19}F splitting, 5
Field-cycling, experiments, 162
Field dependence, 162–163
Field-gradient methods, 64–95, 171
Flip-flop motion, 200, 228
Fluctuation, magnetic field, 149
^{19}F NMR, 201
Frame work, 12, 79, 109
Free model approach, 154
Free rotation, 14
Frequency
  domain, 12
  offset, 13, 119
  passage, 75
Function, spectral density, 17
Fuoss-Kirkwood, distribution, 151

G

Gas, bulk, 174–175
Gauss
  distribution, 220, 253–254
  line shape, 13–14, 58
  multiplication, 30
Generator, 26
Geometry, motional, 142–144, 158

g-Factor, 193
Gibbs-Thomson depression, 160
Glasses, 37, 44, 119; *see also* Silicate glasses
Glass transition, 41
Glassy materials, 157–158
Gradients, 94–95, 116, 136, 172

H

Hahn echo, 31, 75, 257
Half-height width, 13, 229
Hamiltonian, 5, 8, 22, 34, 86, 252
Hartmann-Hahn condition, 77–79, 120
Heteronuclear decoupling, 82
^{199}Hg NMR, 62–63
^{1}H–^{2}H Cross polarization, 80
High power decoupling, 23
^{1}H NMR, 76, 93
Homogeneity, 28–30
Homonuclear coupling, 150
Hydration, 185
Hydrogen bond, 22, 40, 156
Hyperfine coupling constant, 53, 193–194
Hyperfine interactions, 215, 235

I

Imaginary component, 13
Imaging NMR, 174
Immobilized systems, 37, 99, 133, 180, 186
Ineffective excitation, 206
INEPT, 203
Inertia moment, 16
Inhomogeneity, field, 82, 228, 255
^{127}I NMR, 117
Inorganic parts, amorphous, 39; *see also* Materials amorphous
Internal reference, 255
Internuclear distances, 3, 14, 85–89, 109; *see also* Proximity, nuclear
Intramolecular rotations, 19
Inversion-recovery, experiments, 106, 145–147
Irradiation, 3, 29–30
IR spectra, 52
IRS sequence, 91
Isomer shift, 51; *see also* Mossbauer spectra
Isotopic dilution, 102
Isotopic shift, 5
Isotropic chemical shift, 20–21
  reorientations, 13, 16, 152

J

J-coupling, constants, 22–23
  Fermi contact interaction, 23–24
  spin-dipolar mechanism, 23
  spin-orbital mechanism, 23
J(w,tC)-function, 17; *see also* Spectral density function
J-INADEQUITE, 114
Jump
  helical, 158
  reorientations in solids, 143–144
  rotation, 19
  thermally-activated, 143
  two-site, 144

K

Kinetics
  cross polarization, 77, 114, 132 (*see also* CP rate constant; Cross polarization)
    chemical, 16
Knight shifts, 195
Kronecker delta function, 232
^{83}Kr relaxation, 174

L

Lamb, formula, 21
^{139}La NMR, 71
Lanthanides, 205, 207, 232
Larmor frequency, 8–9, 35
Lattice, 12, 14, 17, 45
Lennard-Jones theory, 175
Lignin, 47, 127; *see also* Wood
Line
  broadening, 101, 104
  shape, 19, 24, 57–58, 94, 108
  width, 13, 17
Line-shape analysis, 139–143; *see also* Dynamics
^{6,7}Li NMR, 205
Liquids, bulk, 160
  in pores, 160–161
Local field, 70, 93, 109
Local temperature,141
Longitudinal magnetization, 12, 217; *see also* Relaxation
Lorentz, line shape, 13
Low-dimensionality structures, 41

M

Macroscopic magnetization, 5, 9, 31
Magic angle, 3, 13, 29
Magnetic field, effective, 192–193
  switching, 75, 147, 162
Magnetic quantum number, 7
Magnetogyric ratio, 6, 114
Magnets, 28–29, 74
Mapping experiments, 215–217
Materials amorphous, 48, 58, 66, 85, 151
Mesoporous materials, 51–52, 109, 165, 185
Microscopy, 41, 45

Microwave field, 35, 66
Mixing time, 156
Mixtures, 197, 235
Model, relaxation, 151, 197, 207, 249
Modulation, 193, 239
MOF systems, 39–40
Molecular frame, 61–63
  motions, 13, 15, 154
  motions in polymers, 158–159
Molecules absorbed, 50, 109, 134, 170, 189; *see also* Gas, bulk; Liquids, bulk
Mossbauer spectra, 48, 51
Motion correlation time, 15–17
MREV, 82
Multiple quantum, experiments, 122

N

Nanoparticles, 39, 100, 131, 243
Narrowing line, 19, 29, 134
Negami, model, 151, 197
NMR chromatography, 133, 136
^{15}N NMR, 143
NOESY, 160
Noise effects, 148
Nonexponential relaxation, 146–147, 197, 247–249
Non-selective time T1, 146
^{15}N relaxation, 155
Nuclear overhauser effect (NOE), 3
Nuclear quadrupole coupling constants, 24–25, 116; *see also* Quadrupolar coupling
Nuclei, invisible, 255
Nutation, 81, 109

O

ODESSA, 157
Off-angle spinning, 133
Offset, 119, 192
^{17}O NMR, 120–122, 178
Optical pumping, 34
Orbital interactions, 23, 34
Oscillating field, 10
Overlapping, 176; *see also* Deconvolution
Oxo-tungstates, 232
Oxygen
  coordination type, 44, 112, 178
  in pores, 153, 209

P

Pake pattern, 74, 130
Paramagnetic centers, 53, 76, 149, 153
Paramagnetic relaxation, 191
PASS sequence, 91
Pathways, transfer, 85–86
^{31}P chemical shift calculations, 112, 114–116
Phases, amorphous, 43
Phase separation, 241
Phasing, 141
Phosphates, 47, 111, 194, 205, 216
Phosphonates, 111–112
PISEMA sequence, 109
Planck, constant, 7
Plastic crystals, 13, 68, 130
^{31}P NMR, 38, 66, 134, 205
Point-by-point, method, 201
Point-dipole approximation, 232
Polarization, 28, 34, 81, 88, 92; *see also* Cross polarization
Polarization inversion, 109
Population
  inversion, 192
  transfer, 120
Pore-filling factor, 120
Porosity, measurements, 48, 167, 170–171
Powder
  diffraction, 49–50
  pattern, 49, 123
Power law, 163
Power spectra, 13
Pre-amplifier, 26
Prehistory effects, 153, 241
Principal coordinate system, 59, 61, 86
Principles, Kisman-Armstrong, 30
Progressive saturation, 145
Protocol
  crystallinity determination, 128
  REDOR MAS, 109
  temperature determination, 141
Proton diffusion, 158, 199, 223; *see also* Spin diffusion
Proximity, nuclear 106–108
Pseudo-contact shift, 195, 256
  temperature dependent, 205
Pulse
  adiabatic, 75, 88, 192
  generator, 26
  hard, 146
  selective, 11, 75, 85
PUREX, 157

Q

QCPMG sequence, 117–118
Quadrupolar coupling, 17, 119, 123
  first-order, 70–72, 120, 125
  second-order, 70–72, 123
  tensor, 24, 140, 144
Quadrupole
  echo, 207
  splitting, 84
Quality factor Q, 74
Quantification, 257

Quantum
energy, 8
formalism, 5, 7
transition, 8

R

Radiation damping, 228
Radio-frequency field, 8–9
Raman spectroscopy, 44, 49, 52
Ramsey, theory, 64–65
RAPT, 120
Real component, 13, 27
Receiver, 26, 116, 227
Recoupling-based polarization transfer, 192
Recoupling, experiments, 85, 90, 102–107, 114
Redistribution, 100, 241
REDOR, 86, 90, 109, 125
Refocusing, 192, 228
Regions amorphous, 218
Relaxation
biexponential, 218, 225, 249, 251
curves, 147–148, 150–152, 245, 253
delays, 11, 16
mechanisms, 149–150, 197
times, 30–31, 35
Resolution, digital, 30
^{103}Rh NMR, 202
Rigid deuterons, 160, 188
Ringing, acoustic, 116
Rotating frame, 10, 13, 77, 129, 154, 158
Rotational diffusion tensor, 132
Rotation, double, 105
Rotation, samples, 29
Rotors, MAS, 29, 119, 121
Rotor-synchronized echo, 85

S

Samples amorphous, 58, 153
Satellite transitions, 70–75, 123–125
Saturation, 2, 145, 160, 216
Scalar, 17, 32, 104
Scale
chemical shifts, 62, 117, 127
frequency, 156, 235–236
time, 19, 155, 188
Screening constant, 22
Second moment, 74, 78
SEDOR, 125
Segmental reorientations, 41, 74, 132, 139, 143
Self-assemblies, 40
Sensitivity, 26–28, 34
Shielding constant
diamagnetic contribution, 21
paramagnetic contribution, 21
Shift-compensated quadrupole echo, 207
Sideband patterns, 9, 69, 85, 251–253
Signal-to-noise ratio, 62, 76
increase, 120, 131
Silicate glasses, 44, 76, 119, 178, 241
^{29}Si NMR, 66, 76, 153, 177, 199, 251
^{29}Si spin-diffusion coefficient, 243
Skew, 60
Slow exchange, 144, 156; *see also* Dynamics
motion, 154–157
Small metal particles, 201
Sol/gel materials, 47, 49
Solid echo, 31, 76
Solomon, equations, 199, 252
Solvent-addition effect, 134
Spacer, 135
Span, 60, 72
Spatial resolution, 94
Spectral density function, 149–151
Spin diffusion, 2, 17, 76, 153, 158, 219, 242; *see also* Relaxation
Spin-locking, experiment, 85, 146, 192
Spin operator, 5, 58
Spin-rotation relaxation, 174
Spin-spin coupling, 3, 5, 22
Spin system, 23, 32
Stimulated echo, 156, 173
Stochastic pulses, 216
Surfactant molecules, 50, 141
Susceptibility, magnetic, 53, 136, 195, 252
Synchronous experiments, 192

T

T_2, 151
T1 curves, 148–152
minimum, 17, 150
TEDOR, 109
Temperature calibration, 141
Tensor
electric field gradient, 24
magnetic screening, 60–65, 72
Thomson, theory, 169
Time domain, 12
^{47}Ti NMR, 70, 186
Transfer
charge complexes, 41
magnetization, 35, 114
pathways, 85–86
polarization, 130
Transmitter, 26, 30
Transverse magnetization, antiphase, 204
TRAPDOR, 81, 88, 125
T_2, spinning samples, 246
Tumbling time, 16
Tunneling, 19, 153

U

Ultra-
  centrifugation, 133
  fast spinning, 37, 124
  wide line, 201, 212
Uncoupled nuclei, 8, 13
Unpaired electrons, 34, 70, 197, 200, 210
Unusual relaxation, 149
UV Spectra, 48–49, 52

V

Vanadium-based systems, 123–126
Van Vleck, formula, 74
Variable-amplitude, experiments, 78, 128
Vector model, 9
Visualization, chemical shift tensor, 62–63
$^{50,51}V$ NMR, 124–126
VO_x/ZrO_2 catalyst, 125
V-shaped T_1 curves, 17, 150

W

Wave
  continuous irradiation, 9, 28, 83
  function, 5, 10, 64
Whole echo, 215
Wide line NMR, 73, 212
WISE, 159–160
Wood, morphology, 43

X

XANES, 239
XPS, 51, 53
X-ray, 47–51, 99, 109, 112

Z

Zeeman, splitting, 70
Zeolites, 47–49, 66, 178, 183, 186
Zero-Field experiments, 92–94
Z-filter, 85